W0036788

# Pueraria

# Pueraria

The genus *Pueraria*

*Edited by*

## Wing Ming Keung

*Harvard Medical School, Boston, USA*

## CRC PRESS

Boca Raton   London   New York   Washington, D.C.

## FIRST INDIAN REPRINT, 2012

This book contains information obtained from authentic and highly regarded sources. Reprinted material is quoted with permission, and sources are indicated. A wide variety of references are listed. Reasonable efforts have been made to publish reliable data and information, but the author and the publisher cannot assume responsibility for the validity of all materials or for the consequences of their use.

Neither this book nor any part may be reproduced or transmitted in any form or by any means, electronic or mechanical, including photocopying, microfilming, and recording, or by any information storage or retrieval system, without prior permission in writing from the publisher.

Direct all inquiries to CRC Press LLC, 2000 N.W. Corporate Blvd., Boca Raton, Florida 33431.

© 2002 Taylor & Francis Group, LLC CRC Press is an imprint of Taylor & Francis Group

**Trademark Notice:** Product or corporate names may be trademarks or registered trademarks, and are used only for identification and explanation, without intent to infringe.

### Visit the CRC Press Web site at www.crcpress.com

Printed and bound in India by
Replika Press Pvt. Ltd.

ISBN 10 : 0-415-28492-9
ISBN 13 : 978-0-415-28492-9

### FOR SALE IN SOUTH ASIA ONLY.

# Contents

# Contributors

**Connie Cox Bodner**
Genesee Country Village & Museum
1410 Flint Hill Road
Mumford, NY 14511-0310, USA

**Siang-Shu Chai**
Department of Physiology
Shandong Academy of Medical Sciences
Ji-nan, Shandong 250062
People's Republic of China

**Luc De Cooman**
State University of Gent
Faculty of Pharmaceutical Sciences
Harelbekestr 72
B-9000 Gent, Belgium

**Denis De Keukeleire**
State University of Gent
Faculty of Pharmaceutical Sciences
Harelbekestr 72
B-9000 Gent, Belgium

**Yutaka Ebizuka**
Graduate School of Pharmaceutical
    Sciences
The University of Tokyo
7-3-1 Hongo, Bunkyo-ku
Tokyo 113-0033
Japan

**Guang-Yao Gao**
CBBSM
Harvard Medical School
250 Longwood Avenue
Boston, MA 02115, USA

**Takashi Hakamatsuka**
Faculty of Pharmaceutical Sciences
Science University of Tokyo
12 Funakawara-machi, Ichigaya
Shinjuku-ku, Tokyo 162-0826
Japan

**Theodore Hymowitz**
Department of Crop Sciences
University of Illinois
Urbana, IL 61802, USA

**John L. Ingham**
Department of Food Science &
    Technology
University of Reading
Whiteknights, P.O. Box 226
Reading, RG6 2AP, England

**Anwar Jardine**
Gillette Advanced Technology Center, US
37A Street, Needham
MA 02492-9120, USA

**Wing Ming Keung**
Center for Biochemical and
    Biophysical Sciences and Medicine
Harvard Medical School
250 Longwood Avenue, Boston
MA 02115, USA

**Junei Kinjo**
Faculty of Pharmaceutical Science
Fukuoka University
8-19-1 Nanakuma
Fukuoka 814-0180, Japan

**Kathleen S. Lowney**
Department of Sociology,
Anthropology, and Criminal Justice
Valdosta State University
Valdosta, Georgia 31698
USA

**Scott E. Lukas**
Behavioral Psychopharmacology
    Research Laboratory
East House III
McLean Hospital/Harvard Medical
    School
115 Mill Street, Belmont
MA 02178, USA

**Dr L.J.G. van der Maesen**
Department of Plant Taxonomy
Agricultural University
PO Box 8010
6700 ED Wageningen
The Netherlands

**R. Mathur**
School of Studies in Zoology
Jiwaji University
Gwalior 474 011
India

**Toshihiro Nohara**
Laboratory of Pharmacognosy
Faculty of Pharmaceutical Sciences
Kumamoto University
5-1 Oe-honmachi
Kumamoto 862-0973
Japan

**Keisuke Ohsawa**
Tohoku Pharmaceutical University
4-4-1, Komatsushima, Aoba-ku
Sendai, Miyagi 981-8558
Japan

**Llewellyn J. Parks**
Rhizoma Corporation
120 Rivermont Court, Sheffield
Alabama 35660
USA

**Gerald S. Pope**
c/o, Department of Animal and
    Microbial Sciences
University of Reading
Whiteknights
Reading RG6 6AJ, England

**Ales Prokop**
Department of Chemical Engineering
Box 1604, Station B
Vanderbilt University
Nashville, TN 37235, USA

**Haojing Rong**
State University of Gent
Faculty of Pharmaceutical Sciences
Harelbekestr 72
B-9000 Gent, Belgium

**Sangeeta Shukla**
School of Studies in Zoology
Jiwaji University
Gwalior 474 011
India

**Satoshi Tahara**
Department of Applied Bioscience
Faculty of Agriculture
Hokkaido University
Kita-ku, Sapporo 060-8589
Japan

**Robert D. Tanner**
Department of Chemical Engineering
Box 1604, Station B
Vanderbilt University
Nashville, TN 37235, USA

**Chun-Kowk Wong**
Department of Chemical Pathology
Prince of Wales Hospital
The Chinese University of Hong Kong
Shatin, N.T., Hong Kong

**Takaaki Yasuda**
Tohoku Pharmaceutical University
4-4-1, Komatsushima, Aoba-ku
Sendai, Miyagi 981-8558, Japan

**Ming Zeng**
Department of Pharmacognosy
School of Pharmacy
Second Military Medical University
Shanghai 200433
China

**Han-Ming Zhang**
Department of Pharmacognosy
School of Pharmacy
Second Military Medical University
Shanghai 200433, China

**Ai-ping Zhao**
Department of Physiology
Shandong Academy of Medical Sciences
Ji-nan, Shandong 250062
People's Republic of China

**You-Ping Zhu**
Foundation Hwa To Centre
University of Groningen
Antonius Deusinglaan 1
9713 AV, Groningen
The Netherlands

# Foreword

It is with great pleasure that I have the opportunity to evaluate the manuscript of Dr Keung's book. Kudzu, *Pueraria lobata*, and its relatives are important economic plants. They are valuable resources for medicines, food, fiber and fine chemicals. The earliest written record for the use of Ge (the Chinese name for *Pueraria*) in China dates back to 1000 BC. The long history of human relationship with *Pueraria* can be traced from at least 600 BC via written records from Asia to Europe and America. During the last two decades, we have witnessed an explosion of research on the medicinal and industrial applications of this genus. Thus, a comprehensive book with both the concise description of the age-old knowledge, as well as authoritative scientific information that parallels the rapid increase in our scientific understanding of *Pueraria*, is long overdue.

Wing Ming Keung pioneered the work on the scientific validation and molecular basis of the alcohol craving suppressive activity of *Pueraria lobata*. His thorough and elegant research on this pharmacological activity of *P. lobata*, from the identification of active principle to the elucidation of site and mechanism of action, serve to illustrate a rational process for identifying and evaluating high potential drug candidates from traditional Chinese, and for that matter other indigenous, medicine. This book could not have been put together by a more appropriate scholar than W.M.

This is the first comprehensive work on the genus *Pueraria*. It covers the science, history, chemistry, pharmacology, clinical applications, industrial applications, sociocultural aspects, etc., virtually all that we currently know about the genus. In line with the increasing current interest in phytomedicines, herb-based dietary supplements and health-care products, the recent discoveries of new medical properties of *Pueraria* are also reported.

This book provides detailed accounts on the use of *Pueraria* in traditional Chinese medicine (TCM) and on the new scientific findings that led to its applications in cardiovascular and cerebrovascular diseases in modernized Chinese medicine (MCM). The convergence of ethnopharmacological evidence from a number of Asiatic countries/races such as China, Japan, Korea, Thailand, Burma, and India, strongly speaks for the medical value of Pueraria. The evidence has been amply validated by modern science. The active ingredients responsible for Pueraria's diverse pharmacological activities, including antifertility, antiaging, anticoronary artery and cardiovascular diseases, anticancer and antialcohol abuse, are identified as isoflavonoids and saponins.

In the past, treatment has been too often the sole approach to medicine. But with the world changing, there has been a shift in effort towards the integration of prevention and treatment of diseases. Factors that have propelled such changes are, among others, aging of the world population and the continuously rising health budget with the growing demand of the general public for health care. In this age, a better approach to the eradication of illnesses like cancer and heart disease is by placing more emphasis on prevention and relying less on treatment. The chemo-preventive activities of Pueraria-based medications for cardiovascular diseases and stoke, for osteoporosis and cancer of the breast, prostate and endometrium, and for liver diseases in general, are authoritatively presented in this book. Pueraria-based medications or health food products can be used to prevent these medical problems of the aging population and the associated economic burden of this on society.

This is a truly comprehensive coverage of the genus *Pueraria*. The expertize of contributing authors range widely from taxonomy, botany and cultivation, herbal medicine, sociology, anthropology, zoology, food sciences and technology, physiology, biochemistry, pharmacology, phytochemistry, synthetic chemistry to psychology. Research scientists, graduate students and professionals working in these fields will find this book a valuable resource. In light of the tremendous publicity and misinformation generated around the market place, this book also provides laymen with accurate descriptions of what is really known about the therapeutic efficacy of various medicinal products derived from this genus.

Hin-Wing Yeung, Ph.D.
Director
Institute for the Advancement of Chinese Medicine
Hong Kong Baptist University
Kowloon Tong, Hong Kong

# Preface to the series

Global warming and global travel are among the factors resulting in the spread of such infectious diseases as malaria, tuberculosis, hepatitis B and HIV. All these are not well controlled by the present drug regimes. Antibiotics too are failing because of bacterial resistance. Formerly, less well known tropical diseases are reaching new shores. A whole range of illnesses, for example cancer, occur worldwide. Advances in molecular biology, including methods of *in vitro* testing for a required medical activity give new opportunities to draw judiciously upon the use and research of traditional herbal remedies from around the world. The re-examining of the herbal medicines must be done in a multidisciplinary manner.

Since 1997, 20 volumes have been published in the Book Series *Medicinal and Aromatic Plants – Industrial Profiles*. The series continues. It is characterized by a single plant genus per volume. With the same Series Editor, this new series *Traditional Herbal Medicines for Modern Times*, covers multi genera per volume. It accommodates for example, the Traditional Chinese Medicines (TCM), the Japanese Kampo versions of this and the Ayurvedic formulations of India. Collections of plants are also brought together because they have been re-evaluated for the treatment of specific diseases, such as malaria, tuberculosis, cancer, diabetes, etc. Yet other collections are of the most recent investigations of the endemic medicinal plants of a particular country, e.g. of India, South Africa, Mexico, Brazil (with its vast flora), or of Malaysia with its rainforests said to be the oldest in the world, etc.

Each volume reports on the latest developments and discusses key topics relevant to interdisciplinary health science research by ethnobiologists, taxonomists, conservationists, agronomists, chemists, pharmacologists, clinicians and toxicologists. The Series is relevant to all these scientists and will enable them to guide business, government agencies and commerce in the complexities of these matters. The background to the subject is outlined below.

Over many centuries, the safety and limitations of herbal medicines have been established by their empirical use by the "healers" who also took a holistic approach. The "healers" are aware of the infrequent adverse affects and know how to correct these when they occur. Consequently and ideally, the pre-clinical and clinical studies of a herbal medicine need to be carried out with the full cooperation of the traditional healer. The plant composition of the medicine, the stage of the development of the plant material, when it is to be collected from the wild or when from cultivation, its post-harvest treatment, the preparation of the medicine, the dosage and frequency and much other essential information is required. A consideration of the intellectual property rights and appropriate models of benefit sharing may also be necessary.

Wherever the medicine is being prepared, the first requirement is a well documented reference collection of dried plant material. Such collections are encouraged by organizations like the World Health Organization and the United Nations Industrial Development Organization. The Royal Botanic Gardens at Kew in the United Kingdom is building up its collection of traditional Chinese dried plant material relevant to its purchase and use by those who sell or prescribe TCM in the United Kingdom.

In any country, the control of the quality of plant raw material, of its efficacy and of its safety in use, are essential. The work requires sophisticated laboratory equipment and highly trained personnel. This kind of "control" cannot be applied to the locally produced herbal medicines in the rural areas of many countries, on which millions of people depend. Local traditional knowledge of the "healers" has to suffice.

Conservation and protection of plant habitats is required and breeding for biological diversity is important. Gene systems are being studied for medicinal exploitation. There can never be too many seed conservation "banks" to conserve genetic diversity. Unfortunately such banks are usually dominated by agricultural and horticultural crops with little space for medicinal plants. Developments such as random amplified polymorphic DNA enable the genetic variability of a species to be checked. This can be helpful in deciding whether specimens of close genetic similarity warrant storage.

From ancient times, a great deal of information concerning diagnosis and the use of traditional herbal medicines has been documented in the scripts of China, India and elsewhere. Today, modern formulations of these medicines exist in the form of e.g. powders, granules, capsules and tablets. They are prepared in various institutions e.g. government hospitals in China and Korea, and by companies such as Tsumura Co. of Japan with good quality control. Similarly, products are produced by many other companies in India, the United States and elsewhere with a varying degree of quality control. In the United States, the dietary supplement and Health Education Act of 1994 recognized the class of physiotherapeutic agents derived from medicinal and aromatic plants. Furthermore, under public pressure, the United States Congress set up an Office of Alternative Medicine and this office in 1994 assisted the filing of several Investigational New Drug (IND) applications, required for clinical trials of some Chinese herbal preparations. The significance of these applications was that each Chinese preparation involved several plants and yet was handled as a *single* IND. A demonstration of the contribution to efficacy, of *each* ingredient of *each* plant, was not required. This was a major step forward towards more sensible regulations with regard to phytomedicines.

Something of the subject of western herbal medicines is now being taught again to medical students in Germany and Canada. Throughout Europe, the United States, Australia and other countries' pharmacy and health related schools are increasingly offering training in phytotherapy.

Traditional Chinese medicines clinics are now common outside of China. An Ayurvedic Hospital now exists in London and a degree course in Ayurveda is also available here.

The term "integrated medicine" is now being used which selectively combines traditional herbal medicine with "modern medicine." In Germany there is now a hospital in which TCM is integrated with western medicine. Such co-medication has become common in China, Japan, India, and North America by those educated in both systems. Benefits claimed include improved efficacy, reduction in toxicity and the period of medication, as well as a reduction in the cost of the treatment. New terms such as adjunct therapy, supportive therapy, supplementary medicine, now appear as a consequence of such co-medication. Either medicine may be described as an adjunct to the other depending on the communicator's view.

Great caution is necessary when traditional herbal medicines are used by those doctors not trained in their use and likewise when modern medicines are used by traditional herbal doctors. Possible dangers from drug interactions need to be stressed.

Many thanks are due to the staff of Taylor & Francis who have made this series possible and especially to the volume editors and their chapter contributors for the authoritative information.

Dr Roland Hardman
January 2002

# Preface

Since ancient time, plants of the genus *Pueraria* have been entwined with virtually every Asian culture. Indeed, Pueraria is so much woven into the fabric of Asian societies that it has been used in cooking, weaving, decorating, and treating human ailments for more than two millennia.

The history of *Pueraria lobata* in the United States is short but tale telling. It is by far the best documented ethnobotany of this species. *P. lobata*, brought to Philadelphia from Japan in 1876, rapidly established as an ornamental shade plant. Before long, its excellent nutritional value, remarkable hardiness and growing rate, and elaborate root systems raised it into prominence in the fodder and fertilizer industry, and in soil conservation programs throughout the South. Today, *P. lobata* is widely considered a weed in the South. However, recent ingenuous research efforts may yet again turn this robust plant into a valuable cash crop for the United States' farmers.

*PUERARIA: The Genus Pueraria* is the first monograph to be published on this medicinal and industrial plant genus. It has been assembled with a broad and diverse readers in mind – students, educators, prevention and treatment practitioners, policy-makers, and research scientists in botany, ethnobotany, phytochemistry, and the various disciplines of biological, chemical, and medical sciences. The contents of this book are intended to be comprehensive rather than selective so that readers from all walks of life will find some part of this book useful and/or interesting.

Readers who are interested in botany in general may find the in depth treatment of Pueraria on this subject useful, and the age long human–Pueraria relationship and potential values of Pueraria in food, textile, paper, and pharmaceutical industries fascinating. Others, especially the scientific experts, may be intrigued by the history of *P. lobata* in the United States and alarmed by their power in defining the facts, rules and laws of the natural world for the lay people. For those who use medicinal herbs in the prevention and/or treatment of human ailments, this book provides the most comprehensive, concise, and authoritative guides to Pueraria-based medicinal formulations and preparations.

In the last two decades, the use of medicinal plants in self-directed health care has increased in an alarming rate. This underscores the need for rigorous scientific investigation and validation of all herbal preparations. Therefore, the major objective of this book is to provide thorough reviews on the scientific evaluations of the "therapeutic claims" of Pueraria-based medications through evidence-based pharmacology and medicine. Most of the studies reviewed are preclinical. However, clinical data are also included whenever available. It is hoped that this collection will generate further interests in the research and development of Pueraria-based medicines whose efficacies are based on placebo-controlled, double blind clinical trials: the "gold standard" for the demonstration of clinical utility.

Wing Ming Keung
Boston, Massachusett
May 3, 2001

# The editor

Wing Ming Keung, Ph.D., graduated from The Chinese University of Hong Kong in 1972 with a B.Sc. degree in Chemistry. He then obtained his Master's degree at the same institution and his Ph.D. degree in biochemistry at Colorado State University in Fort Collins, Colorado in 1980. He spent one year as a postdoctoral fellow at Harvard University and then returned to The Chinese University of Hong Kong as a lecturer of biochemistry. In 1987, he rejoined Harvard Medical School's Center for Biochemical and Biophysical Sciences and Medicine as a Senior Research Scientist and has since led the Center's alcohol research program in search of the underlying mechanism that controls/regulates alcohol drinking and novel pharmacological treatments for alcohol abuse and alcoholism. Dr Keung has presented numerous lectures at national and international meetings including the Nobel Symposium at Karolinska Institute and the European Society on Biochemical Research on Alcoholism. He is author of more than 60 research papers and patents. The focus of his current research is on daidzin, the active principle he isolated and identified from a Chinese medicine (*Pueraria lobata*) traditionally used for the treatment of "alcohol addiction."

# 1 *Pueraria*: botanical characteristics

*L.J.G. van der Maesen*

## INTRODUCTION

Kudzu (*P. montana* var. *lobata*), tropical kudzu (*P. phaseoloides*), and their relatives are classified in the genus *Pueraria* that belongs to the subtribe Glycininae of the tribe Phaseoleae in the subfamily Papilionoideae of the legume family, the Leguminosae. This large important family, containing some 18 000 species or 8 per cent of all flowering plant species, is generally well recognizable. The fruit is a legume, a pod, resulting from a superior ovary developing into a bivalved usually dehiscent structure, such as beans and pea pods. The pod is known in a multitude of shapes, from tiny rounded structures containing one seed, to large woody legumes up to 2 m long and 10 to 20 cm wide, the latter in fact the longest fruit structures in the world. The Papilionoideae are characterized by the zygomorph papilionoid flower, a derived structure quite specialized for insect pollination.

Kudzu, as a rather important temperate-zone specialty food and cover crop, and tropical kudzu one of the topmost tropical cover crops, are the only economic species of agricultural use (Figures 1.1, 1.2). *Pueraria tuberosa* is quite widely known as a famine tuber crop or used for animal feed. All other species are either rare or are being gathered from the wild. Medicinal uses of kudzu are age-old. The uses of *P. mirifica* are ancient, but because of its localized occurrence, knowledge about this species has remained limited.

The botany of the genus has been monographed in 1985, and updated in 1994 (van der Maesen, 1985, 1994; Sanjappa, 1992). Lackey, in his synopsis of the Phaseoleae (1977), subdivided the genus into four groups, and these might constitute four separate genera, as recently substantiated by Lee (2000). In her study of Glycinineae she found that on the basis of molecular phylogenetic work *Pueraria* is not monophyletic. Apart from some errors and readjustments of poorly vouchered species, the classical morphological classification has remained rather stable. The rare species are in want of better study.

## ORIGIN AND DISTRIBUTION

The genus *Pueraria* has its species distributed over China and Japan, South and south-east Asia, and parts of Oceania. Japanese poems as early as AD 600 describe the use of kudzu (Shurtleff and Aoyagi, 1977) and the formidable list of vernacular names proves its long standing as a well known plant species. Bodner and Hymowitz (2002) describe the roads the man-dispersed species took.

*Figure 1.1 Pueraria* vine (*P. montana* var. *lobata*) on a grave in Yunnan, China.

Several species have a narrow distribution (Table 1.1), while kudzu is widely spread over China, Japan, and has been introduced ages ago to the highlands of New Guinea and New Caledonia. It became a staple food much before yams (*Dioscorea* spp.) were important, and before sweet potatoes (*Ipomoea batatas*) took over as a main crop there.

*Figure 1.2 Pueraria phaseoloides* var. *phaseoloides* in West Bengal.

Imported into the Americas it first served as an ornamental in the southern United States in 1876. Its use as a pasture became en vogue around 1910, while in the 1930s it was recommended for erosion control. Coverage was 3 million ha in 1950, but due to its rapid growth it smothered trees, houses and telephone poles, and further use as cover was discouraged. In the 1970s, the area had dwindled to 30 000 ha after discouraging notes from the agriculture authorities, and in 1981 kudzu was about to be declared a pest, because it affected e.g. 22 000 ha of commercial forest in the state of Georgia. It is a rather new plant invader in South Africa (Burrows, 1989), but does not act so aggressively in its home countries, or in Argentina, Zanzibar or Switzerland.

*Table 1.1* Geographical distribution of *Pueraria* species

| | |
|---|---|
| *P. alopecuroides* | China, Yunnan; Myanmar; Thailand |
| *P. bella* | India, Arunachal Pradesh; Myanmar |
| *P. calycina* | China, Yunnan |
| *P. candollei* | |
|    var. *candollei* | Bangladesh; India, Assam, Andamans; Myanmar, Thailand |
|    var. *mirifica* | Thailand, Myanmar? |
| *P. edulis* | Bhutan; China, Yunnan; India, Manipur, Sikkim |
| *P. imbricata* | Laos, Thailand |
| *P. lacei* | Myanmar |
| *P. montana* | |
|    var. *chinensis* | Bhutan, China, Hawaii, NE India, Laos, Philippines, Vietnam |
|    var. *lobata* | NE, SE Asia, Australia, Oceania, introduced to many subtropical areas |
|    var. *montana* | NE, E Asia |
| *P. peduncularis* | Bhutan, China, Yunnan, Szechuan, India, E. Himalaya, Khasi hills; Myanmar; Pakistan |
| *P. phaseoloides* | |
|    var. *javanica* | Brunei; Indonesia; Malaysia; Philippines; New Guinea; Solomon Islands; widely introduced in tropical areas |
|    var. *phaseoloides* | E, S, SE Asia; New Guinea, introduced in Africa and S. America |
|    var. *subspicata* | Bangladesh; Bhutan; NE India; Thailand |
| *P. pulcherrima* | E. Indonesia; Philippines; Papua New Guinea; Kei Islands, Solomon Islands |
| *P. sikkimensis* | Bhutan; India, Sikkim, W. Bengal |
| *P. stricta* | China, Yunnan; Myanmar; Thailand |
| *P. tuberosa* | India; Nepal; Pakistan |
| *P. wallichii* | Bangladesh; China, Yunnan; India, E. Himalaya |

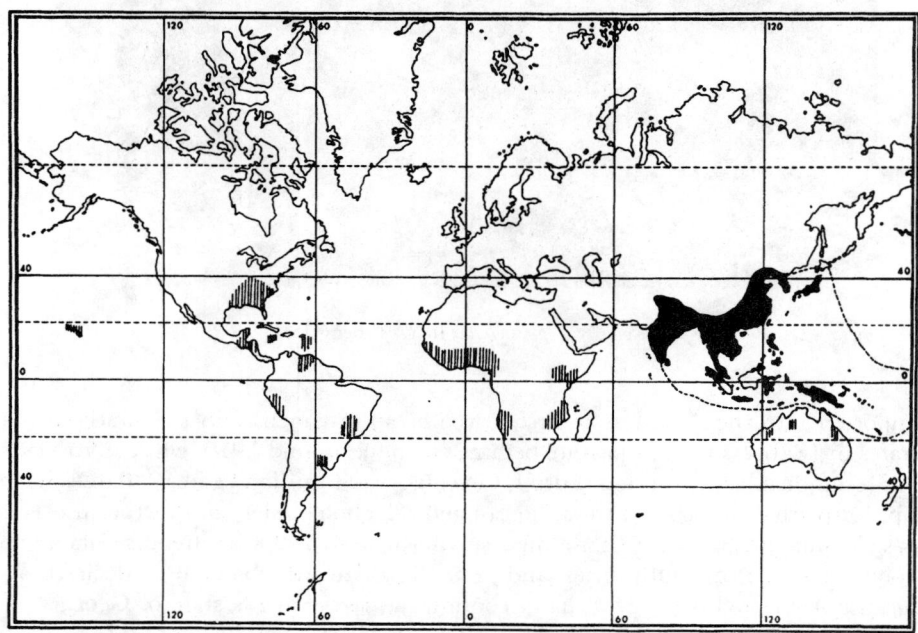

*Figure 1.3* Geographic distribution of *Pueraria* (Original area shown in black; introductions shaded).

Man has spread tropical kudzu all over the tropics, where it is an important cover crop also useful as forage (Figure 1.3). Together with *Calopogonium mucunoides* and *Centrosema pubescens* it forms a trio well appreciated in many areas for soil cover in plantation crops such as oil palm. It is usually the var. *javanica* that is used all around. Peculiarly, just as in kudzu, in new areas flowering and particularly seed set are much less abundant than in the countries of origin, despite its arrival several centuries ago.

## TAXONOMY OF *PUERARIA*

### Generic relationships and species

Relations with the legume family are clear at the tribal level: with the trifoliolate leaves *Pueraria* species are undoubtedly belonging in the tribe Phaseoleae. As far as subtribal level is concerned (Lackey, 1977), Glycininae are a quite natural group and combine species that have flowers adapted to bees, inflorescences not or scarcely nodose (sometimes branched in *Pueraria*) and seeds smooth, granular or shagreened, with short hilum. The genus *Glycine*, containing the soybean *Glycine max* (L.) Merr. and *Neonotonia wightii* (Wight and Arn.) Lackey [=*Glycine wightii* (Wight and Arn.) Verdc.], a cover crop species, are the only two taxa more widely known out of the *c.* 16 genera classified in the Glycininae.

    *Pueraria* now counts 15 species that can be arranged intuitively in three sections: *Pueraria*, with subsections *Pueraria*, subsection *Nonnudiflorae* Maesen and subsection *Pulcherrima* Maesen; the section *Schizophyllon* Baker; and *Breviramulae* Maesen (van der Maesen, 1985). Three species have subspecific taxa: *P. candollei* has two varieties, *P. montana* and *P. phaseoloides* have three varieties each. Zhang and Chen (1995) performed a cladistic analysis. The subsections *Pueraria* and *Pulcherrima* of section *Pueraria* forms a monophyly, the third subsection *Nonnudiflorae* is another monophyly excluding *P. imbricata, P. bella* and *P. alopecurioides*. Zhang and Chen (1995) found the sectional status of *P. phaseoloides* as the only species in sect. *Schizophyllon* unwarranted, in their cladogram this species grouped with *P. lobata* (now *P. montana* var. *lobata*). Lee (2000) detected at least four groups in the genus as hitherto perceived, on the basis of molecular work. Her data can be consolidated with the other ones in the near future.

### Taxonomic history

De Candolle (1825) established the genus *Pueraria* in his Mémoires sur la famille des Légumineuses. The name commemorates M.N. Puerari, a personal friend and Swiss national, who was a Professor at the University of Copenhagen. De Candolle described two species, *Pueraria tuberosum* and *P. wallichii*, and both remain in existence today. In 1831 some species were listed in Wallich's Catalogue, now considered *nomina nuda*, but that were used by Bentham and others. Bentham established the genus *Neustanthus*, to accommodate Roxburgh's *Dolichos phaseoloides*, the tropical kudzu. This new genus differed from *Pueraria* by the non-articulated pods, as *P. tuberosa* was purported to have articulations in the pods. Since the constrictions proved to indicate ovule abortion, Bentham combined the two genera (1867) with together nine species at the time. Several species have been named since, but also many names have fallen into synonymy when more material was collected, better representing the natural ranges of variation within species.

## Morphology

Kudzu and its relatives are usually strong climbers, even lianas, or rarely shrubs. Abundant in growth, most species have short-hairy stems that either ramble over the ground or wind on supports such as shrubs or trees, or man-made structures. Growth of kudzu itself can amount to 30 cm a day, to reach 18 to 30 m in a season. All species are perennial, while not for all species the nature of the roots has been established. Several have tuberous roots, including kudzu, tropical kudzu, *P. tuberosum*, *P. edulis*, and *P. mirifica*.

The trifoliolate leaves are typical for the tribe Phaseoleae, as are the general build of the papilionoid flower and the flat to cylindrical fruit, the pod. Stipules may or may not have basal lobes, which are single, twin-lobed or incised. The peduncles that carry the flowers bear short-shoots or thickened nodes called brachyblasts, that carry 3 (rarely 2) or 4–7 flowers per node. Flowers are mostly without a pedicel, have a bract that drops early, and the calyx has two lateral bracteoles. The calyx has five lobes, but the two uppermost are almost or entirely connate. The length of the calyx lobes vary among each other, the lowest usually the longest. The standard petal is clawed, and often has two auricles near the base of the blade, and a pair of bulges, callosities, is often discernible at the ventral (abaxial) side of the standard. Petal color varies from pure white to blue and purple; the standard often has a green or yellow patch. The stamens are combined into a tube of nine, with the tenth (vexillar) stamen connate to the tube, at least in the middle, in several species the vexillar stamen becomes free when the flowering proceeds. The ovary is elongate, often hairy, sometimes glabrous, with a thread-shaped style, ending in a globular stigma with short papillae, often penicillate below the knob. There are 5–20 ovules. The flowers are fragrant, attracting insects. Pods are most often flat-shaped, either glabrous or hairy, pale brown to black, and carry up to 20 seeds. The brown or black seeds are bean-like, flattened-oblong or barrel-shaped, and the surface is minutely shagreened. The hilum is surrounded by a small strophiole (rim-aril). Germination is epigeal, bringing the cotyledons aboveground. The first two leaves are simple and opposite, the following leaves are trifoliolate and alternate.

Characters that serve to subdivide the species into groups (or genera) are the number of flowers per brachyblast, the shape of stipules and seeds, bracts and pods.

## Habitat

Most species, as usual in climbing Leguminosae, require light, so they are found along forest edges, rivers, roadsides, or in scrub vegetation that allow sufficient insolation. A few, such as *P. peduncularis* tolerate dense shade. Kudzu and tropical kudzu are cultivated and adapted well to ruderal situations. Kudzu is at home in the cooler zones of temperate latitudes or higher altitude zones in the tropics. Most species prefer low-altitude wet evergreen or monsoon semi-deciduous forests. There is apparently no distinct preference for particular soil types, but the labels on the herbarium specimens quote but rarely the substrate. Tropical kudzu is adapted to a wide range of soil types: acidic, rocky mountaneous and fertile lowland soils (Whyte *et al.*, 1953; Skerman, 1977; Duke, 1981). Kudzu prefers a well-drained soil, the pH might range from 5.0 to 7.1 (Duke, 1981). *P. tuberosa* is supported by eroded and exposed soils (van der Maesen, 1985).

## Key to the species and varieties of *Pueraria*

1a. Erect shrubs or rigid stragglers. Flowers 2–10 together on short laterals (brachy-blasts), short-pedicelled, inflorescence sometimes branched ..............................2

1b. Vines, herbaceous to rather lianescent climbers................................................3

2a. Corolla 3–5 times as long as the calyx, calyx lobes short-obtuse, bracts caducous. Erect shrub, rarely straggling branches........................................15. *P. wallichii*

2b. Corolla about twice as long as the calyx, calyx lobes acute, upper ones obtuse; bracts soft, becoming hard and hooked in fruiting stage. Woody climber......................................................................................... 13. *P. stricta*

3a. Corolla 2–3 times as long as the calyx, slender and long-pedicelled. Flowers not crowded, 4–7 per node; pods flat, papery, inflorescences unbranched, 1–2 per axil............................................................................................ 9. *P. peduncularis*

3b. Flower pedicels short, not very slender ....................................................................4

4a. Flowers 2–3 per node, stipules peltate, upper calyx almost or completely united; pods flat ..........................................................................................................................7

4b. Flowers 4 or more per node, stipules not peltate, upper calyx teeth distinct; pods rounded, ca 20 barrel-shaped seeds per pod, papery partitions between the seeds........................................................................... 10. *P. phaseoloides:* 5

5a. Flowers small, corolla 7–13 (–15) mm long; bracts and calyx more or less pubescent, hairs not very long, lateral calyx lobes acute–acuminate, lower calyx lobe acuminate–lanceolate; leaflets entire, lobed or sinuate; fruit 5–9 cm long, 3–4 mm wide .................................................................................10b. var. *phaseoloides*

5b. Flowers larger, corolla 15–23 mm long................................................................6

6a. Bracts and calyx pubescent, lateral calyx lobes obtuse, lower calyx lobe acute; leaflets mostly entire, rarely somewhat lobed; pods 7–11 cm long, 4–5 mm wide ................................................................................................. 10a. var. *javanica*

6b. Bracts and calyx densely long-pubescent, lateral calyx lobes acute–acuminate, lower calyx lobe lanceolate–subulate; leaflets large, entire to usually deeply lobed; pods 7–12.5 cm long, 4–5 mm wide ..................................... 10c. var. *subspicata*

7a. Flowers less than 7 (–12) mm long, fruits hairy mainly on the sutures. Leaflets entire, lanceolate, stipules large, amplexicaul, covering buds, caducous, leaving a line scar................................................................................... 11. *P. pulcherrima*

7b. Flowers more than (10–) 12 mm long; fruits, if hairy, both on sutures and sides. Leaflets often lobed, stipules present, if shed, leaving an oval scar....................8

8a. Pseudoracemes unbranched or with 1–2 branches................................................9

8b. Pseudoracemes branched and/or with stipule-like bracts on the lower portion 13

9a. Stipellae four near petioles of side leaflets; leaflets always prominently lobed; upper calyx lobes minutely distinct ....................................................... 5. *P. edulis*

9b. Stipellae two near petioles of side leaflets, leaflets lobed or not, upper calyx lobes united or minutely distinct ...........................................................................10

10a. Leaflets long-elliptic, not lobed; calyx lobes obtuse ............................2. *P. bella*

10b. Leaflets ovate to orbicular, trilobed or not; calyx lobes acuminate 8. *P. montana*: 11

11a. Leaflets trilobed, sometimes entire, about as wide as long, flowers large, 18 mm or longer, calyx (15–) 17 mm or more, densely grey-pubescent, lobes 2–4 mm wide, imbricate, pods 8–13 × 0.9–1.3 cm.................................. 8a. var. *chinensis*

11b. Leaflets entire or trilobed, flowers smaller......................................................12

12a.   Leaflets entire, often longer than wide, coriaceous, flowers up to 12 (–15) mm long, calyx 8–11 mm long, lobes 1–2 mm wide, short brown pubescent or almost glabrous, pods small, 4–10×0.6–0.9 cm..........................8c. var. *montana*

12b.   Leaflets trilobed, occasionally entire, about as long as wide, membranaceous, pods 5–13×0.7–1.2 cm ................................................................. 8b. var. *lobata*

13a.   Vine densely woolly-pubescent; calyx teeth long .............................................14

13b.   Vine slightly or short-pubescent; calyx teeth at most twice as long as the tube ............................................................................................................15

14a.   Leaflets 5–7 cuspidate; flowers 2 per node; calyx teeth 3–4 times as long as the tube ................................................................................... 3. *P. calycina*

14b.   Leaflets ovate–rhomboid; flowers 3 per node; lower calyx tooth 4–6 times as long as the tube.................................................................................. 7. *P. lacei*

15a.   Inflorescence with few branches, foxtail-like; flowers dense, bracts long, 10–15 mm, and long-hairy, 3 mm, caducous, bracteoles as long as the calyx tube, flowering when in leaves......................................................... 1. *P. alopecuroides*

15b.   Inflorescence more or less copiously branched; bracts up to 10 mm, hairs to 2 mm, caducous, bracteoles half as long as the calyx tube or less15

16a.   Flowering when in leaves, or on old branches without leaves; calyx lobes lanceolate–acute, imbricate ......................................................... 6. *P. imbricata*

16b.   Flowering when leafless; calyx lobes deltoid–obtuse .......................................17

17a.   Inflorescence crowded, rusty pubescent ................................... 12. *P. sikkimensis*

17b.   Inflorescence long and lax, yellow-brown pubescent or glabrous.....................18

18a.   Vexillary stamen free, calyx yellow-brown or greyish pubescent; pods copiously hairy ...................................................................................... 14. *P. tuberosa*

18b.   Vexillary stamen united with staminal tube; calyx grey or brown appressed-pubescent; pods almost glabrous ............................................. 4. *P. candollei* 19

19a.   Inflorescence often more than 30 cm long; calyx purplish green, sparsely pubescent; flowers 12–15 mm long..................................................... 4a. var. *candollei*

19b.   Inflorescence up to 30 cm long; calyx quite densely short-pubescent; flowers 8–10 mm long........................................................................... 4b. var. *mirifica*

## Enumeration of *Pueraria* species

The species are enumerated in alphabetical order, with their synonymy, with the corrections necessary since the 1985 monograph and 1988 and 1994 updates. The main features of the species are highlighted, for a full description refer to van der Maesen (1985). It is usual in taxonomy to place the abbreviated references under each name and synonym, and these are not necessarily repeated in the reference list at the end of the chapter.

### *Pueraria alopecuroides* Craib

Kew Bull. 1910: 276; Craib, Kew Bull. 1911: 40; Craib, Contrib. Fl. Siam, Aberdeen Univ. Stud. 57: 64 (1912); Gagnepain, *Fl. Gén.* Indo–Chine 2: 255 (1916); Craib, Fl. Siam. Enum. 1–3: 448 (1928); Handel-Mazzetti, Symb. Sinicae 582 (1933); Lackey, Synops. Phaseol. 72, 75 (1977); van der Maesen, Agric. Univ. Wageningen Papers 85-1: 15 (1985).

*Lectotype*: Upper Burma, Shan Hills, Gokteik 600 m, *Meebold 8085* (K, holo, designated by Craib).

*Paratype*: China, S. Yunnan, Szemao to Chenlung, Mengwan?, *Bons d'Arty 255* (K).

Strong climber with foxtail-like inflorescences and narrow hairy pods, flowering January–April, fruits in April.

*Distribution*: Upper Mianmar (Burma), Yunnan-China and Thailand.

*Habitat*: Climbing in mixed jungle or among grasses, 100–600 m.


### *Pueraria bella* Prain

*J. Asiatic Soc.* Bengal 67: 288 (1898); Pottinger and Prain, Rec. Bot. Surv. India 1–11: 239 (1898); Burkill, Rec. Bot. Surv. India 10-2: 271 (1925); Lackey, Synops. Phaseol. 73, 76 (1977); van der Maesen, Agric. Univ. Wageningen Papers 85-1: 18–20 (1985).

*Type*: Burma, Kachin Mts nr Myitkyina, *King's collectors (Shaik Mokim) s.n.* (CAL, holo; iso: CAL, K).

Woody climber with glabrous or slightly hairy branches, inflorescences to 35 cm long, with many beautiful white and pale violet flowers, appearing in August, December, pods unknown.

*Distribution*: rare species in Burma and India, Arunachal Pradesh.

*Habitat*: In hills, sprawling over boulders in river bed, 200–1000 m.


### *Pueraria calycina* Franchet

Pl. Delav. 181 (1890); Léveillé, Bull. Soc. Bot. France sér. 4 vol. 8,55: 426 (1908); Gagnepain, Fl. Gén. Indo–Chine 2: 249 (1916); Lackey, Synops. Phaseol. 71, 73 (1977); van der Maesen, Agric. Univ. Wageningen Papers 85-1: 20–22 (1985).

*Type*: China, Yunnan, calcareous hills along Long-teou-chan river above Hee-gni-tang nr Tapintze, *Delavaye 3590* (P, holo; iso: F, K).

*Heterotypic synonym*: *Pueraria forrestii* Evans, Notes Roy. Bot. Gdn Edinburgh 13: 178 (1921); Lackey, Synops. Phaseol. 71, 73 (1977).

*Type*: China, Yunnan, descent to Yangtze river from E boundary of Lichiang valley, *Forrest 10732* (E, holo; iso: BM, E, K).

*Paratypes*: China, Yunnan, Yung-Peh Mts, *Forrest 15312* (CAL, E, K).

Woody pubescent climber, leaves with four stipellae at insertion of the side-leaflets, long-pubescent unbranched inflorescences, broad (8–12 mm) flat papery pods with spreading golden-brown hairs. Flowering and fruiting July–August.

*Distribution*: China, Yunnan, rare.

*Habitat*: Climbing on scrub, calcareous slopes, open dry situations and near rivers, 2000–2600 m.


### *Pueraria candollei* Grah. ex Benth. var. *candollei*

In Miquel, Pl. Jungh. 2: 235 (1852) (as *P. candollei* Wall.); Benth., Bot, J. Linn. Soc., London 9: 123 (1867); Baker in Hooker, Fl. Brit. India 2: 197 (1876); Kurz, J. Asiatic

Soc. Bengal 45-4: 253 (1876); Collett and Hemsley, Bot. J. Linn. Soc., London 28: 47 (1890); Prain, J. Asiatic Soc. Bengal 66-2: 419 (1897); Prain, Bengal Pl. 1: 282 (1903, repr. 1963); Craib, Kew Bull. 1911: 40; Craib, Contrib. Fl. Siam, Aberdeen Univ. Stud. 57: 64 (1912); Gagnepain, Fl. Gén. Indo–Chine 2: 255 (1916); Lackey, Synops. Phaseol. 72, 74 (1977); van der Maesen, Agric. Wageningen Papers 85-1: 23–27 (1985).

*Type*: Myanmar, Pegu, Salween river nr Phanoe, *Graham in Wallich Herb. 5355* (K, holo; iso: BM, G, K).

*Paratypes*: Myanmar, Pegu, *Lobb s.n.* (K); and *McLelland s.n.* (K).

Strong woody climber, sparsely hairy or glabrous, with large leaflets and long inflorescences to 80 cm, many mediumsized flowers and rather glabrous pods of *c.* 8 cm long and 1 cm wide. Tubers have not been found but can be expected. Flowering and fruiting ( January–) February–April, in Thailand also in September–October.

*Distribution*: Bangladesh, India (the Andamans, Assam), Myanmar, Thailand.

*Habitat*: Climbing in deciduous scrub jungle, on limestone rocks, open situations, stream or lake banks, 0–1300 m.

*Vernacular names*: Thailand: Kua ta lan, Kua kao pu (N. Lao), Khampi noi (NE), Kwao khrua (N); Myanmar: Ma U Nwè.


**Pueraria candollei** Grah. ex Benth. var. *mirifica* (A. Shaw and Suvatabandhu) Niyomdham

Nordic J. Bot. 12, 3: 339–346 (1992).

*Basionym*: *P. mirifica* A. Shaw and Suvatabandhu, Kew Bull. 1952: 550; Cain, Nature 198: 774–777 (1960); Lackey, Synops. Phaseol. 72, 74 (1977); Perry, Medic. Pl. SE Asia 225 (1980); van der Maesen, Agric. Univ. Wageningen Papers 85-1: 63–64 (1985).

*Type*: Thailand, Doi Sutep nr Chiengmai, 300–800 m, *Suvatabandhu s.n.* (K, lecto; isolecto: K).

Woody climber similar to *P. candollei* but leaflets with greyish lower surface, inflorescence up to 30 cm, flowers 8–10 mm long and calyx densely pubescent. Flowering February–March, fruiting April.

*Distribution*: Thailand.

*Habitat*: deciduous forest, hill slopes 300–800 m.

*Vernacular names*: Thailand: Kwao keur, Kwao kua (very similar to names of var. *candollei*).

*Uses*: The tubers have been used locally as a rejuvenating medicine (Kerr, 1932; Phya Winit, 1933).

The effect was probably due to the presence of estrogenic substances (Vanijvatama, 1938). Chemical and physical properties were determined in the 1960s by Thai and British scientists (Cain, 1960). The estrogen was called "miroestrol." Its effects were marked, but the side effects were rather disagreeable. In Thailand the local crude preparations of the dried roots are taken with honey.

Some likeness of the old leaves and flowers to those of *Butea superba* Roxb., also a Leguminosae, made at times people refere the medicinal plant to that (much more) common species.

### *Pueraria edulis* Pamp.

Nuov. Giorn. Bot. Ital. n.s. 17: 28 (1910); Gagnepain, Fl. Gén. Indo–Chine 2: 249 (Apr. 1916); Gagnepain, Lecomte Not. Syst. 3: 202, 204 (June 1916); Handel-Mazzetti, Symb. Sinic. 582 (1933); Lackey, Synops. Phaseol. 71, 74 (1977) (as *edulus*); Anon., Icon. Corm. Sinic. (1980); van der Maesen, Agric. Univ. Wageningen Papers 85-1: 27–32 (1985).

*Type*: China, Yunnan, mountain forests near Yunnan-sen, *Maire 100* (FI, holo; iso: P).

*Synonyms*: *Pueraria quadristipellata* Clarke ex W.W. Smith, Rec. Bot. Surv. India 6-2: 36 (1913).

*Type*: India, Sikkim, Yoksun, 2600 m, *C.B. Clarke 25147* (K, holo; iso: BM, CAL, K).

Woody climber with tuberous roots and deeply lobed leaflets, stipels 4, ovate, near side leaflets, 2 near top leaflet, inflorescence to 30 (–75) cm long, and sparsely hairy flattened-oblong pods. Pods are edible, but only Maire and Forrest indicated this use (in Herb.). Flowering between July and October, fruiting up to November.

*Distribution*: Bhutan, China (Yunnan), India (Manipur and Sikkim).

*Habitat*: Climbing over dwarf bushes or oak trees, on hill slopes, in forests, near streams, on schistaceous, sandy and rocky soils, 1300–3300 m.

### *Pueraria imbricata* Maesen

Agric. Univ. Wageningen Papers 85-1: 32–34 (1985).

*Type*: Laos, Sam Neua Prov., between Muong Pun and Xieng Mene, *Poilane 1931* (P, holo; iso: BKF).

*Synonym*: *P. maesenii* Nyomdham, Nord. J. Bot. 12-3: 344–345 (1992).

*Type*: Thailand. Mae Hong Son, Khun Yuam, *Larsen & Larsen 34073* (AAU, holo; iso: BK, K, L; the paratype of *P. imbricata* Maesen).

Climber with golden-brown pubescent leaflets, branched inflorescences up to 47 cm, calyx with imbricate teeth, flowering in September, December. Pods unknown. Related to *Pueraria montana* var. *lobata*, but inforescences branched as in *P. candollei*.

*Distribution*: Laos, Thailand.

*Habitat*: trailing and twining in shrubs, 600–700 m.

### *Pueraria lacei* Craib

Kew Bull. 1915: 399; Lackey, Synops. Phaseol. 72, 75 (1977); van der Maesen, Agric. Univ. Wageningen Papers 85-1: 35–37 (1985).

*Type*: Burma, Shandatgyi, Thayetmyo distr., *Lace 2685* (K, holo; iso: CAL, E).

Woody climber with densely rusty-pubescent hairs, hairy inflorescences of *c.* 20 cm, with stipule-like bracts at the base, branched at the top or not at all, calyx hairy, pods unknown. Flowering in December.

*Distribution*: Myanmar (Burma), *c.* 300 m.

*Habitat*: Not recorded, presumably climbing in shrubs and trees.

### Pueraria montana (Lour.) Merr.

Merr., Trans. Am. Philos. Soc. n.s. 24-2: 210 (1935). For nomenclature and synonymy see the three varieties *montana*, *lobata* and *chinensis* (=*thomsonii*).

### Pueraria montana (Lour.) Merr. var. *montana*

As species: Merrill, Trans. Am. Philos. Soc. n.s. 24-2: 210 (1935); Tanaka's Cyclop. Edible Pl. World 602 (1976); Ohashi, Fl. Taiwan 3: 367 (1977); Lackey, Synops. Phaseol. 72, 74 (1977); Thuan, Fl. Cambodge, Laos, Vietnam 17: 80 (1979); Iconogr. Cormoph. Sinic. 2: 501 (1980); Ohashi *et al.*, Sci. Rep. Tohoku Univ. 4th ser. Biol. 39: 232–236 (1988); Maesen in Proc. 1st Int. Symp. Tuberous Legumes, Guadeloupe, F.W.I, 21–24 April 1992: 66 (1994).

*Basionym*: *Dolichos montanus* Loureiro, Fl. cochinchin. 440 (1790); id. ed. Willd. 536 (1793).

*Type*: Vietnam, Cochinchina, in mountain forests, *Loureiro s.n.* (P, holo).

*Homotypic synonyms*: *Pachyrhizus montanus* (Lour.) DC., Prodr. 2: 402 (1825). *Stizolobium montanum* (Lour.) Spreng., Syst. 3: 352 (1826). *Pueraria lobata* (Willd.) Ohwi var. *montana* (Lour.) Maesen, illegitimate name, Agric. Wageningen Univ. Papers 85-1: 53–58 (1985).

*Heterotypic synonyms*: *Zeydora agrestis* Lour. ex Gomes, Mem. Acad. Sci. Lisb. Pol. Mor. Bel.-Let. n.s. 4-1: 27 (1868), according to Merrill 1935.

*Type*: Not seen.

   *Pueraria tonkinensis* Gagnepain in Lecomte, Not. Syst. 3: 202 (June 1916); id. Fl. Gén. Indo–Chine 2: 250 (April 1916); Merrill, Trans. Am. Soc. Philos. Soc. n.s. 24-2: 210 (1935); Ohwi, Acta Phytotax. Geobot. 5: 63 (1936); Chuang and Huang, Legumin. Taiwan 88 (1966).

*Type*: Vietnam, Tonkin, from Lang Son to Nuoc Binh, *Lecomte & Finet 183* (P, lecto).

   *Pueraria thunbergiana* (Sieb. and Zucc.) Benth. var. *formosana* Hosokawa, J. Soc. Trop. Agr. Taih. 4: 310 (1932).

*Type*: Taiwan, Urai, pref. Taihoku, *Suzuki 3297* (holo: TAI).

   *Pueraria omeiensis* Wang and Tang, *nomen nudum.*, probably in Illust. Guidebook Leguminosae (1955); Iconogr. Cormoph. Sinic. 2: 501 (1980).

### Pueraria montana (Lour.) Merr. var. *lobata* (Willd.) Maesen and Almeida ex Sanjappa and Predeep

Legumes of India: 288 (1992).

*Basionym*: *Dolichos lobatus* Willd., Sp. Pl. 3-2: 1047 (1802).

*Homotypic synonyms*: *Dolichos trilobus* Houtt. non L., Nat. Hist. 10: 153, t.64 f.1 (1779).

*Type*: Japan, "Kudsu," *Thunberg 16757* (UPS, holo). CHECK

*Dolichos hirsutus* Thunb., Trans. Linn. Soc. 2: 339 (1794).

*Pueraria hirsuta* (Thunb.) Matsumura non Kurz, Bot. Mag. Tokyo 16: 33 (1902).

*Basionym*: *Dolichos hirsutus* Thunb.

*Pueraria hirsuta* (Thunb.) Schneid., Illustr. Handbuch Laubholzkunde 2: 114–115 (1912).

*Basionym*: *Dolichos hirsutus* Thunb.

*Pachyrhizus thunbergianus* Siebold and Zuccharini, Abh. Acad. München 4-3: 237 (1846), based on *Dolichos hirsutus* Thunb.

*Pueraria thunbergiana* (Sieb. and Zucc.) Benth., J. Linn. Soc. Bot. 9: 122 (1867); Taubert in Engl. and Prantl, Nat. Pflzfam. 3-3: 371 (1894); Prain, J. Asiatic Soc. Bengal 66: 419 (1897); Matsumura, Tent. Fl. Lutchuensis 426 (1899), J. Coll. Sci. Imp. Univ. Tokyo 12-4 (1900); Léveillé, Bull. Soc. Bot. France sér. 4, 8-55: 425 (1908); Merill, Philipp. J. Sci. Bot. 5: 123 (1910); id., Fl. Manila 253 (1912); Gagnepain, Fl. Gén. Indo–Chine 2: 249 (1916); Degener, Fl Hawaii 2: 10-12 (1934); Merrill, Trans. Philos. Soc. n.s. 24-2: 211 (1935); Ohwi, Acta Phytotax. Geobot. 5: 62 (1936); Merrill, J. Arnold Arb. 19: 348 (1938); Kanjilal *et al*. Fl. Assam 2: 81 (1938); Guillaumin, Fl. Anal. Synopt. Nouv. Calédonie 149 (1950); Burkart, Leguminosas Argentinas ed. 2: 407 (1952); Merrill, Chron. Bot. 14-5/6: 218 (1954); Li, Woody F. Taiwan 359 (1963); Purseglove, Trop. Crops Dicot. 220 (1968); Chun W.Y. and C.C. Chang, Fl. Hainanica 2: 319 (1965).

*Basionym*: *Pachyrhizus thunbergianus* Sieb. and Zucc.

*Pueraria montana* var. *lobata* (Ohwi) Maesen and Almeida, Bombay J. Nat. Hist. Soc. 85-1: 233 (1988), Maesen in Proc. 1st Int. Symp. Tuberous Legumes, Guadeloupe, F.W.I, 21–24 April 1992: 67 (1994).

*Heterotypic synonyms*: *Phaseolus trilobus* (L.) Ait., Hort. Kew. 3: 30 (1789); id. ed. 2, 4: 290 (1812), based on *Dolichos trilobus* L., Sp. Pl. 1021 (1753) and Hort. Kew. cult. ex East Indies, introd. Joseph Banks 1777, no specimen found.

*Neustanthus chinensis* Benth., Fl. Hongkong 86 (1861).

*Type*: Hongkong, Harland (K, holo, not seen).

*Dioclea odorata* Montrouzier, Fl. de l'Ile Art, Mém. Acad. Sci. Bell.-Lett.-Arts Lyon 2: 173–254 (1860) teste Barrau, Ethnology 4: 283 (1965). Herb. *Montrouzier* at Lyon, (LY, not seen).

*Pueraria novo-guineensis* Warburg, Engl. Bot. Jahrb. 13: 235 (1891); Taubert in Engl. and Prantl, Nat. Pflanzenfam. 3-3: 371 (1894); Schumann and Lauterbach, Fl. deutsch. Schutzgeb. Südsee 1: 368 (1901); Baker, Trans. Linn. Soc. Bot. 9: 34 (1916); Hosokawa, Trans. Nat. Hist. Soc. 28: 62 (1938).

*Type*: New Guinea, *Warburg s.n.?* (not seen).

*Paratype*: New Guinea, *Hollrung 231* (not seen).

*Pueraria neo-caledonica* Harms, Engl. Bot. Jahrb. 39: 136 (1906); Guillaumin, Fl. Anal. Synopt. Nouv. Calédonie 149 (1948).

*Type*: New Caledonia, N. area, Mts near Oubatche, *Schlechter 15484* (B, holo? not seen, iso: BM, G, K, W).

*Pueraria argyi* Lévl. and Vaniot, Bull. Soc. Bot. France 55: 426 (1908); Lauener, Notes Roy. Bot. Gdn Edinburgh 30: 239 (1970).

*Type*: China, Kiangsu/Jiangsu, *d'Argy 51* (E, holo; iso: E, *d'Argy 52*).

*Pueraria bodinieri* Lévl. and Vaniot, Bull. Soc. Bot. France 55: 425 (1908); Léveillé, Fl. Kouy-tcheou 214 (1914); Gagnepain in Lecomte Not. Syst. 3: 205 (1916); Lauener, Notes Roy. Bot. Gdn Edinburgh 30: 239 (1970).

*Type*: China, Kouy-tcheou (Kweichow prov.), College Hill, environs of Kweiyang, *Bodinier 2489* (E, holo; iso: P).

*Pueraria caerulea* Lévl. and Vaniot, Bull. Soc. Bot. France 55: 427 (1908); Gagnepain in Lecomte, Not. Syst. 3: 205 (1916); Lauener, Notes Roy. Bot. Gdn Edinburgh 30: 239 (1970).

*Type*: Hongkong, Chay-Ouan (Wanchai?) bay, *Bodinier 239* (E, holo; iso: P).

*Pueraria koten* Lévl. and Vaniot, Bull. Soc. Bot. France 55: 426 (1908); Gagnepain in Lecomte, Not. Syst. 3: 205 (1916); Lauener, Notes Roy. Bot. Gdn Edinburgh 30: 239 (1970).

*Type*: China, Shantung prov., Tche-fou, *Bodinier 239* (E, holo).

*Pueraria harmsii* Rech., Denkschr. Akad. Wiss. Wien, Math.-Nat. 85: 292 (1910).

*Type*: Samoa Islands, Upolu Island, near Motootua, *Rechinger 78* (W, holo).

*Paratype*: Samoa, Apolima Island, *Rechinger 180* (W, holo).

*Pueraria triloba* (Houtt.) Makino in Iinuma, Somoku-Dzusetsu ed. 3 fasc. 3: 954; vol. 13 t. 22 (1912); Verdcourt, Taxon 17: 170–173 (1968). *Pueraria triloba* Makino, Merrill, J. Arn. Arb. 19-4: 348 (1938); Makino, Illustr. Fl. Japan 401 (1949). *Pueraria triloba* sensu Makino non (L.) Makino, Verdcourt, Manual New Guinea Legumes 485 (1979), based on *Dolichos trilobus* Houtt. non L., see above, not a synonym to *Pueraria thomsoni* as quoted by Merrill, 1935 and Mansfeld, 1959.

*Dolichos japonicus* Hort., *nomen nudum*, Bailey, Manual Cult. Pl. 400 (1924) (as *japonica*); Borisov, Fl. USSR 13: 531 (1948 and transl. 1972) as synonym.

*Pueraria triloba* Backer in Heyne, Nuttige Pl. Nederl. Indië 829 (1927), based on *Pachyrhizus trilobus* (Lour.) DC. and *Pueraria thunbergiana* Benth. (should have been (Lour.) Backer).

*Pueraria volkensii* Hosokawa, Trans. Nat. Hist. Soc. Formosa 28: 62 (1938).

*Type*: Yap, at Datyakal, *Hosokawa 8885* (TAI, holo, not seen).

*Pueraria triloba* (Lour.) Makino ex Backer, Fl. Java 1: 632 (1963), based on *P. lobata* (Willd.) Ohwi and *P. thunbergiana* (Sieb. and Zucc.) Benth., obviously citation hybrid of *P. triloba* (L.) Makino 1912 and *Dolichos trilobus* Lour. (1790).

*Pueraria pseudo-hirsuta* Tang and Wang, *nomen nudum*, Hu and Hsun, Native Forage Plants (1955).

*Pueraria lobata* (Willd.) Ohwi subsp. *lobata*, Ohashi, Tateishi and Endo, Sci. Rep. Tohoku Univ. 4th ser. (Biol.) 39: 232–235 (1988).


**Pueraria montana** (Lour.) Merr. var. *chinensis* (Ohwi) Maesen et Almeida
ex Sanjappa and Predeep

Legumes of India: 288 (1992).

*Basionym*: *Pueraria lobata* (Willd.) Ohwi var. *chinensis* Ohwi, Bull. Tokyo Sci. Museum 18: 16 (1947). Based on *Pueraria chinensis* sensu Ohwi, non *Neustanthus chinensis* Benth., Acta Phytotax. Geobot. 5: 63 (1936); Kubo *et al.*, Syoyakugahu Zasshi 31-2: 136–144 (1977); Tahi *et al.*, Syoyakugahu Zasshi 31-2: 145–150 (1977).

*Table 1.2* Nomenclature of Kudzu, correct names for three opinions

| Three species | P. montana<br>(Lour.) Merr. | P. lobata<br>(Willd.) Ohwi | P. thomsonii<br>Benth. |
|---|---|---|---|
| Three varieties<br>(van der Maesen) | P. montana var.<br>montana | P. montana var.<br>lobata (Willd.)<br>Maesen and Almeida<br>ex Sanjappa and Predeep | P. montana var.<br>chinensis (Ohwi)<br>Maesen and Almeida<br>ex Sanjappa and Predeep |
| Two species<br>(Ohashi *et al.*) | P. montana<br>(Lour.) Merr. | P. lobata subsp.<br>lobata | P. lobata subsp.<br>thomsonii (Benth.)<br>Ohashi and Tateishi |

*Type*: China, Hainan, Kwangtung/Guangdong prov., Ngai distr., Wong Kam Shan, *S.K. Lau 552* (KWA, hololecto; isolecto: NY, US, W).

*Homotypic synonym*: *Pueraria montana* (Lour.) Merr. var. *chinensis* (Ohwi) Maesen et Almeida Bombay J. Nat. Hist. Soc. 85-1: 233 (1988), Maesen in Proc. 1st Int. Symp. Tuberous Legumes, Guadeloupe, F.W.I, 21–24 April 1992: 70 (1994).

*Heterotypic synonyms*: *Dolichos trilobus* Lour., Fl. Cochinchin. 439 (1790) not of Linnaeus (1753), according to Merrill (1935).

*Type*: cultivated in Cochinchina and China, *Loureiro* (P?).

> *Pachyrhizus trilobus* (Lour.) DC., Prodr. 2: 402 (1825), based on *Dolichos trilobus* Lour. *Dolichos grandifolius* Grah. ex Wall., *nomen nudum*, Wall. Cat. 5556 (1831).

*Type*: Hongkong, *Harland s.n.* (K, holo, not seen).

> *Pueraria thomsoni* Benth., J. Linn. Soc. Bot. 9: 122 (1867); Baker in Hooker, Fl. Brit. India 2: 198 (1976); Taubert in Engl., Nat. Pflanzenfam. 3-3: 371 (1894); Prain, J. Asiat. Soc. Bengal 66-2: 419–420 (1897); Gagnepain, Fl. Gén. Indo–Chine 2: 251 (1916) and in Lecomte, Not. Syst. 3: 203 (1916); Craib, Fl. Siam. Enum. 1-3: 451 (1928); Merrill, Trans. Am. Philos. Soc. n.s. 24-2: 211 (1935); Wealth of India 8: 315–316 (1969); Tanaka's Cyclop. Edible Pl. World 602 (1979); Iconogr. Cormoph. Sinic. 2: 502 (1980).

*Type*: India, Meghalaya, Khasia regio temp. 5–7000 feet, *Hooker & Thomson s.n.* (K, holo; iso: C, CAL, FI, G, L, NY, OXF, P, U, US, W).

Note: The previous three taxa are here considered as varieties. The characteristics are insufficient to recognize these entities as two or three species. Ohashi *et al.* (1988) specify the distinctive characteristics, that he finds sufficient to classify the taxa at species level. The general facies (habit) of the taxa are quite unambiguous, but insufficient to warrant separation at species level. The intermediate character states, e.g. size of flower parts are linking the varieties rather than separating them. The correct names for the three taxa, depending on the opinions how to rank the taxa, are given in Table 1.2.

### *Pueraria peduncularis* (Grah. ex Benth.) Benth.

J. Linn. Soc. Bot. London 9: 124 (1867); Baker in Hooker, Fl. Brit. India 2: 167 (1876); Taubert in Engler and Prantl, Natürl. Pflanzenfam. 3-3: 371 (1874); Pampanini, Nuovo Giorn. Bot. Ital. n.s. 17: 29 (1910); Smith, Rec. Bot. Surv. India 4-5: 187 (1916); Handel-Mazetti, Symb. Sinicae 581 (1933); Kanjilal *et al.*, Fl. Assam 2:

80 (1938); Rao and Joseph, Bull. Bot. Surv. India: 144 (1965); Biswas, Pl. Darjeeling Sikkim Himalayas 1: 291 (1966); Hara, Fl. E. Himalaya 162 (1966); Thothatri, Rec. Bot. Surv. India 20-2: 81 (1973); Lackey, Synops. Phaseol. 76 (1977); Hara and Williams, Enum. Fl. Pl. Nepal 2: 218 (1979); Thuan, Fl. Cambodge, Laos, Vietnam 17: 84, 86 (1979); Iconogr. Cormoph. Sinic. 2: 503 (1980); van der Maesen, Agric. Univ. Wageningen Papers 85-1: 64–71 (1985).

*Basionym*: *Neustanthus peduncularis* Grah. ex Benth. in Miquel, Pl. Jungh. 2: 235 (1852); Miquel, Fl. Ind. Bat. 1-1: 219 (1855).

*Type*: Nepal, *Graham, Wallich's Herb. 5354* (K, holo; iso: BM, CAL, G, K) as *Pueraria? peduncularis* Grah. ex Wall., *nomen nudum*.

*Heterotypic synonyms*: *Pueraria yunnanensis* Franchet, Pl. Delav. 181 (1890); Gagnepain, Fl. Gén. Indo–Chine 2: 249 (1916); and in Lecomte, Not. Syst. 3: 205 (1916); Handel-Mazzétti, Symb. Sinicae 581 (1933); Lackey, Synops. Phaseol. 76 (1977); Perry, Medicinal Pl. E and SE Asia 225 (1980).

*Type*: China, Yunnan, woods near Tapintze, *Delavaye 506* (P, holo; iso: P).

*Pueraria peduncularis* (Grah. ex Benth.) Benth. var. *violacea* Franchet, Pl. Delav. 182 (1890).

*Type*: China, Yunnan, woods of Hoang-li-pin, above Tapintze, *Delavaye 1983* (P, lecto; iso: K).

*Paratypes*: China, Yunnan, above Chaong-che-teou, near Tapintze, *Delavaye 3588* (P, US); woods of Ta-long-tan, *Delavaye 3567* (K, P, US).

*Derris bonatiana Pampanini*, Nuov. Giorn. Bot. Ital. 17: 8 (1910); Gagnepain, Lecomte Not. Syst. 3: 205 (1916).

*Type*: China, Yunnan-sen, source of the Pe-long-tan river, *Ducloux 377* (F, lecto, not seen).

*Paratype*: China, *Maire 210* (FI, not seen).

Woody climber, ovate to rhomboid leaflets, inflorescence single or paired, to 40 (–68) cm, pedicels slender to 13 mm, flowers white to blue in various tinges, pods papery, 5–7 × 0.5–0.7 (–1) cm. Flowering April to October, fruiting April to November.

*Distribution*: Bhutan, China, India, Myanmar, Pakistan.

*Habitat*: Climbing or pendant on shrubs and trees, in open or shade, medium wet forest, hill slopes or ravines, along jungle edges, bamboo forests, hill tops, on well-drained or sandy soils, scree and rocks, 1200–3600 m.

**Pueraria phaseoloides** (Roxb.) Benth.

J. Linn. Soc. Bot. 9: 125 (1867). For synonyms and typification see varieties *phaseoloides*, *javanica* and *subspicata*.

**Pueraria phaseoloides** (Roxb.) Benth. var. *javanica* (Beneth.) Baker

Benth. var. javanica (Benth.) Baker in Hooker, Fl. Brit. India 2: 199 (1876); Pottinger and Prain, Rec. Bot. Surv. India 1-11: 239 (1898); Verdcourt, Fl. Trop. E. Africa. Leguminosae 4-2: 594–596 (1971); Verdcourt, Manual New Guinea Legumes 487

(1979); Thuan, Fl. Cambodge, Laos, Vietnam 17: 81, 84 (1979); van der Maesen, Agric. Univ. Wageningen Papers 85-1: 75–78 (1985).

*Basionym*: *Neustanthus javanicus* Benth. ex Baker in Miquel, Pl. Jungh. 2: 235 (1852); Miquel, Fl. Ind. Bat. 1-1: 218, t.4 (1855).

*Type*: Java, Merapi, R. Kuning, *Junghuhn s.n.* (K, holo, acc. to Verdcourt, 1971).

*Homotypic synonym*: *Pueraria javanica* (Benth.) Benth., J. Linn. Soc. Bot. London 9: 125 (1867); Koorders, Exkursionsfl. Java 402 (1912); Heyne, Nuttige Pl. Nederl. Indië 829 (1927); Burkart, Rev. Fac. Agron. Univ. La Plata 3-27: 141 (1950); Burkart, Legum. argent. ed. 2: 407 (1952); Verdcourt, Manual New Guinea Legumes 487 (1979).

*Heterotypic synonyms*: *Pachyrhizus mollis* Hasskarl, Flora Bot. Zeitschr. 25-2: Beibl. 74 (1842); Miquel, Fl. Ind. Bat. 1-1: 218 (1855).

*Type*: Java, *Hasskarl s.n.* (L, holo).

*Neustanthus sericans* Miquel, Fl. Ind. Bat. 1-1: 218 (1855).

*Type*: Java, near Surakarta, *Horsfield L 121* (K, holo; iso: U).

The variety most widely spread as green manure by man, recognizable by the relatively lush growth, entire or slightly lobed leaflets, and obtuse lateral calyx lobes. Found in Brunei, Indonesia (Java, Sulawesi), Malaysia (Malaya, N. Borneo), the Philippines, Sri Lanka, New Guinea, Solomon Islands, introduced into Africa and the Americas.

*Habitat*: Growing along rivers, roads, on bushes, in evergreen forest or cleared patches, in grass vegetations, along rice fields, as a soil cover and green manure in perennial crops, such as oil palm, coconut palms, Substrate sandy as well as heavy clay soils. Altitude 0–1100 m.


***Pueraria phaseoloides*** (Roxb.) Benth. var. *phaseoloides*

J. Linn. Soc. Bot. London 9: 125 (1867); Baker in Hooker, Fl. Brit. India 2: 199 (1876); King, J. Asiatic Soc. Bengal 66-1 (1997); Prain, J. Asiatic Soc. Bengal 66-2: 420 (1897); Duthie, Fl. Upper Gangetic Plain 234 (1903), 216 (repr. 1960); Prain, Bengal Pl. 282 (1903, repr. 1963); Perkins, Fragm. Pl. Philipp. 1-3: 212 (1904); Merrill, Enum. Philipp. Legum., Philipp. J. Sci. Bot. 5: 123 (1910); Burkill, Rec. Bot. Surv. India 4-4: 106 (1910); H(a)yata, Icon. Pl. Formos. 1: 198 (1911); Koorders, Exkursionsfl. Java 403 (1912); Merrill, Fl. Manila 254 (1912); Gagnepain, Fl. Gén. Indo–Chine 1: 135 (1917); Merrill, Interpr. Rumph. Herb. Amboin. 282 (1917); Merrill, Sp. Blancoanae 189 (1918); Ridley, Fl. Malay Penins. 1: 571 (1922); Haines, Bot. Bihar Orissa 3: 304 (1922), 2: 295 (repr. 1961); Merrill, Enum. Philipp. Fl. Pl. 2: 311 (1923); Heyne, Nuttige Pl. Nederl. Indië 829 (1927); Craib, Fl. Siam. Enum. 1-3: 450 (1928); Kanjilal *et al.*, Fl. Assam 2: 82 (1938); Amshoff, Fl. Suriname 2-2: 209 (1939); Guillaumin, Fl. Anal. Synopt. Nouv. Calédonie 149 (1948); Backer, Fl. Java 1: 362 (1963); Rao and Joseph, Bull. Bot. Surv. India 7: 144 (1963); Biswas, Pl. Darjeeling Sikkim Himalayas 1: 29 (1966); Verdcourt, Fl. Trop. E. Afr. Legum 4: 596 (1971); Thothatri, Rec. Bot. Surv. India 20-2: 81 (1973); Anon., Fl. Hainanica 2: 320 (1977?); Lackey, Synops. Phaseol. 72, 75 (1977); Ohashi, Fl. Taiwan 3: 370 (1977); Thuan, Fl. Cambodge, Laos, Vietnam 17: 83 (1979); Verdcourt, Manual New Guinea Legum. 485 (1979); Hara and Williams, Enum. Pl. Nepal 2: 218 (1979); Anon., Icon. Cormoph. Sinic. 2: 503 (1980); Perry, Medicinal Pl. E and SE Asia 224–225 (1980); Duke,

Handb. Leg. World Econ. Importance 214 (1981); van der Maesen, Agric. Univ. Wageningen Papers 85-1: 78–84 (1985). Halim, in Prosea 4: 192–195 (1992).

*Basionym*: *Dolichos phaseoloides* Roxb., Fl. Indica 3: 316 (1832).

*Type*: India, Calcutta Bot. Garden, grown from seeds received from Kerr at Canton, China (CAL?). Verdcourt (1971) included as syntype *Roxburgh's drawing no. 1890* (K).

   *Neustanthus phaseoloides* (Roxb.) Benth., in Miquel, Pl. Jungh. 2: 234 (1852); Miquel, Fl. Ind. Bat. 1-1: 219 (1855); Bentham, Fl. Hongkong 88 (1861); Benth., J. Linn. Soc. Bot. London 9: 125 (1867).

*Heterotypic synonyms*: *Dioscorea bolojonica* Blanco, Fl. Filip. 800 (1837); ed. 2 551 (1845); Llanos and Fernandez-Villar, Fl. Filip. ed. 3, 3: 208 (1979); Merrill, Sp. Blancoanae 189 (1918).

*Type*: Philippines, Pasay, Rizal Prov., Luzon, *Merrill Sp. Blancoanae 195* (US, neotype; iso: F, NY, W).

   *Dolichos viridis* Hamilton ex Wallich, *nomen nudum*, based on Wallich Cat. no. 5559 (1831) (K, also BM); Benth. in Miquel, Pl. Jungh. 2: 235 (1852); Benth., J. Linn. Soc. Bot. London 9: 125 (1867); Baker in Hooker, Fl. Brit. India 2: 199 (1876).

   *Phaseolus barbatus* Graham ex Wallich, *nomen nudum*, based on Wallich Cat. no. 5559B (1831), Bangladesh, Sylhet (K, also BM).

   *Phaseolus decurrens* Graham ex Wallich, *nomen nudum*, Wallich Cat. no. 5612 (1831) acc. to Prain. Benth., J. Linn. Soc. Bot. London 9: 125 (1867); Baker in Hooker, Fl. Brit. India 2: 199 (1876); Index Kewensis (1895); Prain, J. Asiatic Soc. Bengal 66-2: 420 (1897), based on Wallich 5612 (K, not seen) from Penang, coll. G. Porter.

   *Pachyrhizus montanus* Blanco (non (Lour.) DC.), Fl. Filip. ed. 2: 406 (1845); Llanos and Fernandez-Villar, Fl. Filip. ed. 3, 2: 380 (1979); Merrill, Enum. Philipp. Legum., Philipp. J. Sci. Bot. 5: 123 (1910); Merrill, Sp. Blancoanae 189 (1918); Merrill, Enum. Philipp. Fl. Pl. 2: 311 (1923).

*Type*: *Blanco*, lost. Description and Tagalog vernacular match *Pueraria phaseoloides*, so Merrill (1918) declared synonymy.

   *Pachyrhizus teres* Blanco, Fl Filip. 580 (1837); Merrill, Enum. Philipp. Legum., Philipp. J. Sci. Bot. 5: 123 (1910); Merrill, Sp. Blancoanae 189 (1918); Merrill, Enum. Philipp. Fl. Pl. 2: 231 (1923).

*Type*: *Blanco*, lost. Description and vernacular match *Pueraria phaseoloides* completely, hence the reduction to synonymy by Merrill.

   Small-flowered variety, with lateral calyx lobe acute–acuminate and leaflets entire, lobed or sinuate.

*Distribution*: Found in China, India, Indonesia, Malaysia, Myanmar, New Guinea, Sri Lanka, Taiwan, Thailand, Vietnam, occasionally introduced e.g. in Liberia, Cameroon, Nigeria, Surinam.

*Habitat*: Trailing in grassland, open grounds in scrub, in mixed forests in shade, bamboo forests, along roadsides, rice fields and rivers, on sandy or clayey soils. Also widely cultivated in plantations. Altitude 0–1600 m.

**Pueraria phaseoloides** (Roxb.) Benth var. *subspicata* (Benth.) Maesen

Agric. Univ. Wageningen Papers 85-1: 84–88 (1985).

*Basionym*: *Neustanthus subspicatus* Benth. in Miquel, Pl. Jungh. 2: 234 (1852); Miquel, Fl. Ind. Bat. 1-1: 219 (1855).

*Type*: Bangladesh, mountains near Sylhet, *Wallich 5557A* (*Dolichos spicatus* Wallich *nomen nudum*) (K, lecto; isolecto: BM, E, G, K, W).

*Paratypes*: India, Goalpara, *Hamilton, Wallich 5557B* (not seen); Bangladesh, Sylhet, *Wallich 5557C* (K, G, NY).

*Homotypic synonym*: *Pueraria subspicata* (Benth.) Benth., J. Linn. Soc. Bot. London 9: 125 (1867); Baker in Hooker, Fl. Brit. India 2: 199 (1976); Kurz, J. Asiatic Soc. Bengal 45-4: 253 (1876); Prain, J. Asiatic Soc. Bengal 66: 420 (1897); Prain, Bengal Pl. 1: 282 (1903, repr. 1963); Craib, Fl. Siam. Enum. 4-3: 455 (1928); Kanjilal *et al.*, Fl. Assam 2: 82 (1938); Thothathri, Rec. Bot. Surv. India 20-2: 81 (1973); Lackey, Synops. Phaseol. 72, 75 (1977). *Dolichos ficifolius* Grah. *nomen nudum*, Wallich Cat. 5563a: based on Burma (Myanmar), Prome 1826 (K, BM, G) and 5563b: Burma, Tavoy *W.G. 447* (K, G).

The large-flowered variety with long-hairy bracts and calyces, deeply lobed leaflets, long pods with dorsal sutures thickened. Found in Bangladesh, India (Assam, Bengal, Meghalaya, Bengal), Myanmar, Thailand from 0–1300 m, in mixed deciduous forest, scrub vegetation, along roads and irrigation tanks.

### *Pueraria pulcherrima* (Kds) Merr.

In Koorders-Schumacher, Syst. Verz. 2: 231 (1914); Merrill, Enum. Philipp. Fl. Pl. 2: 312 (1923); Lackey, Synopsis Phaseol. 71, 75 (1977); Verdcourt, Manual New Guinea Legumes 487 (1979); van der Maesen, Agric. Univ. Wageningen Papers 85-1: 88–93 (1985).

*Basionym*: *Mucuna pulcherrima* Koorders, Meded. Lands Plantentuin 19: 440/630 (1898) (not 1908 as given in 1985).

*Type*: Indonesia, Sulawesi, Minahasa, Menado along Ranoyapo river near Amurang, *Koorders 17699* (BOG, holo, not seen; iso: L).

*Heterotypic synonym*: *Pueraria warburgii* Perkins, Fragm. Fl. Philipp. 1: 87 (1904); Merrill, Enum. Philipp. Legum., Philipp. J. Sci. Bot. 3: 231 (1908), van der Maesen and Almeida, J. Bombay Nat. Hist. Soc. 85-1: 233 (1988).

*Basionym*: *Glycine warburgii* (Perk.) Merr., Philipp. J. Sci. Bot. 3: 231 (1908); Merrill, Philipp. J. Sci. Bot. 5: 124 (1910).

*Type*: Philippines, Mindanao, Taumo, *Warburg 14664* (B, holo, not seen).

*Pueraria novo-guineensis* sensu Pulle non Warburg, Nova Guinea 8-2: 382 (1910); Verdcourt, Manual New Guinea Legumes 487 (1979).

*Type*: Indonesia, Irian Jaya, *Versteeg 1202* (holo: L; iso: U).

*Pueraria pilosissima* Baker f., Trans. Linn. Soc. Bot. 2nd ser. 9: 33 (1916); Verdcourt, Manual New Guinea Legumes 487 (1979).

*Type*: Indonesia, Papua (Irian Jaya), Canoe Camp 150 ft (BM, holo, but with location Utakwa river to Mt. Carstensz, *Boden Kloss*).

*Pueraria sericans* K. Schumann (not of Bentham), Fl. Kaiser Wilhelmsland 99 (1889); Karnbach, Beibl. 37, 5, Engler Bot. Jahrb. 16 (1892).

*Type*: Papua New Guinea, Kuliku-mana near Constantinhafen, *Hollrung 566* (K, lecto, ex B).

*Paratype*: Papua New Guinea, Uassa near Boja river near Finschhafen, *Hollrung 231* (not seen).

*Pueraria textilis* Lauterbach and K. Schumann, in Schum. and Lautb., Fl. deutsch. Schutzgeb. Südsee 368 (1910).

*Type*: Papua New Guinea, Finschhafen, on Bumi river, *Lauterbach 636* (B, holo, not seen; iso: CAL, L).

*Paratypes*: Papua New Guinea, Uassa nr Bonga river, *Hollrung 231* (B, not seen); Oertzen Mts, Nowulja river, *Lauterbach 2072, 2100* (B, not seen); Bismarck Archipelago, New Britain, Gazelle Peninsula, *Warburg s.n.* (B, not seen).

Woody climber with large stipules surrounding the branches, inflorescences crowded with flowers, later crowded with very hairy flat pods of 1.5–4×0.5–0.6 cm. Leaflets rhomboid to lanceolate. Flowering most of the year.

*Distribution*: E. Indonesia, Philippines, Papua New Guinea, Kei and Solomon Islands.

*Habitat*: Climbing in shrubs or trees, open rain forest, *Imperata* grassland, flatlands or on rocks near rivers, either muddy or well-drained soils and coral, 0–1300 m.

### *Pueraria sikkimensis* Prain

J. Asiatic Soc. Bengal 66: 419 (1897); Lackey, Synops. Phaseol. 72, 75 (1977); van der Maesen, Agric. Univ. Wageningen Papers 85-1: 96 (1985).

*Type*: India, Sikkim, Rangeet, *Clarke 27263* (K, lecto; isolecto: BM, CAL, K).

*Paratypes*: India, W. Bengal, Sistra, Darjeeling Terai, *Gamble 2227A* (K); Sikkim Terai, *Anderson s.n., Gamble s.n., Gammie s.n.* (CAL); Tista Valley, *King s.n.* (CAL).

Woody climber with rusty pubescence, 6 mm bracts and dense foxtail-like inflorescences, similar to *P. tuberosa*, and perhaps only an extreme in the diversity of that species. Flowering in March and fruiting in April–May.

*Distribution*: Occurring in Bhutan and India (Sikkim, W. Bengal, NW Himalayas).

*Habitat*: In deciduous forests or scrub, in plains, river valleys, from 330–1600 m.

### *Pueraria stricta* Kurz

J. Asiatic Soc. Bengal 42-2: 254 (1873), and 45-4: 253 (1876); Baker in Hooker, Fl. Brit. India 2: 198 (1876); Prain, J. Asiatic Soc. Bengal 66-2: 420 (1897); Lackey, Synops. Phaseol. 73, 76 (1977); van der Maesen, Agric. Univ. Wageningen Papers 85-1: 100–105 (1985).

*Type*: Burma (Myanmar), Pegu (Yomah), 300–1000 m, *Kurz 2557* (CAL, holo, not seen; iso: K).

*Paratype*: Burma, Martaban Hills, *Kurz s.n.* (CAL).

*Heterotypic synonyms*: *Pueraria brachycarpa* Kurz, J. Asiatic Soc. Bengal 42-2: 232, 254 (1873), and 45-4: 185, 254 (1874); Baker, in Hooker, Fl. Brit. India 2: 199 (1976); Prain, J. Asiatic Soc. Bengal 66-2: 420 (1897); Craib, Fl. Siam. Enum. 1-3: 449 (1928); Lackey, Synops. Phaseol. 73, 76 (1977).

*Type*: Burma, Pegu, *Kurz 2553* (CAL, holo; iso: CAL).

*Pueraria hirsuta* Kurz, J. Asiatic Soc. Bengal 42-2: 254 (1873); Baker in Hooker, Fl. Brit. India 2: 199 (1876); Prain, J. Asiatic Soc. Bengal 66-2: 42 (1897); Craib, Fl. Siam. Enum. 1-3: 450 (1928).

*Type*: Burma, Pegu Yomah, *Kurz 1720* (CAL, lecto).

*Paratype*: Burma, Pegu, *Kurz 2554* (CAL).

*Pueraria collettii* Prain, J. Asiatic Soc. Bengal 66-2: 420 (1897); Craib, Contrib. Fl. Siam, Aberdeen Univ. Stud. 57: 64 (1912); Gagnepain, Fl. Gén. Indo–Chine 2: 254 (1916); Brandis, Indian Trees 228 (1921); Craib, Fl. Siam. Enum. 1-3: 449 (1928); Lackey, Synops. Phaseol. 73-76 (1977); Lackey, Bot. J. Linn. Soc. 74: 170 (1977); Lackey, Phytologia 37: 109 (1977).

*Type*: Burma, Shan Hills at Ywaggyen, *Collett 654* (CAL, lecto; iso: K).

*Paratypes*: Burma, Fort Stedman/Mong Hsawk, *Abdul Huk s.n.* (CAL, K); Taunggyi, *Abdul Khalil s.n.* (CAL, L); Maymyo, *Badal Khan 104* (CAL, G, K); Indine, Saga, *King's collectors*.

*Pueraria siamica* Craib, Kew Bull. 1911: 40; Craib, Contrib. Fl. Siam, Aberdeen Univ. Stud. 57: 65 (1912); Gagnepain, Fl. Gén. Indo–Chine 2: 256 (1916); Craib, Fl. Siam. Enum. 1-3: 449 (1928).

*Type*: Thailand, Chiengmai, Doi Sutep at 420 m, *Kerr 831* (K, holo; iso: BM, CAL, E, K).

*Pueraria collettii* Prain var. *siamica* (Craib) Gagnep., Fl. Gén. Indo–Chine 2: 254 (1916); Craib, Fl. Siam. Enum. 1-3: 449 (1928); based on *P. siamica* Craib.

*Pueraria longicarpa* Thuan, Adansonia sér. 2, 16-4: 509 (1977); Thuan, Fl Cambodge, Laos, Vietnam 17: 82, 91 (1979); van der Maesen, Agric. Univ. Wageningen Papers 85-1: 100 (1985).

*Type*: Laos, Pak Leun 1500 m, near Xieng Khouang, *Poilane 16897* (P, holo).

Shrub, sometimes straggling, short grey-pubescent, leaflets rhomboid to ovate, bracts persistent, 8 mm small flowers and pods 3.5–6×0.5–0.7 cm. Flowering May, July, October–November (–January), fruiting November to March.

*Distribution*: Found in Burma, China (Yunnan) and Thailand.

*Habitat*: in open grassy jungle, deciduous or dry evergreen forest with bamboo or oak, hilsides, dry ridges, sandston or calcareous soil from 400–1700 m.


**Pueraria tuberosa** (Roxb. ex Willd.) DC.

Ann. Sc. Nat. sér 1-4: 97 (1825) and Mém. Lég. 254 (1825) and Prodr. 2: 240 (1825); Wight and Arnott, Prodr. Fl. Penins. Ind. Or. 205, 449 (1834); Wight, Icon. Pl. Ind. Or. 2-1 t. 412 (1843); Bentham in Miq., Pl. Jungh. 2: 235 (1852); Bentham, J. Linn. Soc. Bot. London 9: 123 (1867); Kurz, J. Asiat. Soc. Bengal 45-4: 253 (1876); Baker in Hooker, Fl. Brit. India 2: 197 (1876); Taubert, in Engler and Prantl, Nat. Pflanzenfam. 3-3: 370 (1894); Collett, Fl. Siml. 139 (1902); Wood, Rec. Bot. Surv. India 2: 64, 97 (1902); Cooke, Fl. Bombay 1-2: 374 (1902) and 1: 399 (repr. 1958); Duthie, Fl. Upper Gang. Plain 1-1: 233 (1903) and 215 (repr. 1960); Prain, Bengal Pl. 1: 282 (1903, repr. 1963); Gagnepain, Fl. Gén. Indo–Chine 2: 250 (1916); Gamble, Fl. Madras 1-2: 360 (1918) and 1: 245 (repr. 1967); Brandis, Indian Trees and Timbers 245 (1922); Haines, Bot. Bihar Orissa 3: 294–295 (1922) and 2: 294 (repr. 1961); Kanjilal *et al.*, Fl. Assam 2: 79 (1938); Barrau, Ethnology 4: 285 (1966); Ahuja, Medicinal Pl. Saharanpur 62 (1965); Vartak, Enum. Pl. Gomantak 43 (1966); Patel, Forest Fl. Melghat 116

(1968); Wealth of India 8: 313–317 (1969); Lackey, Synops. Phaseol. 72, 74 (1977); Ali, Fl. W. Pakistan 100: 99 (1978); Shah, Fl. Gujarat 1: 237 (1978); Hara and Williams, Enum. Pl. Nepal 2: 128 (1979); van der Maesen, Agric. Univ. Wageningen Papers 85-1: 105–109 (1985).

*Basionym*: *Hedysarum tuberosum* Roxb. ex Willd., Sp. Pl. 3-2: 1197 (1803); Wight and Arnott, Prodr. Fl. Penins. Ind. Or. 205, 449 (1834); Roxburgh, Fl. Indica 3: 363 (1832).

*Type*: India, Bengal, *Roxburgh s.n.* (B-Willd., holo; iso: G, K).

Woody climber, with large tubers to 35 kg and up to 0.75 cm circumference, large rounded–ovate leaflets to 32 cm long, conspicuously yellow in dry season before shedding, long inflorescences to 50 cm, pods few and often poorly fertilized, then constricted, *c.* 7 × 1.1 cm, with dense golden brown hairs. Flowering after leaf fall from December to February.

*Distribution*: Widely distributed in India, Nepal and Pakistan.

*Habitat*: In hill forest and deciduous vegetation, in exposed and eroded areas, covering ground, bushes and trees (0–1300 m).

*Uses*: The enormous tubers are used as a famine food or to feed cattle, and have medicinal properties against fevers, are applied as cataplasms to cure swellings of joints, and as a lactagogue. Some genotypes appear to be used as fish poison.

### *Pueraria wallichii* DC.

Ann. Sci. Nat. Sér. 1-4: 97 (1825); DC., Mém. Lég. 254, t. 43 (1825); DC., Prodr. 2: 240 (1825); Benth. in Miquel, Pl. Jungh. 2: 235 (1852); Benth., J. Linn. Soc. Bot. London 9: 124 (1867); Baker in Hooker, Fl. Brit. India 2: 198 (1876); Kurz, J. Asiatic Soc. Bengal 45-4: 253 (1976); Collett and Hemsley, L. Linn. Soc. Bot. London 28: 47 (1890); Prain, J. Asiatic Soc. Bengal 66-2: 419 (1897); Duthie, Fl. Upper Gang. Plain 234 (1903); and 216 (repr. 1960); Craib, Kew Bull. 1911: 41; Craib, Contrib. Fl. Siam, Aberdeen Univ. Stud. 57: 65 (1912); Gagnepain, Fl. Gén. Indo–Chine 2: 252, 257 (1916); Brandis, Indian Trees 228 (1921); Craib, Fl. Siam. Enum. 3: 452 (1928); Kanjilal *et al.*, Fl. Assam 2: 80 (1938); Hara, Fl. E. Himalaya 162 (1966); Lackey, Synops. Phaseol. 77 (1977); Hara and Williams, Enum. Fl. Nepal 2: 218 (1979).

*Type*: Nepal, *Wallich s.n. (1821)* (G, holo; iso: BM, C, K: *Wallich 5353 a/c*).

*Homotypic synonym*: *Neustanthus wallichii* (DC.) Benth. in Miquel, Pl. Jungh. 2: 234 (1852).

*Heterotypic synonyms*: *Pueraria wallichii* DC. var. *composita* (Grah. ex Wall.) Benth., J. Linn. Soc. Bot. London 9: 124 (1867); Kurz, J. Asiatic Soc. Bengal 45-4: 253 (1876).

*Type*: Burma, Taong Dong, *Wallich 5570* (K, holo; is: BM, G, K), based on *Pueraria composita* Grah. ex Wall., *nomen nudum*, Cat. Herb. Ind. 5570.

*Dolichos frutescens* Ham. in Don, Prodr. 240 (1825); Prain, J. Asiatic Soc. Bengal 66-2: 419 (1897).

*Type*: Nepal, *Hamilton s.n.* (CAL, holo).

Shrub, sometimes straggling, sparsely pubescent and glabrous with age, rhomboid-elliptic leaflets, rather long inflorescences to 30 (–52) cm, 5–8 flowers per short lateral, pods somewhat s-shaped, 6–7-seeded, 6–10 × 0.7–1.1 cm wide.

*Distribution*: Found in Bangladesh, China (Yunnan), N. India, Myanmar, Nepal, Thailand from 180–2300 m.

*Habitat*: In or near dry evergreen forests, associated with *Shorea robusta* Gaertn., *Pinus*, *Dipterocarpus, Quercus*; in open grassy vegetations, on slopes and along rivers, erect or straggling in shrubs. Flowering October–January or earlier, fruiting into February.

*Use*: One mention is made of *Pueraria wallichii* as a hedge in the Khasi Hills of Meghalaya, India.


## Fossil species

*Pueraria tanaii* Ozaki, sci. Rep. Yokohama Nat. Univ. Sect. 2 Biol. Geol. Sci 21: 15 (1974).

*Type*: Japan, Izumi distr., Inkyoyama hill, N. of Toki, *GYNU-CMP* 1029, Geol. Inst. Yokohama Nat. Univ.

*Paratypes*: same location, *GYNU-CMP* 1028, 1030. The single-leaf fossils resemble *P. montana* var. *lobata* leaflets.

*Distribution*: Japan.

*Habitat*: Warm-temperate to transitional forests of the Miocene era.


## Excluded Species

Several legume species have once been described in *Pueraria* because of initial likenesses and erroneous placement in this genus. For notes and typifcation see van der Maesen (1985).

*Pueraria anabaptista* Kurz, *J. Asiatic Soc. Bengal* 45-2: 253 (1876). *Shuteria hirsuta* Baker in Hooker, *Fl Brit. India* 2: 182 (1876).

*Pueraria barbata* Craib, Kew Bull. 1927: 379. *Teyleria barbata* (Craib) Lackey ex Maesen, Agric. Univ. Wageningen Papers 85-1: 117 (1985).

*Pueraria chaneti* Lévl., Bull. Acad. Geogr. 17 suppl iii (1907). *Phaseolus chaneti* Lévl., Bull. Soc. Bot. France 55: 427 (1908)

*Pueraria elegans* Wang and Tang. Reference obtained at the closure of this manuscript, the protologue and voucher material have not been seen. Probably described *c.* 1955, but unknown in literature.

*Pueraria ferruginea* Kurz, J. Asiatic Soc. Bengal 42-2: 232 (1873). *Shuteria hirsuta* Baker; Lackey (1977).

*Pueraria ficifolia* (Benth.) L. Bolus, Ann. Bolus Herb. 1: 189 (1915). *Neorautanenia ficifolius* (Benth.) C.A. Smith in Burtt Davy, Fl. Pl. Ferns Transvaal 2: XXVII and 417 (1932); Verdcourt, Kew Bull. 24: 306 (1970).

*Pueraria garhwalensis* L.R. Danwal and D.S. Rawat, J. Bombay Nat. Hist. Soc. 93: 570–572 (1996) is probably a *Shuteria*, as the type specimens have only been compared to *Pueraria ferruginea* Kurz, a species now relegated to the synonymy of *Shuteria hirsuta* Baker. I have not yet seen the type material.

*Pueraria hochstetteri* Chiov., Ann. Inst. Bot. Roma 8: 434 (1908). *Neorautanenia mitis* (A. Rich.) Verdc., Kew Bull. 24: 306 (1970).

*Pueraria maclurei* (Metcalf) F.J. Hermann, Techn. Bull. USDA 1268: 46 (1962). Basionym *Glycine maclurei* Metcalf, Lingnan Sci. J. 19: 557 (1940). *Sinodolichos lagopus* (Dunn.) Verdc., Kew Bull. 24: 398 (1970). *Dolichos lagopus* Dunn, J. Linn. Soc. Bot. 35: 490 (1903).

*Pueraria rigens* Craib, Kew Bull. 1927: 380. *Craspedolobium schochii* Harms, see Phan Ke Loc, Bot. Zhurnal. 83-6: 118–122 (1998).

*Pueraria rogersii* L. Bolus, Ann. Bolus Herb. 1: 189 (1915). *Neorautanenia amboensis* Schinz, Bull. Herb. Boiss. 7: 35 (1899).

*Pueraria stracheyi* Baker in Hooker, Fl. Brit. India 2: 198 (1876). *Shuteria sp.?* Cf. Lackey, Synops. Phaseol. 77 (1977).

*Pueraria strobilifera* Kurz ex Prain, J. Asiatic Soc. Bengal 66: 2–403 (1897). *Shuteria hirsuta* Baker, Fl. Brit. India 2: 182 (1876).

*Pueraria tetragona* Merr., Philipp. J. Sci. 5: 122 (1910). *Teyleria tetragona* (Merr.) Lackey ex Maesen, Agric. Univ. Wageningen Papers 85-1: 117 (1985).

## Biosystematics

The anatomy of the Glycininae is not very particular (Lackey, 1977). Paraveinal mesophyll is common, but in *P. peduncularis* and *P. wallichii* these structures are lacking, confirming their status within a section apart from other *Pueraria* species.

Cytotaxonomically the genus has been poorly researched. The basic number of chromosomes in *Glycininae* is usually $x=10$ and $x=11$ (Lackey, 1977). Indeed chromosome counts in *Pueraria* either give $2n=22$ and $2n=24$ for both *P. montana* var. *lobata* and *P. phaseoloides*. *P. collettii* reportedly has $2n=20$ or $4n=44$ chromosomes. Numbers other than those divisible by 11 seem less probable, and may be accounted for by interpretation differences or the occasional presence of aneuploids. Wu *et al.* (1994) published some data on Chinese *Pueraria*.

Chemical data useful for chemotaxonomy have been inadequate. With the attention paid in this volume to medicinal uses, chemical properties of at least the commoner species are highlighted into much greater detail than was possible to glance from the literature previously (van der Maesen, 1985). The list of various constituents is rather impressive despite the findings having been based on a limited number of species. The value of these chemical data with regard to relationships within the genus and within the tribe has so far been very limited.

## GENETIC RESOURCES AND BREEDING

A few germplasm collections hold accessions of *Pueraria* species. CIAT, the Centro Internacional de Agronomia Tropical, in Cali, Colombia, has no less than 288 accessions. ILRI maintains five accessions (IPGRI Internet information). In Japan and China collections most likely contain *Pueraria*.

## CONCLUSIONS

The taxonomy and nomenclature of *Pueraria* are relatively well established. Sectional classification may need further improvement after phylogenetic analysis. At present 15 species are recognized, some with two or three varieties. If strict monophyletic groups were to be distinguished, a number of genera could be established, but as a whole the genus as presently understood forms a recognizable group within subtribe Glycininae of the papilionoid legumes. Apart from the three economically used and widely spread taxa, *P. montana* var. *lobata*, *P. phaseoloides* and *P. tuberosa*, the other 13 species are rather restricted in their distribution. More efforts are required to acquire further collections in order to complete the some times unknown characteristics, such as shape and size of pods and tubers, or even the presence and absence of tuberous roots, and add to the scantily known distribution areas. The valorization of the species can be improved, as will be shown in the next chapters.

## REFERENCES

Bodner, C. and Hymowitz, T. (2002, this volume) Ethnobotany of *Pueraria* species.

Burrows, J.E. (1989) Kudzu vine – a new plant invader of South Africa. *Veld & Flora* 75(4), 116–117.

De Candolle, A.P. (1925) Mémories sur la famille des Légumineuses, 252–255.

Duke, J.A. (1981) *Handbook of legumes of world economic importance*. Plenum Press, New York.

Lackey, J.A. (1977) *A synopsis of the Phaseoleae (Leguminosae, Papilionoideae)*. Ph.D. dissertation, Dept. Bot. Pl. Path. Iowa State Univ., Ames, pp. 293.

Lee, J. (2000) *Molecular genetics on the genus Glycine: I. Phylogenetic study of subtribe Glycininae. II Development of a universal soybean genetic map*. Ph.D. Thesis, Department of Crop Sciences, University of Illinois, Urbana, Illinois, USA.

Maesen, L.J.G. van der (1985) Revision of the genus *Pueraria* DC. with some notes on *Teyleria* Backer (Leguminosae). *Agric. Univ. Wageningen Pap.* 85(1), 1–132.

Maesen, L.J.G. van der (1994) *Pueraria*, the Kudzu and its relatives, an update of the taxonomy. In *Proc. 1st Int. Symp. on Tuberous Legumes*, Guadeloupe, F.W.I., 21–24 April 1992: 55–86.

Maesen, L.J.G. van der and Almeida, S.M. (1988) Two corrections to the nomenclature in the revision of *Pueraria* DC. *J. Bombay Nat. Hist. Soc.* 85(1), 233.

Niyomdham, C. (1992) Notes on Thai and Indo–Chinese Phaseoleae (Leguminosae-Papilion-oideae). *Nordic J. Bot.* 12(3), 339–346.

Ohashi, H., Tateishi, Y., Nemoto, T. and Endo, Y. (1988) Taxonomic studies on the Leguminosae of Taiwan III. *Sci. Rep. Tohoku Univ. ser.* 4, 39, 191–248.

Sanjappa, M. (1992) *Legumes of India*. Bishen Singh Mahendra Pal Singh, Dehra Dun. pp. 338.

Shurtleff, W. and Aoyagi, A. (1977) *The book of Kudzu. A culinary and healing guide*. Autumn Press, pp. 102.

Skerman, P.J. (1977) *Tropical Forage Legumes*. FAO Plant Production and Protection Series no. 2, Rome.

Whyte, R.O., Nilsson-Lessner, G. and Trumble, H.C. (1953) *Legumes in Agriculture*. FAO Agricultural Studies No. 21, Rome.

Wu, T., Chen, Z. and Huang, X. (1994) A study of Chinese *Pueraria. J. Trop. Subtrop. Bot.* 2(3), 35–40.

Zhang, D. and Chen, Z. (1995) A cladistic analysis of *Pueraria* DC. (Leguminosae). *J. Trop. Subtrop. Bot.* 3(1), 35–40.

## INDEX

*Dolichos frutescens* Ham.
*Dolichos grandifolius* Grah. ex Wall. nom. nud.
*Dolichos hirsutus* Thunb.
*Dolichos japonicus* Hort. nom. nud.

*Dolichos lagopus* Dunn
*Dolichos lobatus* Willd.
*Dolichos montanus* Lour.
*Dolichos phaseoloides* Roxb.
*Dolichos spicatus* Wall. nom. nud.
*Dolichos tuberosus* Lam.
*Dolichos trilobus* L.

**Dolichos trilobatus L.**
*Dolichos trilobus* Houtt. non L.
*Dolichos trilobus* Lour.
*Dolichos viridis* Ham. ex Wall. nom. nud.
*Glycine warburgii* (Perk.) Merr.
*Hedysarum tuberosum* Roxb. ex Willd.
*Millettia rigens* (Craib) Niyomdham
*Mucuna pulcherrima* Koorders.
**Neorautanenia amboensis Schinz**
**Neorautanenia ficifolius (Benth.) C.A.Smith**
**Neorautanenia mitis (A.Rich.) Verdc.**
*Neustanthus* Benth.
*Neustanthus chinensis* Benth.
*Neustanthus javanicus* Benth.
*Neustanthus phaseoloides* Grah. ex Benth.
*Neustanthus sericans* Miquel
*Neustanthus subspicatus* Benth.
*Neustanthus wallichii* (DC.) Benth.
*Pachyrhizus mollis* Hassk.
*Pachyrhizus montanus* Blanco
*Pachyrhizus montanus* (Lour.) DC.
*Pachyrhizus teres* Blanco
*Pachyrhizus thunbergianus* Sieb. and Zucc.
*Pachyrhizus trilobus* (Lour.) DC.
*Phaseolus barbatus* Grah. nom. nud.
*Phaseolus chaneti* (Lévl.) Lévl.
*Phaseolus decurrens* Grah. ex Wall.
*Phaseolus trilobus* (L.) Ait.

**Pueraria DC.**
**Pueraria alopecuroides Craib**
*Pueraria anabaptista* Kurz
*Pueraria argyi* Lévl. and Vaniot
*Pueraria barbata* Craib

**Pueraria bella Prain**
*Pueraria bicalcarata* Gagn.
*Pueraria bodinieri* Lévl. and Vaniot
*Pueraria brachycarpa* Kurz
*Pueraria caerulea* Lévl. and Vaniot

**Pueraria calycina Franch.**
**Pueraria candollei Grah. ex Benth.**

*Pueraria chaneti* Lévl.
*Pueraria chinensis* (non Benth.) Ohwi
*Pueraria collettii* Prain
　　var. *siamica* (Craib.) Gagn.
*Pueraria composita* Grah. ex Wall. nom. nud.
*Pueraria decurrens* Grah. ex Wall. nom. nud.

**Pueraria edulis Pamp.**
*Pueraria elegans* Wang and Tang
*Pueraria ferruginea* Kurz
*Pueraria ficifolia* (Benth.) L. Bolus
*Pueraria forrestii* Evans
*Pueraria harmsii* Rech.
*Pueraria hirsuta* Kurz
*Pueraria hirsuta* (Thunb.) Matsum.
*Pueraria hirsuta* (Thunb.) Schneid.
*Pueraria hochstetteri* Chiov.

**Pueraria imbricata Maesen**
*Pueraria javanica* (Benth.) Benth.
*Pueraria koten* Lévl. and Vaniot

**Pueraria lacei Craib**
*Pueraria lobata* (Willd.) Ohwi
　　var. *chinensis* (non Benth.) Ohwi
　　var. *lobata*
　　var. *montana* (Lour.) Maesen
　　var. *thomsoni* (Benth.) Maesen
*Pueraria longicarpa* Thuan
*Pueraria maclurei* (Metcalf) F.J. Hermann
*Pueraria maesenii* Niyomdham
*Pueraria mirifica* Shaw and Suvatabandhu

**Pueraria montana (Lour.) Merr.**
　　**var. chinensis (Ohwi) Maesen and Almeida ex Sanjappa and Predeep**
　　**var. lobata (Willd.) Maesen and Almeida ex Sanjappa and Predeep**
　　**var. montana**
*Pueraria neo-caledonica* Harms
*Pueraria novo-guineensis* Warb.
*Pueraria novo-guineensis* sensu Pulle non Warb.
*Pueraria omeiensis* Wang and Tang
**Pueraria peduncularis Grah. ex Benth.**
var. *violacea* Franch.
**Pueraria phaseoloides (Roxb.) Benth.**
　　**var. javanica (Benth.) Bak.**
　　**var. phaseoloides**
　　**var. subspicata (Benth.) Maesen**
*Pueraria pilosissima* Bak. f.
*Pueraria pseudo-hirsuta* Tang and Wang
**Pueraria pulcherrima (Koorders) Merr.**
*Pueraria quadristipellata* Clarke ex W.W. Smith
*Pueraria rigens* Craib
*Pueraria rogersii* L. Bolus
*Pueraria seguini* Lévl. and Vaniot
*Pueraria sericans* K. Schum.

*Pueraria siamica* Craib
**Pueraria sikkimensis Prain**
*Pueraria stracheyi* Bak.
**Pueraria stricta Kurz**
*Pueraria strobilifera* Kurz ex Prain
*Pueraria subspicata* (Benth.) Benth.
**Pueraria tanaii Ozaki (fossil)**
*Pueraria tetragona* Merr.
*Pueraria textilis* Lauterb. and K. Schum.
*Pueraria thomsoni(i)* Benth.
*Pueraria thunbergiana* (Sieb. and Zucc.) Benth.
    var. *formosana* Hosokawa
*Pueraria tonkinensis* Gagn.
*Pueraria triloba* (Houtt.) Mak.
*Pueraria triloba* (Lour.) Backer
*Pueraria triloba* (Lour.) Mak. ex Backer
**Pueraria tuberosa (Roxb. ex Willd.) DC.**
*Pueraria volkensii* Hosokawa
**Pueraria wallichii DC.**
*Pueraria warburgii* Perkins
*Pueraria yunnanensis* Franch.
*Shuteria anabaptista* (Kurz) Wu
*Shuteria ferruginea* (Kurz) Bak.
**Shuteria hirsuta Bak.**
**Sinodolichos lagopus (Dunn) Verdc.**
*Stizolobium montanum* (Lour.) Spreng.
**Vigna radiata (L.) Wilczek**
*Zeydora agrestis* Lour. ex Gomes

# 2 Ethnobotany of *Pueraria* species

*Connie Cox Bodner and Theodore Hymowitz*

## INTRODUCTION

Jones (1941) defined *Ethnobotany* as "the study of the interrelations of... man and plants," urging that this study not be limited solely to human use of plants, but instead expanded to include all aspects of contact between plants and humans. He characterized the discipline as bridging the plant sciences and anthropology, drawing upon both for its approach, methods, and data. He then identified the task of the ethnobotanist as that of correlating the data on problems of interest to researchers in these disciplines and then presenting the results in a form that is useful to the plant scientist, the anthropologist, or both.

We have adopted such an approach for a cross-cultural investigation of human interrelations with the 16 member species of the Asiatic genus *Pueraria* DC. (Leguminosae). Data were gleaned from sources in botany, agronomy, anthropology, chemistry, medicine, and history. *Pueraria* of East and south-east Asia, and Oceania are emphasized, but that of other areas where it is particularly well documented and has specific relevance is also considered.

*Pueraria* DC. was delimited as a genus by de Candolle in 1825. With 16 species it is the second largest genus in the subtribe *Glycininae* Benth. (Lackey, 1977). Information on distribution, ecology, and habitat is scanty or simply unavailable for many areas where *Pueraria* occurs. The cytology, chemistry, and medicinal properties of many species have never been studied. The historical picture is incomplete, and ethnographic, linguistic, and archaeological data are sorely lacking for most species and geographical areas. This is surprising, considering that humans have exploited *Pueraria* for food, medicine, fiber, forage, and ground cover on nearly every continent of the world.

Three species have been cultivated for a long time over a large geographical area, but none was ever domesticated, a fact that holds import for considering the nature of domestication processes. Two species have been transported by humans from an original mainland Asia center of diversity to China, Japan, India, south-east Asia, and Oceania into Europe, Africa, and the Americas. Much of the Asian/Oceanic spread probably took place in prehistoric times, and certain *Pueraria* species may have been among the earliest food crops in south-east Asia and Oceania. As such, they hold an important potential for aiding the understanding of early agriculture in that area. Lastly, the study of *Pueraria*–human relationships offers a unique opportunity to examine the nature, significance, and consequences of shifts in human attitudes toward plants. An analysis of how and why such changes occur can be beneficial to plant scientists and anthropologists alike who are interested in the dynamics of human–plant relationships.

BOTANY

## Morphology

Members of the genus *Pueraria* are robust climbing or trailing perennials, and some have tuberous roots. Stems are often elongated, up to 20 m in length. Leaves are often large (15–18 cm, long, 10–15 cm wide) with three entire or sinuately lobed leaflets arranged pinnately. The stipules are sometimes produced below the point of insertion. Inflorescences are axillary, raceme-like, or paniculate, and are often very long (30–40 cm). The flowers vary from white to blue and purple and are often clustered on reduced side branches along the rachis; bracts usually are obvious. Corollas are of small- to medium-size. Fruits are 3–11 cm long, linear, compressed, many-seeded, and usually pubescent (Verdcourt, 1979; Duke, 1981).

## Taxonomy

See Maesen, this volume.

## Cytology

Chromosome counts have been reported for only four of the sixteen *Pueraria* species (Table 2.1). Gametophytic numbers of 10 and 11 and sporophytic numbers of 20 and 22 correspond to those of other members of subtribe *Glycininae*. The counts of $n = 12$ and $2n = 24$ are probably incorrect (Lackey, 1977).

## Chemical composition and nutritional value

Analyses of chemical composition and nutritional value are available for only two species, *P. montana* (Lour.) Merr. var. *lobata* (Willd.) and *P. phaseoloides* (Roxb.) Benth., which are economically important as forage and fodder crops. These data are presented in

*Table 2.1* Chromosome numbers reported for *Pueraria* species

| Species | $n$ | $2n$ | References |
|---|---|---|---|
| P. stricta | | | |
|   as P. colletti | – | 20 | Lackey (1977) |
| P. montana var. lobata | – | 24 | Duke (1981) |
|   as P. hirsuta | – | 22, 24 | Lackey (1977) |
|   as P. thunbergiana | – | 24 | Lackey (1977) |
| | 11, 12 | – | Darlington and Wylie (1955) |
| P. phaseoloides | – | 22 | Duke (1981) |
| | – | 20, 22 | Lackey (1977) |
| | 11, 12 | – | Darlington and Wylie (1955) |
|   as P. javanica | – | 22, 24 | Lackey (1977) |
| | 11, 12 | – | Darlington and Wylie (1955) |
| P. tuberosa | 11 | – | Lackey (1977) |

Table 2.2 Chemical constituents and nutritional value of *Pueraria* species and selected tropical food crops

| Species | Plant part analyzed | Moisture % | Protein % | Fat % | Carbohydrate % | Fiber % | Ash % | Other % | References |
|---|---|---|---|---|---|---|---|---|---|
| *P. montana* | Raw tuber | 68.6 | 2.1 | 0.1 | 27.8 | 0.7 | 1.4 | – | Duke, 1981: 211–212 |
| var. *lobata* | Fresh leaves | 76.9 | 4.0 | 0.6 | 10.2 | 6.8 | 1.5 | – | Shurtleff and Aoyagi, 1977: 87 |
| | Tuber starch | 16.5 | 0.2 | 0.1 | 83.1 | – | 0.1 | – | Duke, 1981: 212 |
| *P. phaseoloides* | Fresh leaves, stems | 80.9 | 3.8 | 0.4 | 7.9 | 5.5 | 1.5 | – | Duke, 1981: 214 |
| *Colocasia esculenta* | Corm | 63–85 | 1.4–3.0 | 0.2–0.4 | 13–29 | 0.6–1.2 | 0.6–1.3 | – | Coursey, 1968: 27 |
| | Fresh leaves | 87.2 | 3.0 | 0.8 | 6.0 | 1.4 | 1.6 | – | Purseglove, 1972: 63–64 |
| *Dioscorea* spp. | Tuber | 65–75 | 1.0–2.5 | 0.05–0.20 | 15–25 | 0.5–1.5 | 0.7–2.0 | – | Purseglove, 1972: 111–112 |
| *Ipomoea batatas* | Tuber | 70.0 | 1.5–2.0 | 0.2 | 27.0 | 1.0 | – | – | Purseglove, 1968: 85 |
| | Fresh leaves | 86.0 | 3.2 | 0.8 | 8.5 | – | – | 1.5 | Purseglove, 1968: 85 |
| *Manihot esculenta* | Tuber | 62.0 | 1.0 | 0.3 | 35.0 | – | – | 1.7 | Purseglove, 1968: 282 |
| *Pachyrhizus erosus* | Tuber | 87.1 | 1.2 | 0.1 | 10.6 | 0.7 | 0.3 | – | Purseglove, 1968: 177 |
| *Xanthosoma* spp. | Tuber | 70–77 | 1.3–1.7 | 0.2–0.4 | 17–26 | 0.6–1.9 | 0.6–1.3 | – | Coursey, 1968: 27 |

Table 2.2 along with comparative data for other selected tropical crops. These are not definitive comparisons however, due to differences in analytical techniques. The overall chemical compositions of *Pueraria* species seem to most closely approximate those of *Ipomoea batatas* and *Dioscorea* spp.

Considerable research effort also has been directed toward the isolation and characterization of chemical compounds present in the tubers, flowers, and leaves of *Pueraria* spp. These compounds include oils, amino acids, flavonoids, steroids, and anthocyanins (see Rong *et al.* and Parks *et al.*, this volume).

## Plant growth and life cycle

Members of the *Pueraria* genus are perennials. *P. montana* var. *lobata* can be propagated from seed, cuttings, or crowns. Seeds planted in the spring usually develop four to six true leaves and at least one root approximately 1.3 cm in diameter and 15 cm in length in four months. These plants can grow to a length of 14 m in one year (Duke, 1981). During the growing season, early spring to late fall, a dense cover of broad-leafed foliage is produced. Leaves are dropped after the first frost, often numbering 250–350 per square meter, and thereby leaving a substantial layer of soil-enriching organic material (Whyte *et al.*, 1953; Kumar, 1977).

Growth resumes the next spring, and new plants are established at the nodes where vines are in contact with the soil. Roots of the new plants enlarge and form new crowns. Each crown supports three to five vines, which spread rapidly. Young vines are ≈1.3 cm in diameter, while one- to two-year-old vines are ≈2.5 cm and woody. Older vines can measure 10 cm or more across. By the end of the second growing season, the vine connecting the crowns of the old and new plants dies, with each crown producing new vines the following year (Dickens, 1974).

The root system of *P. montana* var. *lobata* spreads horizontally to an average depth of 1 m, although depths of 2.5 m have been reported in the south-eastern United States. Individual roots swell and form large fleshy tuberous structures in which starch is stored over the winter to support plant growth in the spring. Such roots range in size from 3.8 cm in diameter for young vines to 18 cm in diameter and 2.5–3 m in length for older vines. Roots 45–50 cm in diameter weighing 100–180 kg have also been reported.

Flowers and seeds are produced in the late summer and early fall of the third year, with production increasing each year thereafter. *P. montana* var. *lobata* is reported to be cross-pollinated by bees.

*P. phaseoloides* grows throughout the year whereas *P. montana* var. *lobata* has a dormant period. Seeds are planted at the beginning of the rainy season. Growth is slow for the first few months, but the plant usually covers the ground in about 6 months. *P. phaseoloides* and *P. montana* var. *lobata* produce runners that root at both the nodes and internodes. With sufficient light and water, the plants will continue to spread. The root system of *P. phaseoloides* is deep (1.5 m) and widely branched with starchy tuberous structures formed at various points. *P. phaseoloides* sets more seed than *P. montana* var. *lobata*, and seed production is greater when plants are allowed to climb (Whyte *et al.*, 1953; Duke, 1981). Both *P. montana* var. *lobata* and *P. phaseoloides* produce runners. They can grow successfully and spread even in areas that are not conducive to flowering and seed setting.

Unfortunately, similar information for other species of the genus is unavailable.

## Ecology

Most *Pueraria* species are adapted to monsoon forests with medium rainfall and are found in open areas, forest edges, scrub jungle, and disturbed areas, usually climbing in shrubs and trees. *P. phaseoloides* requires humid climate with moderate to high rainfall and temperatures. An extensive root system allows the species to withstand short periods of drought and tolerate waterlogging and short periods of flooding. *P. phaseoloides* grows successfully from 22.1–27.4 °C. Cultivation is difficult below 18.3 °C (Whyte *et al.*, 1953; Purseglove, 1968; Kumar, 1977; Skerman, 1977; Duke, 1981).

*P. phaseoloides* grows in a wide range of soil types, including acidic, rocky upland and fertile lowland soils. Soil pH suitable for supporting *P. phaseoloides* ranges from 4.3 to 8.0. This species grows well in full sun or moderate shade, but it cannot tolerate dense shade. It is a low elevation species, generally found below 600 m, although it does occur at higher elevations in some regions, up to about 1600 m. It grows along rivers, roads, in rice fields, in evergreen forests, and in open grassland (see also Verdcourt, 1979; Maesen, 1985).

In contrast to *P. phaseoloides*, *P. montana* var. *lobata* is not adapted to the tropics, preferring instead the climate of the humid subtropics and warm temperate regions. *P. montana* var. *lobata* also is deep-rooted and therefore drought-resistant, but it cannot tolerate waterlogging. Aerial ground parts of the plant are killed by frost, and deep-freezing kills the entire plant.

*P. montana* var. *lobata* also grows on a wide range of soil types, but prefers well-drained loams of good fertility. Soil pH suitable for supporting *P. montana* var. *lobata* ranges from 5.0 to 7.1 (Duke, 1981). This species grows at considerably higher altitudes than *P. phaseoloides*, from near sea level on Fiji (Parham, 1943) and Niue Island (Yuncker, 1943), 1000 m in Japan (Shurtleff and Aoyagi, 1977), to 2000 m in the Philippines (Merrill, 1923) and New Guinea (Verdcourt, 1979). It grows in forests, thickets, hedges, swamps, and along riversides and roadsides.

*P. tuberosa* (Roxb. ex Willd.) DC. is a tropical species, but it grows well throughout India and Upper Burma except where climatic conditions are very wet or arid (Baker, 1897; Rao, 1958; Krishnamurthi, 1969; Arora and Chandel, 1972). It is common in the deciduous hill forests (Ahuja, 1965; Bentham, 1867; Gamble, 1967) and grows well in exposed and eroded areas. Collections have been made from near sea level to 1300 m (Watt, 1892; Collett, 1971; Maesen, 1985).

Information regarding the ecology of other *Pueraria* species is very meager. In general species occur at low to medium elevations, up to 1500 m. Exceptions are *P. calycina*, *P. edulis* and *P. peduncularis* which grow at altitudes up to 2600 m, 3300 m and 3600 m, respectively.

## Diseases and pests

*Pueraria* species are affected little by insects or diseases. Leaves of *P. montana* var. *lobata* are attacked by velvet bean caterpillars (*Anticarsia gemmatilis*), and roots are damaged by several nematodes (*Meloidogyne hapla, M. incognita acrita, M. javanica, M. thamesi*, and *Rotylenchulus reniformis*). Bacterial blight and halo blight are caused by *Pseudomonas syringae* pv. *phaseolicola* and *Ps. syringae* pv. *syringae*, respectively (Völksch and Wingart, 1997). Fungi that may cause damage are *Alternaria* spp. (leaf-spot), *Colletotrichum lindenuthianium* (anthracnose), *Fusarium* spp. (stem rot), *Macrohomina phaseolina* (charcoal rot),

*Mycosphaerella puericola* (angular leaf-spot), and *Rhizoctonia solani* (damping-off) (Duke, 1981). Powell (1974) reported that *Mycovellosiella* spp. (yellow mold) and *Synchytrium minitrum* (false rot) are problematic in New Guinea.

*P. phaseoloides* is equally resistant to attack. Reported pests include leaf-eating caterpillars, which cause damage in ungrazed plots, and pod-borers, which interfere with seed production (Skerman, 1977). Blackbirds and pigeons reportedly eat recently planted seeds in Puerto Rico (Duke, 1981).

## GEOGRAPHY

Data from herbarium collections, floras, ecological surveys, and other botanical studies were combined to produce distribution maps of *Pueraria* species (Maesen, this volume). Such results are tenuous at best, however. Precise distribution of the genus awaits systematic collections throughout the Indian subcontinent, East and south-east Asia, and the Pacific. All but one of the sixteen species have been collected within an area comprising north-east India, Upper Burma, northern Thailand, and southern China, an apparent geographical center of diversity for the genus.

Using Harlan's (1992) definitions of the terms domesticated, cultivated, escaped, and wild, thirteen *Pueraria* species can be classified as wild. They are not tended or cared for by humans, although one species, *P. mirifica*, is exploited for medicinal purposes (Kashemsanta *et al.*, 1952; Perry, 1980). The remaining three species, *P. montana* var. *lobata*, *P. phaseoloides*, and *P. tuberosa*, are each known in both cultivated and wild forms. Cytology data (Table 2.1) suggest no genetic difference between the wild and cultivated forms for any of these species, nor is there any indication of reduced sexual fertility in the cultivated forms when the ecological requirements of the species are met. They are therefore not considered domesticated species.

In several instances, cultivated *Pueraria* species have escaped and returned to the wild. Perhaps the best publicized is *P. montana* var. *lobata* (kudzu) in the south-eastern United States (see Lowney, this volume). Originally introduced from Japan as an ornamental plant, kudzu eventually found use as a source of fodder, ground cover, and a means of enriching nitrogen-depleted soils. However, its growth is so vigorous and uncontrollable, by 1970, it was officially declared a noxious weed (Focht, 1972; Winberry and Jones, 1973). Kudzu has now spread to New York City (Frankel, 1989) and Connecticut (Stewart, 2000).

*P. montana* var. *lobata* and *P. phaseoloides* returned to the wild state after cultivation in the islands of western and southern Pacific, including Japan (Ohwi, 1965); Okinawa, Kume Island, Miyako, Ishigaki, Iriomote (Walker, 1976); the Philippines (Merrill, 1918, 1923); Indonesia (Merrill, 1917; Backer and Bakhuizen van den Brink, 1963); New Guinea, New Britain, New Ireland, Bougainville, Guadalcanal (Powell, 1974; Verdcourt, 1979); Niue Island (Yuncker, 1943); New Caledonia, the New Hebrides, Fiji, Tonga, and Samoa (Leenhardt, 1930; Christophersen, 1935; Parham, 1943; Yuncker, 1959; Haudricourt, 1964; Barrau, 1965; Roberts, 1970; Herklots, 1972). The apparently extensive distribution of wild *P. tuberosa* throughout Indian subcontinent suggests that this species too may have escaped cultivation several times in the past.

That *Pueraria* species never became domesticated is not surprising when the nature of both the plants and their exploitation is considered. In each case it is the tuber or the leaves and stems of *Pueraria* that humans find useful. Small, young tubers usually are preferred because large ones can be tough, woody, and difficult to harvest and process. Consequently, there is no selection for larger roots or stems. Although the portion of the *P. montana* var. *lobata* root that grows above ground has been regarded as somewhat poisonous (Smith, 1969), none of the *Pueraria* species is considered toxic. In addition and in contrast to certain other tropical root crops (e.g. *Dioscorea* spp., *Manihot* spp.), none has any other natural defense system that poses serious problems to their easy exploitation by humans. Consequently, there has been no need for selection against toxicity, spiny growths, or a deep-burying habit. Because these species can reproduce by cuttings and crowns as well as or better than by seed, there is little need to select against shattering seed cases or for increased seed production.

The question of the geographic origin of *Pueraria* cultivation remains. The evidence indicates that a center of genetic diversity for the genus may be identified in the area comprising north-east India, Burma, Thailand, and southern China, and although three *Pueraria* species have been widely cultivated, none has ever been domesticated. Further, the nature of the exploitation of these species by humans is such that little if any selection pressure is exerted. Not surprisingly, there are no secondary centers of diversity for *Pueraria* species; the geographic center of diversity and the geographic origin of cultivation are one and the same (Li, 1970). The evolution pattern best fits Harlan's (1992) monocentric class, which consists of crops with definable centers of origin and wide dispersals without secondary centers of diversity.

## HISTORY

The history of human relationships with *Pueraria* can be traced via written records from Asia proper to Oceania, Europe, North America, South America, and Africa, beginning in at least the sixth century BC. The prehistoric picture is harder to piece together because direct archaeological evidence pertaining to the early exploitation of *Pueraria* is lacking. It is unlikely that carbonized *Pueraria* seeds will be recovered in an archaeological context since it was most probably the tubers, stems, and leaves that were most frequently utilized, and these materials are rarely preserved under tropical conditions. However, recent advances in detecting the presence of diagnostic starch grains on stone tools relating to the cultivation of manioc, yams, arrowroot, and maize in Panama (Piperno *et al.*, 2000) suggest that data relating to *Pueraria* may be recoverable from like contexts. Its prehistory, for now however, must remain a matter of speculation.

### *Pueraria montana* var. *lobata*

#### *China*

The earliest written reference to the use of *Pueraria montana* var. *lobata* in China is in the classic *Shih Ching*, a collection of 305 folk songs, odes, sacrificial psalms, and poems composed by numerous authors between 1000 BC and 500 BC (Keng, 1974). *P. montana* var. *lobata*, or *ko*, is mentioned in nine poems, and in addition to *Cannabis sativa* L. (hemp) and *Boehmeria nivea* (L.) Gaud. (ramie), is identified by Keng as one of three fibrous plants used to manufacture summer cloth in northern China at that time.

Bretschneider (1881) ascertained that *P. montana* var. *lobata* (his *P. thunbergiana*) was among the plants listed in the *Shen Nung Pen Ts'ao*, which is often referred to as the earliest treatise on Chinese pharmacopoeia and the basis for all subsequent herbal studies. Traditionally, this work is attributed to the Emperor Shen Nung, who is said to have reigned from 2838 to 2698 BC. However, Wong and Wu (1936), among others, have suggested that the book was actually written during the first century BC.

*P. montana* var. *lobata* also is mentioned in the classic medical book *Shang Han Lun*, or *Treatise on Fevers*, written in the second century AD. *Ge-gen Tang*, a decoction prepared from the root, is suggested as treatment for cases of neck stiffness, lack of perspiration, and aversion to air drafts (Fang, 1980; also Zhu, this volume).

Reference to the plant is made again in the *C'hi Min Yao Shu*, or *Essential Ways for Living of the Common People* (Chia Ssu-hsieh, *c.* AD 540). Intended as a practical guide for the general improvement of rural life, this book consists of excerpts from classics, contemporary books, proverbs, and folk songs as well as experts' opinion and personal experience (Shih, 1974). *Pueraria* is discussed as one of several wild plants (e.g. *Zizania, Andropogon*) used as sources of fiber but never cultivated as crops in northern China.

Lastly, *P. montana* var. *lobata* is included in a discussion of cultivated textile plants in the *Nung Cheng Ts'uan Shu*, or *Complete Treatise on Agriculture*. Published in 1640, this 60-volume work was written by Su Kuang K'i (1562–1633) of Shanghai (Bretschneider, 1881).

Martini reported a textile plant found in the province of "quei cheu" (Guizhou) and referred to it by the proper Chinese name for *P. montana* var. *lobata*, *Co* (Martini, 1655). Le Comte also identified the plant in the English translation of *Nouveaux Mémoires sur l'Etat de la Chine*, noting that "the Linnen which is the most valued, and is to be found no where else, is called *Coupou*; because it is made of a Plant that the People of the Country call *Co*, found in the Province of *Fokien* [Fujian]" (Le Comte, 1698).

The earliest reference by a European to the use of what probably was *P. montana* var. *lobata* is the mention of the Chinese manufacture of "stuffs of the bark of certain trees which form very fine summer clothing" by Polo (1932) in his account of his late thirteenth-century travels in China.

*P. montana* var. *lobata* was collected by several European missionaries, administrators, and naturalists stationed in China during the eighteenth and nineteenth centuries. The first was the English physician James Cunningham. Among his plant collection was *co*, described by botanist Leonard Plunkenet in his *Amaltheum botanicum* of 1703–1704. Quoting probably from Cunningham's original description, Plunkenet indicated that the plant was edible and was used in making *co-pou*, a wearable cloth (Bretschneider, 1880). In 1793, Sir George Leonard Staunton collected the plant as *Dolichos hirsutus* near Beijing and as *Pachyrhizus thunbergianus* in Zhejiang. In 1824, French missionary Joseph Etienne Polycarpe Voisin obtained seeds of the plant *ko*, which were relayed through Stanislas Julien to the Museum d'Histoire Naturelle in Paris. Plants grown from these seeds were later examined by Prof. de Jussieu, who characterized them as a *Phaseolus* cousin of *Dolichos* and particularly *Dolichos bulbosus* (Julien and Champion, 1869). One of these plants was then sent to the herbarium of the Botanical Gardens of St Petersburg, where it was labeled *Pachyrhizus trilobus*, and later recognized as *P. montana* var. *lobata* (Bretschneider, 1898).

In 1867, William Carles of the British Consular Service produced an unpublished account of the manufacture of fiber from *P. montana* var. *lobata* by the Chinese, and at about the same time, Father Armand David, a French priest and naturalist stationed in northern Jiangxi, reported seeing Chinese men carrying loads of stems from the "haricot-chanvre" (hemp bean), a wild plant esteemed for its textile fibers which were used in making cloth (referred to as *Pachyrhizus thunbergianus*) (Bretschneider, 1898). Bretschneider himself collected the plant in the plain and mountains surrounding Beijing and sent seeds to the Société d'Acclimatation in 1881. Lastly, in 1885, William Marsh Cooper, an interpreter and consul for the British in China, filed an unpublished report on the fiber of *P. montana* var. *lobata* and the cloth made from it in Ningpo.

### Japan

Because the Chinese character for *ge* (Figure 2.1) is identical to the Japanese character for kudzu, the Japanese name for *P. montana* var. *lobata*, it has been widely suggested that the plant was used first in China, and later taken to Japan. This correlates with evidence from the distribution and geography of the genus discussed earlier. Shurtleff and Aoyagi (1977) have speculated that *P. montana* var. *lobata* was used in manufacturing fences, baskets, and rough cloth as early as the Jomon period (*c*. 10 000 BC–400 BC), and that techniques for weaving the fibers were quite refined by the end of the Yayoi period which continued until AD 400 in Japan. There is no archaeological evidence for weaving in Japan before the Yayoi, although it is agreed that the early cloth was woven on primitive looms using vegetable fibers of some type (Chard, 1974).

The first written reference to *P. montana* var. *lobata* in Japan is perhaps in the *Manyoshu*, a collection of poems compiled around AD 600 (Shurtleff and Aoyagi, 1977). In the ninth century, the use of the leaves as a wild vegetable was described in the *Wamyosho*. Mention of the plant also is made in the ninth-century *Gi-shiki* account of the ritual *Ohotono Hogahi*, or "Luck-Wishing of the Great Palace." Kuzu (*P. montana* var. *lobata*) is listed as one of the climbing plants from which cords were made in the description of the palace of the Japanese sovereign. The framework of the palace was tied together using these cords (Satow, 1906).

Shurtleff and Aoyagi suggested that by AD 1200 farmers of central and southern Japan were extracting starch for cooking and for medicinal use. Cloth woven from *P. montana* var. *lobata* fiber was in production by the thirteenth century. Commercial production of starch powder began in 1610 in Nara Prefecture, and at roughly the same

*Figure 2.1* Ge, the Chinese character for *Pueraria montana* var. *lobata*.

time kudzu cloth became well-known throughout Japan for its durability, resistance to tearing, and the protection afforded the wearer.

*P. montana* var. *lobata* was included in the Japanese pharmacopoeia during the Edo period (AD 1600–1867). A white powder extracted from the root and dried cubes of unprocessed root were used in preparing medicines for treating a wide variety of ailments and disorders.

*P. montana* var. *lobata* root also became established in Japanese cooking as evidenced by the inclusion of recipes for "kudzu mochi" and "grilled kudzu mochi" in the *Ryori Monogatari* (*The Story of Cooking*) which dates to *c.* 1620.

Since 1868, commercial demand for kudzu cloth has decreased as western influences became stronger. However, new looms were developed to produce grasscloth from *P. montana* var. *lobata* fiber and opened new markets in Japan and abroad, particularly the United States. This industry flourished until the 1960s when Korea began its own kudzu industry and became the center of kudzu cloth production. Kudzu cloth is still woven in Japan as a small-scale cottage industry (Shurtleff and Aoyagi, 1977).

### Europe

*P. montana* var. *lobata* was introduced to France from Japan and China, as discussed above. In 1878, it was sent to Paul de Mortillet, who described it to E.A. Carrière. Carrière (1891) published an account of the plant and outlined its value as an ornamental, a source of starch, and as a possible raw material for paper production. From France, the plant was introduced to Germany, where it was enthusiastically welcomed as an ornamental vine (Wittmack, 1896). Eventually, *P. montana* var. *lobata* became acclimatized in Switzerland, the Mediterranean region, the Crimea and Caucasus regions, Australia, and South America.

### Oceania

*P. montana* var. *lobata* is distributed throughout much of Oceania, both as a wild and cultivated species. This has been used to fuel one of the major controversies in Oceania culture history, namely, the geographic and cultural origins of the Polynesians. However, reconstruction of the history of *P. montana* var. *lobata* in this region has been fraught with difficulties due to a lack of early written records and pertinent archaeological remains, and a number of misunderstandings involving early erroneous botanical identifications of the plant.

Early botanical collections were made throughout the Pacific as part of Captain James Cook's voyages. A substantial collection was made by Parkinson on Cook's 1769 voyage to Tahiti, and Merrill (1954) believed that *Pueraria* was among the several unclassified and unnamed economic plants included. Unfortunately, these particular plants were not described, but on the basis of the local names cited by Parkinson, Merrill tentatively identified one as *Pueraria*, one as *Dioscorea*, and another as *Cyrtosperma*.

Seemann (1865) collected *P. montana* var. *lobata* in Fiji, the New Hebrides, and New Caledonia but misidentified the material as the American *Pachyrhizus trilobus* DC. Some 40 years later, Guppy (1906) included *Pachyrhizus trilobus*, but again meant *P. montana* var. *lobata*, in his list of food plants of the pre-Polynesians, the earliest inhabitants of the Pacific (Barrau, 1956, 1958, 1965). Seemann's error was corrected by Smith (1942), but the correction went unnoticed by many (Merrill, 1954; Barrau, 1956). As a result, it became an accepted truth to some that the American species *Pachyrhizus trilobus* (now *Pachyrhizus erosus* [L.] Urb.) was widely distributed throughout Oceania in pre-contact times. Heyerdahl (1952) used Seemann's original misidentification of *P. montana* var. *lobata*, Guppy's classification of it as an early food plant, and various ethnographic reports of the plant's current use as a famine food and as a source of fiber in Oceania to support his contention that Amerindians did indeed engage in transpacific voyaging in prehistoric times and were very likely responsible for settling Polynesia. There should be no question now, however, that the plant collected by Seemann and discussed by Guppy was *P. montana* var. *lobata*, an Asiatic, not American, species.

How and when *P. montana* var. *lobata* was brought to the Pacific Islands remain uncertain. Powell (1976) suggested a hypothetical reconstruction of the vegetation and environment for New Guinea at the time of the arrival of humans at 11 000 B.P. She has suggested that the montane forests were rich in fruit and nut trees, including *Pandanus, Elaeocarpus, Castanopsis*, and *Sterculia*, and that *P. montana* var. *lobata*, edible ferns, herbs, and woody species with edible leaves also may have been available. In light of the natural distribution of *Pueraria*, its center of diversity, and the origin of cultivation, we believe it is more probable that *P. montana* var. *lobata* was part of the floral assemblage introduced by the new settlers. When and where this took place and whether it was a single event or a recurring introduction are not known.

A number of researchers have speculated that regardless of exactly when and where *P. montana* var. *lobata* was introduced, it became an important staple crop for at least some Oceania populations in prehistoric times (Bulmer and Bulmer, 1964; Watson, 1968; Golson, 1976; White and Allen, 1980). The argument that crops of presently marginal importance and/or those that are utilized only in times of shortage are remnants of an ancient system of cultivation has been used not only for *P. montana* var. *lobata*, but for *Cordyline fructicosa, Dioscorea nummularia*, and other crops. However, Bellwood (1979) and others have pointed out that while such an argument may provide an attractive means of explaining the apparent decline of a geographically widespread introduced plant, there is no *a priori* reason why this must be the case.

### United States

The history of *P. montana* var. *lobata* in the United States is by far the best documented (Ahlgren, 1949; Stephens, 1953; Stevens, 1976; Stewart, 2000) and as a result is both informative and enlightening in terms of human relations with not only this species but with plants in general (see Lowney, this volume; Winberry and Jones, 1973).

### India

*P. montana* var. *lobata* was introduced to India from the United States in 1925–1926, where it was grown experimentally at the Imperial Agricultural Research Institute at Pusa, Bihar (Kumar, 1977). It has produced favorable results in some areas (e.g. Bihar, Uttar Pradesh, Delhi, Assam) but has performed poorly in others (e.g. Mysore, Coorg, South Kanara, Poona) (Krishnamurthi, 1969; Narayanan and Dabadghao, 1972). It is most often planted as green manure, a fodder crop, and as a cover crop on land that is otherwise unsuitable for cultivation.

## *Pueraria phaseoloides*

The first reference to the movement of *P. phaseoloides* is Roxburgh's (1874) note of having received seeds of *Dolichos phaseoloides* from Mr Kerr of Canton at the Botanic Garden of Calcutta in 1804. There, the plants thrived, blossomed, and set seed. In 1919, seeds were received by the United States Office of Foreign Seed and Plant Introduction from Darjeeling, Bengal, India (USDA, 1922). From the United States, *P. phaseoloides* was introduced to Puerto Rico as a forage crop in 1943. Although it was not well received at first, eventually the plant was found useful as a means of halting erosion and as a nitrogen-providing ground cover for banana and plantain acreage as well as forage for dairy cattle (Smith and Chandler, 1951).

In the same year, *Pueraria phaseoloides* was introduced to Fiji and was distributed widely as a fodder plant (Parham, 1949). While it is still used for feeding livestock, it has a greater value as a ground cover under developing tree crops (e.g. coconuts, oil palm) (Roberts, 1970).

*P. phaseoloides* was also introduced to New Caledonia, West and East Africa, and the West Indies, where it is valued as a ground cover and as green manure (Whyte *et al.*, 1953; Guillaumin, 1954; Duke, 1981).


## ANTHROPOLOGY

### Economic uses

The results of a survey of the literature relating to the economic uses of *Pueraria* species are presented in Table 2.3. In brief, *P. montana* var. *lobata* is used as a source of food for humans and livestock, as fiber for the manufacture of a wide variety of items, and as medicine for treating various ailments and disorders. It is also valued as an ornamental or shade plant, as a ground cover, and it holds considerable potential as a raw material for fuel production via anaerobic fermentation. The root of *P. candollei* var. *mirifica* is traditionally used in Thailand to make a rejuvenating tonic, and that of *P. peduncularis* is used as an insecticide in China.

*P. phaseoloides* is employed widely in the tropics as a ground cover and green manure, and like *P. montana* var. *lobata*, it is valued for its soil-enriching properties. Parts of the plants also are used medicinally, as food for humans and livestock, as tying materials, and in at least one instance as a source of magic. Lastly, *P. tuberosa* also is exploited as food for humans and livestock and as medicine to treat a number of ailments.


### Indigenous cultivation, harvesting, and preparation techniques

Fragmentary data pertaining to agricultural practices associated with *Pueraria* species are available for New Guinea, New Caledonia, Fiji, China, Japan, Vietnam, Indonesia, and the Philippines, although it is only in New Guinea that specific research on *Pueraria* cultivation has been conducted.

In his description of the structure of gardens in the New Guinea Highlands, Barrau (1958) noted that several plant species were grown on earthen banks inside garden fences. These fences were constructed to exclude pigs and were themselves protected from the elements by their own thatched roofs. Drainage ditches were dug along the outsides of the fences, with the earth piled inside providing garden areas. Plants grown

Table 2.3 Economic uses of *Pueraria* species

| Species | Plant part | Economic use | Geographical area | References |
|---|---|---|---|---|
| *Pueraria montana* var. *lobata* | Whole plant | To treat skin rashes | China | Smith, 1969 |
| | | Used in unspecified medicines | Tonga | Yuncker, 1959 |
| | | Ritual plant associated with myth | New Guinea | Girard, 1957 |
| | | Used in unspecified ceremonies | Fiji | Parham, 1943; Seemann, 1865 |
| | | Thickets provide source of preferred honey | Japan | Shurtleff and Aoyagi, 1977 |
| | | To prevent and/or halt soil erosion | United States | Duke, 1981; Bailey, 1939 |
| | | Forage for livestock | Japan, United States, tropics, subtropics | Siebold, 1827; Vieillard, 1862; Alexander, 1930; Porterfield, 1938; Guillaumin, 1954; Yuncker, 1959; Herklots, 1972; Shurtleff and Aoyagi, 1977; Duke, 1981 |
| | | Fuel production via anaerobic fermentation | United States | Tanner *et al.*, 1979; Wolverton and McDonald, 1981 |
| | | Soil enrichment | United States, subtropics | Duke, 1981; Winberry and Jones, 1973 |
| | | Ornamental, shade plant | United States | Winberry and Jones, 1973 |
| | Seeds | To treat dysentery, alcoholic excess | China | Smith, 1969 |
| | Shoots | To treat insufficient secretion of milk, incipient boils, and aphthous sore mouth | China | Smith, 1969 |
| | Flowers | To treat alcoholic excess Diaphoretic, febrifuge | China, south-east Asia | Smith, 1969; Perry, 1980; |
| | Leaves | To treat dysentery | Malaya | Hooper, 1929 |
| | | | East and south-east Asia | Perry, 1980 |
| | | Styptic | China | Smith, 1969 |
| | | In anthropophagic preparations | New Caledonia | Guillaumin, 1954 |
| | | Decoction ingested as a tonic, cough remedy | East and south-east Asia | Perry, 1980 |
| | Stems | To make baskets, trunks | Japan, French Polynesia | Maclet and Barrau, 1959; Shurtleff and Aoyagi, 1977; |
| | | Fishing line | Japan | Shurtleff and Aoyagi, 1977 |
| | | To make fishing nets | New Caledonia, Fiji | Vieillard, 1862; Seemann, 1865; Parham, 1943; Guillaumin, 1954; |
| | | Tying temporary bundles | Fiji, Philippines | Smith, 1942; Bodner, 1986 |
| | | Rope | New Guinea, Japan | Siebold, 1827; Powell, 1974, 1976; |
| | | To make net bags | New Guinea | Straatsman, 1967 |

Table 2.3 (Continued)

| Species | Plant part | Economic use | Geographical area | References |
|---|---|---|---|---|
| | | In house construction (reinforcing agent in mud/clay walls) | New Guinea, Japan | Satow, 1906; Porterfield, 1938; Powell, 1976; Siebold, 1827; Shurtleff and Aoyagi, 1977 |
| | | As weft thread in weaving cloth | China, Japan | Julien and Champion, 1869; Siebold, 1827; Shurtleff and Aoyagi, 1977 |
| | | To make woven cloth for dressing the dead | Philippines | Bodner, 1986 |
| | | To make paper | Japan | Shurtleff and Aoyagi, 1977 |
| | | To stuff cushions, chairs, beds | Japan | Shurtleff and Aoyagi, 1977 |
| | | Burned as mosquito repellent | Japan | Shurtleff and Aoyagi, 1977 |
| | Tuber | Valued for edible grubs found in stems | Philippines | H.C. Conklin, 1982, personal communication |
| | | Food | Vietnam, New Guinea, Philippines | Bowers, 1964; Watson, 1964, 1968; Strathern, 1969; Powell, 1974, 1976, 1977; Tanaka, 1976; Bodner, 1986 |
| | | | New Caledonia, Fiji, Tonga, Niue Island | Vieillard, 1862; Seemann, 1865; Smith, 1942; Parham, 1943; Yuncker, 1943, 1959; Haudricourt, 1964 |
| | | To treat colds, fever, alcoholic excess, dysentery, chicken pox, measles, skin rash, snake and insect bites, influenza, diarrhea, gastritis, enteritis, diabetes, typhoid fever, flatulence, and retching | East and south-east Asia | Crevost and Lemarie, 1917; Hooper, 1929; Roi, 1955; Smith, 1969; Tseng et al., 1975; Shurtleff and Aoyagi, 1977; Perry, 1980 |
| | | To treat sudden deafness | China | Anon, 1975 |
| | | To relieve symptoms associated with hypertension | China | Tseng et al., 1975; Fang, 1980 |
| | | To counteract croton oil and other poisonous drugs | China | Smith, 1969; Perry, 1980 |
| | | Externally applied to dog bites | China | Smith, 1969 |
| Pueraria mirifica | Tuber | As a tonic for its rejuvenating powers | Thailand | Kashemsanta et al., 1952; Perry, 1980 |
| Pueraria peduncularis | Tuber (?) | Insecticide | China | Perry, 1980 |
| Pueraria phaseoloides | Whole plant | Cultivated as fertilizer, green manure, soil-enricher | Tropical Asia, Africa, America | Burkill, 1935; Smith and Chandler, 1951; Backer and van den Brink, 1963; Purseglove, 1968; Duke, 1981 |

*Table 2.3* (Continued)

| Species | Plant part | Economic use | Geographical area | References |
|---|---|---|---|---|
| | | Livestock forage | Tropical Asia, Africa, America | Smith and Chandler, 1951; Purseglove, 1968; Duke, 1981 |
| | | To prevent and/or halt soil erosion | Tropical Asia, Africa, America | Smith and Chandler, 1951; Purseglove, 1969; Duke, 1981 |
| | Leaves | Chewed as an intoxicant | New Britain | Powell, 1976 |
| | | Rubbed on arrows, bows, and guns for luck in hunting | Philippines | Fox, 1953 |
| | Bark | Extract of crushed bark taken to aid childbirth | New Britain | Powell, 1976 |
| | Stems | Cordage | Indochina, Malaya | Burkill, 1935 |
| | Tuber | Food | Vietnam | Crevost and Lemarie, 1917 |
| | Unspecified | Decoction taken internally to treat foul-smelling ulcers; poultice applied to treat boils in children | Malaya | Burkill and Haniff, 1930; Perry, 1980 |
| *Pueraria tuberosa* | Whole plant | Livestock fodder | India | Rao, 1958; Krishnamurthi, 1969; Collett, 1971; Kay, 1973 |
| | Tuber | Food | India | Watt, 1892; Gamble, 1967; Patel, 1968; Krishnamurthi, 1969; Collett, 1971; Tanaka, 1976 |
| | | Tonic, emetic, galactagogue, cooling medicine, demulcent, and poultice | India | Dymock, 1885/1886; Watt, 1892; Ahuja, 1965; Patel, 1968; Kay, 1973 |
| | | To treat gravel, low fever, ulcers in the nose, and menorrhagia | India | Bodding, 1927 |

in these areas included *Setaria palmiflora, Amaranthus hybridus, Solanum nigrum*, Acanthaceae (*Rungia?*), a sterile Malvaceae (*Hibiscus manihot?*), a *Colocasia, Dolichos lablab*, and a legume with edible roots, which Barrau noted was most likely *Pueraria*.

Watson (1964) described the cultivation of *P. montana* var. *lobata* in gardens in the Kainantu subdistrict of New Guinea. The plants climbed on poles 3–4 m high in gardens, and was propagated via layered vine cuttings 30–60 cm long, tied in a loose knot, and sometimes with the ends split to encourage sprouting. He noted that cultivation seemed easy, that insects were not a problem, and that the plant bore reliably, particularly in bush gardens. Tubers were harvested over a period of years, with each planting yielding between five and ten tubers each, with some tubers weighing as much as 36 kg and measuring over 1.2 m long by 0.5 m in diameter.

Watson also recorded variation in cultivation practices between grassland and bushland groups. Grassland villagers occasionally gathered wild *P. montana* var. *lobata* specimens and sometimes replanted and/or tended them. Informants noted that unless the plants were allowed to grow on poles, however, the tubers were scattered and did not attain the size of well-tended ones. As a crop, *P. montana* var. *lobata* was planted on the edge of the garden in the earth dug out of the trenches while the center was devoted to the cultivation of another crop, often sweet potatoes. In bush gardens, *P. montana* var. *lobata* was not limited to garden edges, although it was not planted in orderly rows as were yams. It was often planted near the house close to a small tree for the vine to climb as well as in the garden proper (Watson, 1965). Bushland and grassland villagers alike acknowledged that the crop grew better in bush gardens.

Watson indicated that the roots could be roasted, cooked in an earth oven with pork and other foodstuffs for consumption the next day, or boiled with greens. Even after a long period of cooking, *P. montana* var. *lobata* root was often hard and woody with no distinctive flavor.

In a more detailed paper, Watson (1968) presented data collected during the course of two surveys in 1963–1964 in the Highlands and one in 1967 in the Dani (Balim) area of West Irian. Cultivation techniques varied, but in all cases, propagation was by vine cuttings and poles were usually supplied to support the plants unless a small tree was handy. Some informants indicated that *P. montana* var. *lobata* plants were grown singly rather than in stands and that ant hills were good places to grow it. Plantings were made in gardens and near houses, but not always in both contexts by a given group. Compost was sometimes added.

Tubers were left in the ground for three to five years and sometimes reached a size so large that they required two or more men to carry them. In addition to the roasting, boiling, and earth-oven preparations, *Pueraria* tubers could also be grated and cooked in a bamboo tube. Sometimes the tubers were cooked twice, the first time to remove the skin. The keeping quality of cooked *Pueraria* was often mentioned as a virtue as this made it suitable food to take on a journey.

Strathern (1969) described *P. montana* var. *lobata* cultivation in the Central and Northern Melpa regions of the New Guinea Highlands. Grown in small mixed-vegetable patches near settlements as well as in the larger mixed gardens cut from forest or from woodland fallow, *Pueraria* was planted among taro, yams, sugar cane, banana, and a variety of green vegetables. *Pueraria* plantings were often made after early ripening crops (e.g. greens, cucumbers, corn) had been harvested, and harvesting *Pueraria* was delayed until after that of most other crops since the digging of its roots disturbed large areas of the garden. Here, too, poles were provided for the vine to climb. In contrast to the harvesting practices recorded by Watson (1964, 1968), Strathern noted that Hageners did not leave the tubers in the ground for any appreciable period of time. Plants were considered mature when they no longer produced new shoots and the leaves began to fall, although tubers occasionally were dug when plants were green and flourishing. Tubers were dug between 9 and 18 months after planting, and the main root was sometimes replanted in its original spot so the plant would regenerate.

Elsewhere in Oceania, *P. montana* var. *lobata* is now considered a wild plant, and no information on its cultivation is available. Haudricourt (1964) reported that in New Caledonia, tubers were gathered for food every August. Guillaumin (1954) summarized several techniques for preparing stems for use in making fishing nets. One involved heating the stems on hot stones, removing the bark, and separating out the fibers to make thread. An alternate method was to steep the stems, presumably in water, and then separate out the fibers. A third technique entailed chewing in order to loosen and remove the starch between its fibers. Guillaumin also noted that both the leaves and tubers were edible, the latter resembling licorice in its taste. Barrau (1956) described New Caledonian *P. montana* var. *lobata* tubers seasoned with a turmeric sauce as agreeable in taste but very fibrous, requiring a long period of chewing and then spitting out the fibers. Parham (1943) characterized the Fijian method of cooking the tuber in coconut milk and mashing them with a bit of sugar as producing quite palatable results.

Heyne (1927) described the method of *P. montana* var. *lobata* cultivation in the Kangean Archipelago of Indonesia. Propagation was achieved by planting cuttings of one-year-old vines at the beginning of the rainy season. Cuttings were allowed to dry for 24 h and were then planted almost horizontally with poles or trees for support. One tilling and moderate watering were apparently all that was necessary for starting the plants, after which growth was luxurious and yields were abundant.

Less detail was provided by Crevost and Lemarie (1917) in their account of *P. montana* var. *lobata* cultivation in Vietnam, but there too propagation was by cuttings or slips which were planted in January, and harvested 12 months later, although some tubers were left in the ground for 2 to 3 years.

In his work on vegetables in south-east Asia, Herklots (1972) included a discussion of *P. montana* var. *lobata* cultivation and harvesting by the Chinese. The crop was often grown in beds with taro and ginger and was propagated via cuttings taken from the first 30–50 cm of one-year-old vines. Taro corms and pieces of ginger rhizomes were planted in February, and rooted cuttings of *P. montana* var. *lobata* were interspersed among these plants in March. The vines were trained up poles, and tubers were harvested in December or January.

Because *P. montana* var. *lobata* is considered a wild plant through much of China and Japan, little information pertaining to its cultivation there is available. Shurtleff and Aoyagi (1977) emphasized its uses in Japan and detailed the procedures for processing the roots and stems of wild plants. Roots of wild *P. montana* var. *lobata* are gathered between early December and late March, cut into 1-m lengths, and are taken to shops for processing. There the roots are crushed, and the resulting pulp is mixed with water to form a slurry which is filtered through screens to remove the fiber. The water mixture is then put in tanks where the starch settles out in layers. The top and bottom layers are removed, and the middle layer is again washed and allowed to settle. This process continues for 16 days, and the resulting clay-like material is then broken into small pieces, dried for 6 weeks, and packaged. The final product, kudzu powder, is used as a thickener similar to cornstarch or arrowroot, as a coating for fried foods, and as a jelling agent in desserts. Alternatively, wild roots are washed, peeled, cubed, and allowed to dry in the sun. The resulting cubes are used with other herbal preparations in making medicinal teas.

Stems are gathered from early June until late July. The best vines are one year old, green, pliant, straight, growing directly out of the crown, and 2–5 m in length. Leaves are stripped by hand, and 20–30 vines are tied together and wound into coils. The coils are then cooked in boiling water until the outer portion of the vine begins to separate from the rest. Next, the coils are taken to a stream and soaked, after which they are drained, placed in a straw- and grass-lined pit, covered over, and left for two days. They are then subjected to a series of washings, soakings, and strippings until the filaments are separated out and split to the desired fineness. Filaments are tied together, end-to-end, to a length of 75 m, and finally are wound into bobbins, each weighing about 11 g.

Kudzu thread is used as the weft with cotton or linen as the warp in making cloth. Grasscloth is made from a wide fiber in a loose weave and is sold as wallpaper. Netting is constructed with twisted fiber in a net weave and is used to make fish nets and pressing sacks.

A limited amount of information about indigenous techniques for preparing *Pueraria tuberosa* is available. In Saharanpur, India, tubers are collected during and after the rainy season and are used in the Ayurvedic medicinal preparation, *chyavanprash* (Ahuja, 1965). Watt (1892) reported that in parts of India the tubers were pounded and made into a poultice for application to swollen joints. In Nepal, they were prepared with milk and given as a tonic and/or an emetic. *P. tuberosa* has been used traditionally by the Santals to treat low fever, syphilitic ulcers in the nose, gravel, and menorrhagia (Bodding, 1927).

### Naming and classification systems

Comparative philology of vernacular plant names provides valuable information in identifying genera, species, and even varieties, and in understanding the origins, histories, and both temporal and geographical limits of dissemination of plants by humans (Merrill, 1946). A survey of the literature has resulted in the compilation of over 200 vernacular names for *Pueraria* species and an additional 17 names for *P. montana* var. *lobata* parts or products. A detailed analysis of the *Pueraria* data for either of these purposes is beyond the scope of this book and outside the limits of either author's capabilities. A further problem lies in the fact that these names were collected by a variety of observers, and few were trained linguists. The resulting orthographic problems very nearly preclude any but the most rudimentary analysis.

Nevertheless, several names have shed light on the plant's habit, its relationship to other plants, and/or its usefulness to humans. For example, the viny, climbing character of *P. montana* var. *lobata* is reflected in the Chinese *mao-man-do* ("hairy creeping bean"), the Japanese *tsuru mame* ("vine bean"), the Fijian *va-yaka* or *wa-yaka*, and the Dutch *slingerboon* ("creeping bean"). Its similarity to wisteria is clear in the Japanese names *kudzu-fuji* and *kusho* ("kudzu wisteria"). The speed with which it grows and spreads is expressed in the south-eastern United States through terms such as "foot-a-night-vine," "mile-a-minute vine," and "Jack and the Bean Stalk." *P. montana* var. *lobata*, the fiber plant, is called *haricot-chanvre* in French, or "hemp bean," and *kopou-bohne* in German, referring to the Chinese cloth made from the plant's fibers. Its usefulness as a shade plant is reflected in the English term "porch vine," and lastly, its promise as a cure for agricultural ills in the south-eastern United States is attested to by names such as "miracle vine" and "wondervine."

For *P. phaseoloides*, a commonly expressed characteristic in vernacular names is its wildness. The Chinese *ye luk tau* ("wild green bean"), *ye sha ge* ("wild sandy vine"), and the Malay *kachang hijau butan* ("woodland green bean") are examples. Its climbing habit is disclosed in the Malay *suloh*. That the plant is known as a source of fiber is manifested in the Malay name *tampong urat* ("fiber") and the Lao classification of *piet* as *po* ("fiber plant").

Perhaps even more relevant to ethnobotanical considerations of *Pueraria* species are local classification systems that reveal *Pueraria*'s standing in the plant world. Such classification systems are often multifaceted and can provide insights into cultivation practices of the past, the history of certain plants in given areas, the cultural bases of taboos, and present-day attitudes toward both old and new or introduced species.

A commonly expressed distinction in these systems is between wild and cultivated plant forms. In discussing the popular nomenclature of food plants in New Caledonia, Leenhardt (1930) noted that cultivated plants, such as bananas, taro, and some species of yams, were contrasted with wild food plants, most of which were tuber-bearing climbers. These wild plants were classed as the food of *hou*, a caterpillar totem. Traditionally, altars were established throughout the bush where these plants abounded, and occasionally, the plants in the immediate vicinity of the altars were tended. At the time of Leenhardt's study, these plants were no longer used except in times of food shortage. *P. montana* var. *lobata* was included in the *hou* group.

Watson (1964, 1968) recorded several instances of a probable wild/cultivated dichotomy in plant classification in the New Guinea Highlands, noting that while it was unclear whether clones of *P. montana* var. *lobata* were distinguished in the same manner in all areas, it was possible that "wild" constituted one class and "cultivated" and "edible" the other. He also reported that some gardeners saw the use of supports as the difference between wild and cultivated forms.

In a consideration of attitudes toward *P. montana* var. *lobata* and *Ipomoea batatas* among the Melpa Hageners of the Highlands, Strathern (1969) presented a more thorough sketch of their plant and animal classification system. Again, a distinction was made between wild and cultivated forms, here marked linguistically by a prefix or suffix of *mbo*, meaning "planted" or "domestic," or *romi/rakra*, meaning "uncultivated" or "wild." Several species were reported to have both wild and cultivated forms. *P. montana* var. *lobata* was classed as a cultivated plant, and opinion was divided as to the existence of a wild counterpart.

A second means of distinguishing among plant forms is by use. Barrau (1965) pointed out that in both the New Guinea Highlands and New Caledonia, islanders discriminated between two varieties of *P. montana* var. *lobata* – one that yielded edible tubers and one that was used as tying material, even though the latter might also have an edible tuber. Vidal (1958–1959, 1963) outlined the system of popular botanical nomenclature in Laos and determined that *P. phaseoloides* was classified as both *po*, a category comprising fiber plants, and as *khua*, encompassing all vines.

Watson (1968) documented a third criterion by which *P. montana* var. *lobata* was separated from certain other plants in the New Guinea Highlands. Specifically, this was *Pueraria*'s "strong" nature, a notion applied in at least two different senses. It was a "strong" food in that it could be kept for a long time without spoiling, even when cooked. This quality made *P. montana* var. *lobata* tubers an excellent food for journeys.

It was also considered "strong" in the sense that it was not a suitable food for children. Several informants indicated that it was at that time, or had been in the past, tabooed to infants and young children, and in some instances it was considered food for old people only.

A fourth and also widespread means of classifying plants is by gender. Gender can be reflected in either the sex of those persons who are allowed to tend, harvest, and/or eat the plant, or in the sex ascribed to the plant itself. Sillitoe (1981, 1983) published the results of his investigations of crop gender associations among the Wola, who live in the Southern Highlands of New Guinea. He determined that the Wola used a classificatory continuum from "Male Only" to "Predominantly Male" to "Both Sexes," through "Predominantly Female," to "Female Only," indicating who was allowed to plant and tend the crop. *P. montana* var. *lobata* was classified as "Male Only." Gender was also expressed through a choice of "to be" verb forms used in association with a plant. The form *wiy* had connotations of things in a recumbent state, things growing horizontal to the ground, weakness, and femaleness, while *hae* corresponded to things in an erect state, things growing vertically, strength, and maleness. The *hae* form was used with *P. montana* var. *lobata*.

Watson's (1964, 1968) data also support the notion of *P. montana* var. *lobata* as a "male" plant. Both planting and harvesting were male concerns, although a few informants indicated that women were sometimes involved. Similarly, Strathern (1969) reported that all the *Pueraria* plants she observed belonged to men, but some informants said that women occasionally planted it or assisted in the harvesting.

The last aspect of indigenous plant classification to be considered here is the relationship of *P. montana* var. *lobata* to *Ipomoea batatas*, or sweet potato. In both form and habit, *P. montana* var. *lobata* resembles sweet potato much more than it does either yams or taro. The overall leaf shapes of the two plants are similar, both can be propagated by vine cuttings, and are capable of rejuvenating themselves, producing tubers that can be harvested in sequence over long periods time (Strathern, 1969; Watson, 1968). Classificatory connections between these two crops were recorded by Watson (1964, 1968), Bowers (1964), and Strathern (1969) for the New Guinea Highlands; however, the details of the specific associations varied from case to case.

According to Watson (1964, 1968), the Dani of West Irian's Balim Valley termed *P. montana* var. *lobata* "the mother of sweet potatoes," and planted it in the gardens and near certain men's houses to insure large yields of the sweet potato crop. His informants in the Mt. Hagen area characterized *P. montana* var. *lobata* as a "strong" sweet potato or "stronger" than sweet potato and often commented that it was more nourishing. *P. montana* var. *lobata* was also sometimes mentioned as a variety of sweet potato.

Strathern's (1969) informants distinguished *P. montana* var. *lobata* from sweet potato because *P. montana* var. *lobata*: (1) was very filling and quickly satisfied hunger, (2) lasted a long time in the ground after harvesting and even after cooking, and (3) was available even in times of food shortage. Nevertheless, it was considered an inferior crop. Some said it tasted bad, and most found it coarse, stringy, and difficult to harvest.

Among the Melpa Hageners, *P. montana* var. *lobata*, the sweet potato, and related vines were grouped together as *oka* (Strathern, 1969). This grouping included neither yams nor taro, both of which were covered by other terms of the same order as *oka*. Such terms could be considered primary taxa, and there was no generic term for "tuber." *Oka* alone referred invariably to sweet potato, whereas *oka* modified by *mapumb* denoted *P. montana* var. *lobata*.

Bowers (1964) reported a slightly different classification used by the Kakoli people. Their term *ngga* referred to "tuber" as a swollen storage root and included more than 40 varieties of sweet potato, manioc, Irish potato, and *P. montana* var. *lobata*. Mutually exclusive to this category was that covered by the term *me*, which corresponded to "corm" and included certain taros and taro-like plants.

Strathern (1969) further determined that *P. montana* var. *lobata* and sweet potato were viewed by at least some Hageners as *porman*, meaning "companions" or "equals." This was not the same as pairing the two in the sense that yams and taro were paired, since pairing was done only between two taxa of the same order. Rather, *porman* status indicated a significantly closer relationship. In sum, *P. montana* var. *lobata* was more like sweet potato than yams were like taro or than yams were like sweet potato. Yet, *P. montana* var. *lobata* was also accorded an independent status complementary to that of sweet potato.

## Myths and Rituals

Unfortunately, only a few fragmented bits of information pertaining to myth and ritual associations of *Pueraria* species are available. Even when abundant data on such topics are accessible, it is rarely clear why certain plants are featured in mythic contexts, and it is often equally puzzling why some plants are used on ritual occasions to the exclusion of others. To be sure, in cases where the data are disjointed and incomplete, speculation regarding their significance is ineffectual and can be dangerous. Let the reader beware.

Girard (1957) discussed *P. montana* var. *lobata* as a ritual plant among the Buang of the Morobe District of New Guinea and noted that it was associated with a myth, which was reported as follows:

A mother had to flee from her sons because they had killed their father. Along the path, however, she left her milk on the leaves of *Pueraria montana* var. *lobata* for her youngest son, who was still nursing.

Taken totally out of context, this story imparts little information. We do not know what special significance is attached to the father's murder by his sons or to the mother's leaving her sons because of their crime, let alone the reason for choosing *Pueraria* over other plants as a repository for the milk. It might be that the mother's act of leaving milk on the *Pueraria* leaves for her still-nursing son signifies the plant's present-day or past importance as a food crop and perhaps even as a staple food crop.

Strathern (1969) reported that among the Melpa Hageners there were no stories about *P. montana* var. *lobata* itself or its antiquity. Informants indicated that both *oka* (sweet potato) and *oka mapumb* (*P. montana* var. *lobata*) had always been with them. There was, however, the following story about sweet potatoes which Strathern felt was pertinent.

In the past everyone grew sweet potatoes up sticks; the sweet potato vine was trained up sticks, although its tuber grew underground. Once, an old woman neglected her garden; she did not go near it for six months. Then when she looked at her garden, she saw that some of the sticks had fallen down, and when she went to lift the sweet potato plants up, she discovered that they had sent shoots into the ground and could not be moved. So she dug for some tubers, and found they were good ones. Then she broke down all the other stakes in the garden and let the plants fall to earth. People who saw this said she must be mad and was ruining her garden; they demanded [to know] why. But she said they should hold their tongues; she could ruin her own garden if she wanted to. A while passed, then she sent word to other women to come and watch her harvest her sweet potato garden. As they watched, she dug out some good tubers. And they exclaimed at how good they were. They were delighted and said they would follow suit. So they no longer trained the sweet potato up sticks but let the vine go underground, which is how we do it now.

Strathern suggested the story could be interpreted as: (1) describing cultivation methods of the past when sweet potatoes were grown in the same way as *P. montana* var. *lobata*, (2) as reflecting a time when *P. montana* var. *lobata* was widely grown and when *oka* referred to it rather than to sweet potato, or (3) as simply a comment on the current status of *oka* as a taxon. The fact that *oka* is applied to two crops, one which grows on the ground and one which climbs up poles, is ambiguous and perhaps puzzling, but by separating the opposing elements in time, this story alleviates the problem by implying that *oka* changed from one type to another. The myth also validates the present-day dominance of ground-growing *oka* (sweet potato) over pole-climbing *oka* (*P. montana* var. *lobata*).

Fox (1953) described *P. phaseoloides* as a plant having supernatural powers according to the Pinatubo Negritos of the Zambales Range, the Philippines. Leaves were rubbed on bows and arrows, and more recently, on guns for luck in hunting.

Watson's (1968) informants in New Guinea also indicated that there were ritual associations of several kinds for *P. montana* var. *lobata*. There was a common feeling that *P. montana* var. *lobata* as a food should be shared, particularly on special occasions, and that it was not a food for routine individual or family meals. Relevant comments included:

If friends come, we like to share it with them. When we have it, we invite them and share it – not everybody, just kinsmen.

When we gather food for another group, we hang the tubers in trees where the dance will take place.

It is used in dance feasts as a food.

Watson (1968) also mentioned its ritual importance among Kamano peoples as evidenced by the prominent display of *P. montana* var. *lobata* tubers on walls and speakers' stands at the inauguration of new Kamano Council buildings in 1963.

Bowers (1964) reported that *P. montana* var. *lobata* as a food is forbidden to most Kakoli clans of the New Guinea Highlands for religious reasons. Other such foods included dog meat, red pandanus, and an unidentified type of bean. If an individual failed to observe the food taboo, he could become seriously ill and the climate could worsen as a result of his violation.

Lastly, Strathern (1969) recorded a series of spells and precautions to be taken when planting *P. montana* var. *lobata* among the Melpa Hageners of the New Guinea Highlands. It was recommended that planting be done in the evening, since if the planter had a day's activities or a journey still ahead of him, the roots of the plant would be likely to spread out too far. It was also advised that the gardener should squat with his arms and legs close to his side so as to prevent the underground roots from spreading and making the tubers difficult to find.

Each of the three spells having to do with *P. montana* var. *lobata* emphasized that the tuber should be soft and "should crumble away as easily as a lizard's tail falls away at a touch." One sought tubers which were "tender and full of grease [i.e. juicy] like little marsupials with their fragile bones," and another compared *oka mapumb* (*P. montana* var. *lobata*) with *oka* (sweet potato) in the hope that the *Pueraria* too would be full of grease, plump, and tender.

## SUMMARY AND CONCLUSIONS

In the foregoing pages we have considered the botany, geography, history, and anthropology of *Pueraria*. It has been our aim to assemble data from widespread sources for the mutual benefit of anthropologists and plant scientists alike. In so doing, we have determined that of 16 species in the genus, 13 are wild and three are cultivated, although none has ever been domesticated. We have attributed this non-domesticated status to the manner in which humans have exploited these species. By focusing largely on the young stems and tuberous roots, humans have exerted little if any selection pressure, which has resulted in little or no genetic change through time.

After considering species distribution, ecological requirements, and flowering and fruiting patterns, we have postulated that the area comprising north-east India, Burma, Thailand, and southern China is most likely the geographic center of genetic diversity for *Pueraria* and have speculated that this was also the area of the earliest cultivation of *P. montana* var. *lobata*, *P. phaseoloides*, and *P. tuberosa*. There does not appear to be a secondary center of diversity for any of these species; consequently, the crop evolutionary pattern conforms to that of Harlan's monocentric class.

From a mainland Asia homeland, *Pueraria* species have been transported by humans throughout Asia proper and the Pacific to Europe, Africa, and the Americas. It is inferred that much of the Asian and Oceanic transport of the genus by humans took place in prehistoric times, although there is no direct archaeological evidence to support this.

The earliest written records of the use of *Pueraria* by humans date to at least the sixth and possibly to the tenth century BC and refer to *P. montana* var. *lobata* as a source of fiber for summer cloth in north China. Species of *Pueraria* have been used for food, medicine, fiber, shade, ground cover, and fuel by humans on nearly every continent and at every socioeconomic level.

We have also contemplated indigenous agricultural practices and preparation techniques, vernacular names, local classification systems, and myths and rituals associated with *Pueraria* species. As a result, a number of issues have been raised pertaining to human relationships with *Pueraria* species, not without relevance to their relationships with other plants in general. These issues are worthy of further discussion here.

A common theme that runs throughout our consideration of human interaction with *Pueraria* has been that of the plant's change in status as perceived by people. *P. montana* var. *lobata* is by no means the only plant to have fallen into disfavor (Li, 1969), but its case has been particularly well-documented in a number of situations.

The suggestion has been made that *P. montana* var. *lobata* was once an important and possibly staple food crop in New Guinea before the arrival of the sweet potato. We have posited in this paper that the species was most likely introduced to New Guinea by humans at some time in prehistory, and Watson (1964) has noted that it has no distinctive Pidgin name, as do other recently introduced crops (e.g. Irish potato, manioc). Further, Watson's (1968) informants often described *Pueraria* as ancient – a food of the ancestors, and a crop planted by their forefathers. In combination with the linguistic, myth, and ritual data supplied by Bowers (1964), Watson (1964, 1965, 1968) and Strathern (1969), this can be interpreted as evidence that *P. montana* var. *lobata* was once more important to subsistence in the New Guinea Highlands than it is now.

The question then arises, why did the plant undergo a decline in status from an important food crop to a marginally cultivated and sometimes unrecognized plant? Strathern (1969) has postulated that *P. montana* var. *lobata* may have been cultivated in at least the recent past as a complement to the sweet potato. Its hardiness and keeping qualities would have been particularly valuable during war and on long hazardous journeys. The decline may be a result of changing conditions which have made its positive qualities less in demand since European contact. Provisioning for war and for long journeys is no longer an important consideration; hence, the niche once filled by *P. montana* var. *lobata* no longer exists.

Watson (1968) has argued for a greater antiquity of the crop and contends that it was used in New Guinea not only before sweet potatoes but before taro and yams as well. He sees its continued use after the introduction of taro and yams as indicative of strong cultural preferences (perhaps ritual related) for its retention. The rapid decline of *P. montana* var. *lobata* since the introduction of the sweet potato is attributed to the intensive use of the latter, the logic being that if *P. montana* var. *lobata* were primarily cultivated for subsistence, then sweet potato production would have lessened the need for it (Watson, 1968).

Golson (1976) also acknowledged that the "now unimportant legume" *P. montana* var. *lobata* may have been an important food in the past. He also noted that it, taro, and yams all take longer to mature than does sweet potato, which could have been an important factor in the development of the supremacy of the sweet potato and a concomitant decline in the cultivation of *P. montana* var. *lobata*.

We have also seen that *P. montana* var. *lobata* has undergone a decline in use in Japan. The production of cloth for garments, starch powder for cooking and medicinal purposes, and grasscloth for export has degenerated to a cottage-industry scale. Shurtleff and Aoyagi (1977) have discussed several reasons for this, including a diminishing market for such products, the labor-intensive methods required for harvesting the wild plants, and the establishment of a competing and thriving kudzu industry in Korea.

Lastly, *P. montana* var. *lobata* has experienced a drastic shift in status from a miraculous agricultural cure-all to a destructive, menacing threat in the south-eastern United States. Once the object of widespread government-sponsored planting campaigns, it is now officially classified as a noxious weed, and considerable sums of money are expended each year in efforts to eradicate it. Winberry and Jones (1973) proposed that this change was the result of: (1) early unreasonably high expectations of the plant, (2) an initial glossing over of its negative aspects, and (3) significant changes in Southern agriculture from 1930 to the present.

While the plant does grow easily, covers well, halts soil erosion, enriches the soil with nitrogen, and provides a nutritious hay for livestock, it does not have a high carrying capacity as pasturage, its grazing must be carefully controlled, and as a hay, the woody stems, which can be up to half the hay crop by weight, make *P. montana* var. *lobata* difficult to cut and rake. Perhaps, even more important have been the changes in Southern agriculture over the last several decades. In traditional Depression-era farming, *P. montana* var. *lobata* was indeed valuable. Requiring little capital investment, no fertilizers, and next to no attention once it was planted, *P. montana* var. *lobata* still produced forage. As an added bonus, it rejuvenated soils depleted by cotton and halted soil erosion. After the Depression, however, a growing interest in beef and milk production on the part of Southern farmers resulted in a concern for more specialized pasture. Grasses such as coastal bermuda, bahia, and fescues were preferred because of their higher nutritional values and the greater ease with which they could be harvested. They were more expensive than *P. montana* var. *lobata* in terms of both time and money, but farmers now had the resources and incentives for investment.

Similarly, substantial acreage has been turned over to forest production. The aggressive nature of *P. montana* var. *lobata* has made it a detriment insofar as it chokes out saplings and even mature trees and completely prevents establishment of new forests. Even as a conservation crop, *P. montana* var. *lobata* is now at a disadvantage since it must be planted by hand, whereas other crops can be started from seed and therefore can be planted mechanically.

In this study, we have conformed to Jones' original concept of ethnobotany as the study of the interrelations between plants and humans and have extended it to include complex and state-level societies. The basic principles are the same, and data from each case informs our understanding of the whole. We have seen examples of a plant falling into disfavor as the result of: (1) the disappearance of a niche in the subsistence system, (2) replacement by a "better" crop, (3) diminishing markets for products derived from the plant, (4) an evaluation of the necessary harvesting techniques as too labor intensive, (5) competition from neighboring areas for the same market, (6) unreasonable expectations of the plant's potential, (7) a failure to properly evaluate the positive and negative aspects of the plant, and (8) changing agricultural perceptions, goals, and practices.

These issues are of direct relevant to understanding the nature of specific as well as general plant–human interrelations. They are important to researchers whose aims may be reconstructing histories of plant movement and use by people for whatever purpose and are also vital to those who may be involved in introducing new crop plants into already existing agricultural systems.

## REFERENCES

Ahlgren, G.H. (1949) *Forage Crops*. McGraw-Hill Book Company, Inc., New York, p. 102.

Ahuja, B.S. (1965) *Medicinal Plants of Saharanpur*. Central Council of Ayurvedic Research, Hardwar, India, pp. 62–63.

Alexander, E.J. (1930) *Pueraria thunbergiana. Addisonia*, 15, 57–58.

Anonymous (1975) Treatment of sudden deafness with *Pueraria* root: a comparative analysis of therapeutic results in 294 cases. *Chin. Med. J.* (English edition), 1(5), 343–349.

Arora, R.K. and Chendel, K.P.S. (1972) Botanical source areas of wild herbage legumes in India. *Trop. Grasslands*, 6, 213–221.

Backer, C.A. and Bakhuizen van den Brink, R.C. (1963) *Flora of Java*, Vol. 1. N.V.P., Noordhoff, Groningen, p. 632.

Bailey, R.V. (1939) Kudzu for erosion control. *USDA Farmers' Bulletin*, **1840**, 1–31.

Baker, J.G. (1897) Leguminosae. In J.D. Hooker (ed.), *The Flora of British India*, Vol. II. L. Reeve and Company Ltd., Ashford, p. 197.

Barrau, J. (1956) Les légumineuses à tubercules alimentaires de la Melanesie. *La Terre et la Vie*, **103**, 11–16, 40.

Barrau, J. (1958) *Subsistence Agriculture in Melanesia*. Bernice P. Bishop Museum Bulletin 219, pp. 29, 64.

Barrau, J. (1965) Witnesses of the past: notes on some food plants of Oceania, *Ethnology*, **4**, 282–294.

Bellwood, P. (1979) *Man's Conquest of the Pacific*. Oxford University Press, New York, p. 140.

Bentham, G. (1867) Notes on *Pueraria*, DC. correctly referred by the author to Phaseoleae. *J. Linnean Soc. London, Botany*, **9**, 121–125.

Bodding, P.O. (1927) Studies in Santal medicine and connected folklore. Part II. *Memoirs Asiatic Soc. Bengal*, **10**(2), 133–426.

Bodner, C. Cox. (1986) *On the Evolution of Agriculture in Central Bontoc*. Unpublished Ph.D. Dissertation, Department of Anthropology, University of Missouri, Columbia.

Bowers, N. (1964) A further note on a recently reported root crop from the New Guinea Highlands. *J. Polynesian Soc.*, **73**, 333–335.

Bretschneider, E. (1880) Early European researches into the flora of China. *J. N.-China Branch R. Asiatic Soc.*, n.s., **15**, 1–194.

Bretschneider, E. (1881) Botanicon Sinicum: Notes on Chinese botany from native and western sources. *J. N.-China Branch R. Asiatic Soc.*, n.s., **16**, 18–230.

Bretschneider, E. (1898) History of European botanical discoveries in China, 2 vols., Zentral-Antiquariat, Leipzig (Reprinted in 1962.).

Bulmer, S. and Bulmer, R. (1964) The prehistory of the Australian New Guinea Highlands. *Am. Anthropol.*, **66**(4), 39–76.

Burkill, I.H. (1935) *A Dictionary of the Economic Products of the Malay Peninsula*, 2 vols., Crown Agents for the Colonies, London.

Burkill, I.H. and Haniff, M. (1930) Malay village medicine. *The Gardens' Bulletin Straits Settlements*, **6**, 165–321.

Carrière, E.A. (1891) *Pueraria thunbergiana. Revue Horticole* (Paris), **63**, 31–32.

Chard, C.S. (1974) *Northeast Asia in Prehistory*. University of Wisconsin Press, Madison, pp. 167–172.

Christophersen, E. (1935) *Flowering Plants of Samoa*. Bernice P. Bishop Museum Bulletin 128, p. 104.

Collett, H. (1971) *Flora simlensis*. Bishen Singh Mahendra Pal Singh, Dehradun, India, pp. 139–140.

Coursey, D.G. (1968) The edible aroids. *World Crops*, **20**(4), 25–30.

Crevost, C. and Lemarie, C. (1917) *Catalogue des Produits de l'Indochine*, Vol. 1., Produits Alimentaires et Plantes Fourrageres, Gouvernement General de l'Indochine, pp. 133–134.

Darlington, C.D. and Wylie, A.P. (1955) *Chromosome Atlas of Flowering Plants*. George Allen & Unwin Ltd., London.

de Candolle, A. (1825) Notice sur quelques genres et espèces nouvelles de légumineuses, extraite de divers mémoires présentés à la Société d'Histoire Naturelle de Genève, pendant le cours des années 1823 et 1824. *Ann. des Sci. Natur.*, **4**, 90–103.

Dickens, R. (1974) Kudzu: friend or foe? *Weeds Today*, **5**(3), 9.

Duke, J.A. (1981) *Handbook of Legumes of World Economic Importance*. Plenum Press, New York, pp. 212–215, 319, 323.

Dymock, W. (1885/1886) *The Vegetable Materia Medica of Western India*, Second edition, Educational Society Press, Bombay.

Fang, Q. (1980) Some current study and research approaches relating to the use of plants in the traditional Chinese medicine. *J. Ethnopharmacol.*, **2**, 57–63.

Focht, J. (1972) Additions to the flora of Rockland County. *Bull. Torrey Bot. Club*, **99**(5), 249–250, New York.

Fox, R.B. (1953) The *Pinatubo negritos*: their useful plants and material culture. *Philippine J. Sci.*, 81(3–4), 173–414.

Frankel, E. (1989) Distribution of *Pueraria lobata* in and around New York City. *Bull. Torrey Bot. Club*, 116(4), 390–394.

Gamble, J.S. (1967) *Flora of the Presidency of Madras*, Vol. 1., Botanical Survey of India, Calcutta, p. 254.

Girard, F. (1957) Quelques plantes alimentaires et rituelles en usage chez les Buang. *J. d'Agriculture Tropicale et de Botanique Appliquée*, 4, 212–227.

Golson, J. (1976) Archaeology and agricultural history in the New Guinea Highlands. In G. de Sieveking, I.H. Longworth and K.E. Wilson (eds), *Problems in Economic and Social Archaeology*, Westview Press, Boulder, pp. 201–220.

Guillaumin, A. (1954) Baite, magnagna et kudzu: légumineuses alimentaires de Nouvelle Calédonie. *La Terre et la Vie*, 101, 174–176.

Guppy, H.B. (1906) *Observations of a Naturalist in the Pacific between 1896 and 1899*, Vol. 2., Macmillan and Company Ltd., London, pp. 412–413.

Harlan, J.R. (1992) *Crops and Man*, Second edition, American Society of Agronomy, Inc. and Crop Science Society of America, Inc., Madison.

Haudricourt, A.G. (1964) Nature et culture dans la civilisation de l'igname: l'origine des clones et des clans. *L'Homme*, 4, 93–104.

Herklots, G.A.C. (1972) *Vegetables in South-East Asia*. George Allen & Unwin, London, pp. 468, 470.

Heyerdahl, T. (1952) *American Indians in the Pacific*. George Allen & Unwin, London, pp. 474–475.

Heyne, K. (1927) *De Nuttige Planten van Nederlandsch–Indie*. Ruygrok & Company, Batavia, p. 829.

Hooper, D. (1929) On Chinese medicine: drugs of Chinese pharmacies in Malaya. *The Gardens' Bulletin Straits Settlements*, 6, 1–163.

Jones, V.H. (1941) The nature and status of ethnobotany. *Chron. Bot.*, 6(10), 219–221.

Julien, S. and Champion, P. (1869) *Industries Anciennes et Modernes de l'Empire Chinois*. Eugene Lacroix, Paris, p. 171.

Kashemsanta, M.C., Lakshnakara, S.K. and Airy Shaw, H.K. (1952) A new species of *Pueraria* (Leguminosae) from Thailand, yielding an estrogenic principle. *Kew Bull.*, 7, 549–551.

Keng, H. (1974) Economic plants of ancient north China as mentioned in *Shih Ching* (Book of). *Econ. Bot.*, 28, 391–410.

Krishnamurthi, A. (ed.) (1969) *The Wealth of India. Raw Materials*, Vol. VIII, Council of Scientific and Industrial Research, New Dehli, pp. 313–317.

Kumar, R. (1977) Kudzu – a perennial fodder legume vine. *Indian Farm.*, 27(7), 17–19.

Lackey, J.A. (1977) *A Synopsis of the Phaseoleae (Leguminosae, Papilionoideae)*. Unpublished Ph.D. Dissertation, Departments of Botany and Plant Pathology, Iowa State University, Ames.

Le Comte, L. (1698) *Memoirs and Observations Topographical, Physical, Mathematical, Mechanical, Natural, Civil, and Ecclesiastical*, Translated from the Paris edition, Printed for B. Tooke and to be sold by G. Huddleston, London, p. 141.

Leenhardt, M. (1930) Notes d'ethnologie Néo-Calédonienne. *Travaux et Mémoires d'Institut d'Ethnologie*, 8, 1–340.

Li, H.-L. (1969) The vegetables of ancient China. *Econ. Bot.*, 23, 253–260.

Li, H.-L. (1970) The origin of cultivated plants in Southeast Asia. *Econ. Bot.*, 24, 3–19.

Maclet, J-N. and Barrau, J. (1959) Catalogue des plantes utiles aujourd'hui présentes en Polynésie Frantaise. *Journal d'Agriculture Tropicale et de Botanique Appliquée*, 6, 161–184.

Maesen, L.J.G. van der (1985) Revision of the genus *Pueraria* DC. with some notes on *Teyleria* Backer (Leguminosae). *Agric. Univ. Wageningen Pap.*, 85-1, 1–132.

Maesen, L.J.G. van der (this vol.) *Pueraria*: botanical characteristics.

Martini, M. (1655) *Novus Atlas Sinensis, a Martino Martinio Descriptus*. Seste Deel van de Nieuwe Atlas, Oft, Toonneel des Aer Drijex, Uytgegeven door Joan Blaeu, Amsterdam, pp. 186–189.

Merrill, E.D. (1917) *An Interpretation of Rumphius's Herbarium Amboinense*. Department of Agriculture and Natural Resources Bureau of Science Publication 9, Manila, p. 287.

Merrill, E.D. (1918) *Species Blancoanae: A Critical Revision of the Philippine Species of Plants Described by Blanco and Llanos*. Bureau of Printing, Manila, p. 89.

Merrill, E.D. (1923) *An Enumeration of Philippine Flowering Plants*, Vol. 2. Bureau of Printing, Manila, p. 312.

Merrill, E.D. (1946) On the significance of certain oriental plant names in relation to introduced species. *Chron. Bot.*, **10**, 295–315.

Merrill, E.D. (1954) The botany of Cook's voyages. *Chron. Bot.*, **14**, 161–384.

Narayanan, T.R. and Dabadghao, P.M. (1972) *Forage Crops of India*. Indian Council of Agricultural Research, New Delhi, pp. 77–79.

Ohwi, J. (1965) *Flora of Japan*. In Frederick G. Meyer and Egbert H. Walker (eds), Smithsonian Institution, Washington, DC, p. 570.

Parham, H.B.R. (1943) *Fiji Native Plants*. The Polynesian Society Memoir, 16, p. 136.

Parham, B.E.V. (1949) List of introduced leguminous plants recorded in Fiji. *Fiji Agricultural Journal*, **10**, 21–25.

Patel, R.I. (1968) *Forest Flora of Melghat*. Bishen Singh Mahendra Pal Singh, Dehradun.

Perry, L.M. (1980) *Medicinal Plants of East and Southeast Asia-Attributed Properties*. Massachusetts Institute of Technology Press, Cambridge, p. 225.

Piperno, D.R., Ranere, A.J., Holst, I. and Hansell, P. (2000) Starch grains reveal early root crop horticulture in the Panamanian tropical forest. *Nature*, **407**(6806), 894–897.

Polo, M. (1932) *The Travels of Marco Polo*. E.P. Dutton & Co., Inc., New York.

Porterfield, W.M. (1938) Ko, the kudzu vine, provides food, shade, clothes, and medicine. *J. New York Bot. Garden*, **39**, 203–205.

Powell, J.M. (1974) Traditional legumes of New Guinea Highlands. *Sci. N. G.*, **2**, 48–62.

Powell, J.M. (1976) Ethnobotany. In K. Paijmans (ed.), *New Guinea Vegetation*, Elsevier Scientific Publishing Company, New York, pp. 106–183.

Powell, J.M. (1977) Plants, man and environment in the island of New Guinea. In J.H. Winslow (ed.), *The Melanesian Environment*. Australian National University Press, Canberra, pp. 11–20.

Purseglove, J.W. (1968) *Tropical Crops: Dicotyledons*. Longman, London, p. 218, 220.

Purseglove, J.W. (1972) *Tropical Crops: Monocotyledons*. Longman, London.

Rao, P.S. (1958) Non-cereal foods: *Pueraria tuberosa* as a food and fodder. *Indian Forest*, **84**(5), 281–283.

Roberts, O.T. (1970) A review of pasture species in Fiji. II. Legumes. *Trop. Grasslands*, 4(3), 213–222.

Roi, J. (1955) Traité des plantes médicinales Chinoises. *Encyclopédie Biol.*, 47, 1–500.

Roxburgh, W. (1874) *Flora Indica*. Thacker, Spink & Co., Calcutta.

Satow, E. (1906) Ancient Japanese rituals, part III. *Transactions of the Asiatic Society of Japan*, 9, 183–211.

Seemann, B. (1865) *Flora Vitiensis*. L. Reeve and Company, London. (Reprinted 1977, *Historiae Naturalis Classica*, Vol. 103.)

Shibata, S., Katsuyama, A. and Noguchi, M. (1978) On the constituents of an essential oil of kudzu. *Agric. Biol. Che.* (Tokyo), 42(1), 195–198.

Shih, S.-H. (1974) A Preliminary Survey of the Book Ch'i Min Yao-Shu. *An Agricultural Encyclopaedia of the 6th Century*, Second edition, Science Press, Peking, pp. 1–3.

Shurtleff, W. and Aoyagi, A. (1977) *The Book of Kudzu, a Culinary and Healing Guide*. Autumn Press, Brookline, Massachusetts, pp. 10, 72–73, 79.

Siebold, P.F. von (1827) *Synopsis Plantarum Oeconomicarum Universi Regni Japonici*, Batavia.

Sillitoe, P. (1981) The gender of crops in the Papua New Guinea Highlands. *Ethnol.*, 20, 1–14.

Sillitoe, P. (1983) *Roots of the Earth: Crops in the Highlands of Papua New Guinea*. Manchester University Press, Manchester.

Skerman, P.J. (1977) *Tropical Forage Legumes*. FAO Plant Production and Protection Series No. 2, Rome, pp. 317, 363.

Smith, A.C. (1942) Fijian plant studies, II. *Sargentia* I. Arnold Arboretum, Jamaica Plain, Massachusetts, p. 39.

Smith, F.P. (1969) *Chinese Materia Medica: Vegetable Kingdom*, Ku T'ing Book House, Taipei (Extensively revised from F. Porter Smith's work by G.A. Stuart, 1911.), p. 299.

Smith, R.M. and Chandler, J.V. (1951) Tropical kudzu moves into Puerto Rico. *Crops Soils*, 3(6), 12–14.

Stephens, R.W. (1953) Kudzu: versatile wonder-bean. *Org. Farmer*, 4(6), 32–35.

Stevens, L. (1976) King Kong kudzu, menace to the South. *Smithsonian Mag.*, 7(9), 93–98.

Stewart, D. (2000) Kudzu: love it – or run. *Smithsonian Mag.*, 31(7), 65–70.

Straatsman, W. (1967) Ethnobotany of New Guinea in its ecological perspective. *J. d'Agriculture Tropicale et de Botanique Appliquée*, 14, 1–20.

Strathern, A.M. (1969) Why is the *Pueraria* a sweet potato? *Ethnol.*, 8, 189–198.

Tanaka, T. (1976) *Tanaka's Cyclopedia of Edible Plants of the World*, Sasuke Nakao (ed.), Keigaku Publishing Company, Tokyo.

Tanner, R.D., Hussain, S.S., Hamilton, L.A. and Wolf, F.T. (1979) Kudzu (*Pueraria lobata*): potential agricultural and industrial resource. *Econ. Bot.*, 33, 400–412.

Tseng, K.-U., Chou, Y.-P., Chang, L.-Y. and Fan, L.-L. (1975) Pharmacologic studies on radix Puerariae. I. Effects of Puerariae on blood pressure, vascular reactivity, cerebral and peripheral circulation in dogs. *Chin. Med. J.* (English edition), 1, 335–342.

United States Department of Agriculture (1922) *Inventory of Seeds and Plants Imported by the Office of Foreign Seed and Plant Introduction during the Period from April 1 to June 30, 1919*, No. 59, Bureau of Plant Industry, p. 67.

Verdcourt, B. (1979) *A Manual of New Guinea Legumes*. Botany Bulletin 11, Office of Forests, Division of Botany, Lae, pp. 483–488.

Vidal, J. (1958–59) Noms vernaculaires de plantes (Lao, Méo, Kha) en usage au Laos. *Bulletin de l'Ecole Francaise d'Extrême Orient*, 49, 425–608.

Vidal, J. (1963) Systematique, nomenclature et phytonymie botanique populaire au Laos. *J. d'Agriculture Tropicale et de Botanique AppliquTe*, 10, 438–448.

Vieillard, M.E. (1862) Plantes utiles de la Nouvelle Calédonie. *Annales des Sciences Naturelles, Botanique*, ser. 4, 16, 28–76.

Völksch, B. and Wingart, H. (1997) Comparison of ethylene-producing *Pseudomonas syringae* strains isolated from kudzu (*Pueraria lobata*) with *Pseudomonas syringae* pv. *phaseolicola* and *Pseudomonas syringae* pv. *glycinea*. *European Journal of Plant Pathology*, 103(9), 795–802.

Walker, E.H. (1976) *Flora of Okinawa and the Southern Ryukyu Islands*. Smithsonian Institution Press, Washington, DC, 588–589.

Watson, J.B. (1964) A previously unreported root crop from the New Guinea Highlands. *Ethnology*, 3, 1–5.

Watson, J.B. (1965) From hunting to horticulture in the New Guinea Highlands. *Ethnol.*, 4, 295–309.

Watson, J.B. (1968) *Pueraria*: names and traditions of a lesser crop of the central highlands, New Guinea. *Ethnol.*, 7, 268–279.

Watt, G. (1892) *A Dictionary of the Economic Products of India*, Vol. 6, Part 1, W.H. Allen and Company, London, p. 363.

White, J.P. and Allen, J. (1980) Melanesian prehistory: some recent advances. *Science*, 207, 728–734.

Whyte, R.O., Nilsson-Leissner, G. and Trumble, H.C. (1953) *Legumes Agric.*, FAO Agricultural Studies No. 21, Rome, pp. 318–320.

Winberry, J.J. and Jones, D.M. (1973) Rise and decline of the "miracle vine": kudzu in the southern landscape. *Southeast. Geographer*, 13, 61–70.

Wittmack, L. (1896) *Pueraria thunbergiana* (Sieb. et Zucc.). Benth. Eine für Deutschland neue ausdauernde schlingpflanze. *Gartenflora*, 45, 401–404.

Wolverton, B.C. and McDonald, R.C. (1981) Energy from vascular plant wastewater treatment systems. *Econ. Bot.*, **35**, 224–232.

Wong, K.C. and Wu, L.-T. (1936) *History of Chinese Medicine*. Tientsin Press Ltd., Tientsin, p. 5.

Yuncker, T.G. (1943) *The Flora of Niue Island*. Bernice P. Bishop Museum Bulletin, 178, p. 65.

Yuncker, T.G. (1959) *Plants of Tonga*. Bernice P. Bishop Museum Bulletin, 200, pp. 147–148.

# 3 *Pueraria* (Ge) in traditional Chinese herbal medicine

*You-Ping Zhu, Han-Ming Zhang, and Ming Zeng*

## INTRODUCTION

The root of kudzu (Ge Gen) is one of the most common ingredients used in traditional Chinese medicine. The stem, leaf, flower and seed of Ge have also been used in traditional Chinese medicine. Only the root and flower are commercially available nowadays. The medical record of kudzu (Ge, *Pueraria*) dates back nearly 2000 years in China's first medicinal work, Shen Nong Ben Cao Jing (Divine Husbandman's Classic of Materia Medica, Anonymous, *c*. AD 25–200), in which Ge is listed as a middle grade medicinal plant[1], and Ge Gen is used for the relief of "thirsting and wasting symptoms," fever, vomiting, and non-specific intoxication (Jiangsu College of New Medicine, 1977). The stem is used for the treatment of carbuncles and sore throat. The leaf is used for wound bleeding. The flower was first recommended for alcohol intoxication (Sun Si-Miao, 581–682) and later for alcohol abuse in Li Dong-Yuan's (1180–1251) Pi Wei Lun (Discussion of the Spleen and Stomach, 1249). In Shen Nong Ben Cao Jing, the seed is described as an antidysenteric agent but in Ben Cao Gang Mu (Compendium of Materia Medica, 1596) it is recommended as an anti-alcohol intoxication agent by Li Shi-Zhen (Zeng *et al.*, 2000).

There are disputes over the classification of Chinese *Pueraria* species (van der Maesen, 1985; Zeng, 1999). In this chapter, the authors adopt the species names used in Chinese literatures, which include *P. lobata* (*P. thunbergiana*, *P. hirsute*, *P. pseudo-hirsuta*), *P. thomsonii* (*P. chinesis*), *P. omeiensis*, *P. phaseoloides*, *P. montata*, *P. edulis*, *P. alopecuroides*, *P. calycina*, *P. peduncularis* and *P. wallichii*. *P. lobata* can be found growing wild all over the country except in Tibet, Xinjiang and Qinghai provinces while *P. thomsonii* is mostly cultivated or growing wild in the southern provinces of Guangxi, Guangdong, Fujian and Yunnan. Other *Pueraria* species are only found in southern part of China (Lian *et al.*, 1995; Zeng, 1999).

Based on the historical record of traditional Chinese medicine, researchers believe that the original plant Ge used for medical purposes were *Pueraria lobata* and *P. thomsonii* although a few other *Pueraria* species were reported to be used in some areas (Lian *et al.*, 1995; Zeng, 1999).

A nationwide market survey in the 1980s revealed that *Pueraria lobata* was the main commercial source of Ge Gen. Ge Gen derived from *P. thomsonii* is being used in some

---

1 In *Shen Nong Ben Cao Jing* herbs are classified into three basic categories: upper, middle and lower according to their applications. The upper grade nourishes life, the middle grade nourishes constitutional types and the lower grade expels disease.

southern provinces such as Guangxi, Guangdong and Sichuan, and that of *P. omeiensis* is only used in some remote areas of Guizhou province (Lian *et al.*, 1995). *P. edulis* was found in use in southern Chinese provinces of Yunnan, Guizhou and Sichuan, *P. omeiensis* in Guizhou and Sichuan, and *P. phaseoloides* in eastern province of Zhejiang (Lian *et al.*, 1989). In the late 1990s, only *P. lobata* and *P. thomsonii* were found to be used as the commercial sources of Ge Gen (Zeng, 1999). Chinese Pharmacopoeia includes the root of *P. lobata* and *P. thomsonii* as the only legitimate sources of Ge Gen (Pharmacopoeia Commission of the Ministry of Health of PRC, 1995).

## PROPERTIES AND INDICATIONS OF GE GEN IN TRADITIONAL CHINESE MEDICINE

In traditional Chinese medicine, properties of herbs are described in terms of temperature and taste characteristics. There are five temperature properties (warm, cold, hot, cool and neutral) and five basic tastes (sour, sweet, bitter, pungent and salty). These properties do not necessarily refer to the literal meanings of these designations but reflect their functions and therapeutic indications. "Hot diseases must be cooled, cold diseases must be warmed." Thus, an herb able to cure heat or warm symptoms is of cool or cold property, and an herb for cold symptoms has a hot or warm property. The tastes are correlated with their functions. Pungent herbs disperse exopathogens from superficies and facilitates the move of Qi and blood; sweet herbs nourish, replenish, tonify and harmonize; bitter herbs clear heat and fire, send down adverse flow of Qi and dry dampness; sour herbs arrest discharges; salty herbs purge and soften hard mass (Zhu, 1998).

Ge Gen is an herb with pungent and cold properties and hence, according to the principles of traditional Chinese medicine, it belongs to a group of medicinal materials that cure heat and warm symptoms and disperse pathogenic factors from the superficies of the body and relieve exterior syndromes. According to Chinese medical theory, exterior symptoms appear when the exogenous causes of diseases invade the body and lodge in the superficies. The main function of Ge Gen is to dispel pathogenic factors from the superficial muscles to allay fever, headache and stiffness of the nape. It also promotes the rash of measles to surface to hasten recovery from measles with incomplete eruption of the rash. In addition, Ge Gen also promotes production of body fluid to alleviate thirst and diabetes. It is also used to arrest diarrhea in acute dysentery and diarrhea (Zhu, 1998).

Modern scientific research has led to the use of Ge Gen for coronary disease and angina pectoris. This new use is based on the research on the effects of Ge Gen on smooth muscle, cerebrovascular and cardiovascular systems (see Chapter 8 for details). Ge Gen is also recommended for stiff neck and pain from hypertension based on lengthy clinical experience of using Ge Gen for neck stiffness and pain from externally contracted disorders (Bensky and Gamble, 1986).

## COMMON GE GEN FORMULATIONS: INDICATIONS AND EFFICACIES

Rare is the case in traditional Chinese medicine that an herb is used alone. Herbs are often combined in order to produce optimal therapeutic effects, to accommodate complex clinical conditions, and/or to minimize toxicity and side effects (Zhu, 1998).

For cold or flu with headache, neck stiffness and pain, Ge Gen is often used with Gui Zhi (*Caulis cinnamomi*) and other herbs as in Ge Gen Tang (Kudzu decoction) to relieve exterior symptoms. For measles in the early stages where the rash has not yet appeared, Ge Gen is used together with Sheng Ma (*Radix cimicifugae*) as in Sheng Ma Ge Gen Tang (Cimicifuga and kudzu decoction). For hot diarrhea and dysentery-like disorders caused by damp heat, Ge Gen is used together with Huang Lian (*Rhizoma coptidis*) and often with Huang Qin (*Radix scutellariae*) as well, as in Ge Gen Qin Lian Tang (Kudzu, scutellaria and coptis decoction). With Tian Hua Fen (*Radix trichosanthis*) and Mai Meng Dong (Tuber *Ophiopogonis japonici*), Ge Gen is used to relieve thirst in what is called "thirsting and wasting syndrome" in traditional Chinese medicine, which often refers to diabetes (Wang, 1985).

There are many Chinese herbal formulas containing Ge Gen as the major ingredient and the most renowned ones are Ge Gen Tang (Kudzu decoction) and Ge Gen Qin Lian Tang (Kudzu, scutellaria and coptis decoction).

Ge Gen Tang (Kudzu decoction) is composed of Ge Gen (20 per cent), Ma Huang (*Herba ephedrae*, 15 per cent), Gui Zhi (*Ramulus cinnamomi*, 10 per cent), Shao Yao (*Radix paeoniae*, 10 per cent), Sheng Jiang (*Rhizoma Zingiberis Officinalis Recens*, 15 per cent), Da Zhao (*Fructus Zizyphi jujubae*, 20 per cent) and Gan Cao (*Radix glycyrrhizae*, 10 per cent). This formula is indicated for fever and chills without sweating, and for stiff and rigid neck and upper back. This is the type of externally-contracted symptoms caused by pathogenic factors and is defined in Chinese medicine as wind-cold, which often refers to upper respiratory infection and other disorders with upper respiratory manifestations such as rhinitis and sinusitis.

In this formula, Ge Gen is the principal ingredient. It acts to relieve the muscle layer (especially of the upper back and neck) by drawing fluids to the affected area, and release the exterior. Gui Zhi helps the principal to relieve the exterior and Ma Huang induces sweating (Bensky and Barolet, 1990).

Pharmacological studies showed that Ge Gen Tang has antimicrobial, antipyretic, diaphoretic and natural killer (NK) cell stimulating activities. This decoction and its ingredient were also found to have anti-allergic activities (Deng, 1990).

The primary indications of Ge Gen Tang are various acute infectious diseases including respiratory infection, enteritis and dysentery, and suppurative skin infections such as carbuncles and sores. In Japan, Ge Gen Tang is one of the most commonly used formulas for cold and flu. According to a Japanese survey comparing effects of various cold formulas, Ge Gen Tang is the most effective. Subjective symptom improvement, clinical efficacy and safety in cold patients reached 93 per cent, 90 per cent and 100 per cent, respectively. It is particularly effective in children (Deng, 1990).

Another indication of the formula is cervical spondylopathy, neck and shoulder problems. In addition, the formula has also been used in the treatment of allergic diseases including urticaria and eczema (Deng, 1990).

Ge Gen Qin Lian Tang (Kudzu, scutellaria and coptis decoction), consisting of Ge Gen (50 per cent), Huang Qin (*Radix scutellariae baicalensis*, 18.75 per cent), Huang Lian (*Rhizoma coptidis*, 18.75 per cent) and Gan Cao (*Radix glycyrrhizae*, 12.5 per cent), is indicated for acute gastroenteritis, acute enteritis, early-stage poliomyelitis and bacillary dysentery. The importance of Ge Gen is reflected in its relatively large dosage. It releases the exterior, clears heat and treats dysenteric diarrhea. Huang Lian, as the deputy, stops diarrhea and Huang Qin assists the deputy in stopping diarrhea. Irrespective of the prior treatment history, the use of this formula has been expanded to include any early-stage dysenteric disorder characterized by fever, foul-smelling stools and a burning sensation in the anus. With the appropriate presentation, it may be used in treating such biomedically-defined diseases as acute gastroenteritis, acute enteritis, measles, early-stage poliomyelitis and bacillary dysentery (Bensky and Barolet, 1990).

Pharmacological activities of the formula relavent to its applications include anti-microbial, antipyretic, anti-inflammatory and spasmolytic actions. Anti-arrythmic actions of the formula have also been reported. Clinically this formula is mainly used in cases of acute intestinal infection such as acute bacillary dysentery, enteritis and diarrhea (Deng, 1990).

This formula has been made into tablet and pill forms and they are included in the Chinese Pharmacopoeia. The tablet (Ge Gen Qin Lian Pian) is indicated for dysenteric diarrhea, and the pill (Ge Gen Qin Lian Wei Wan) is indicated for dysenteric diarrhea, bacillary dysentery and enteritis (Pharmacopoeia Commission of the Ministry of Health of PRC, 1995).

## GE GEN AS A DIETARY SUPPLEMENT

In addition to their medical applications, Ge Gen is also a nutritional plant component used as food and dietary supplements. As a dietary supplement, Ge Gen is often ground with water to a powder known as Ge Fen (Kudzu starch, see Chapter 17) (Wang, 1992).

Ge Gen Cha (Kudzu tea): Tea made from 30 g of Ge Gen can be used daily to improve hypertension and chronic headache. It can also be ground to powders together with Huai Hua (*Flos sophorae*) to make Ge Gen Huai Hua tea which is beneficial to people with mild form of hypertension.

Ge Gen Jiu (Kudzu wine): The juice of fresh Ge Gen or powders of kudzu flower in wine can be used to prevent and relieve drunkenness.

Ge Gen Zhou (Kudzu and rice soup): There are three modifications to the soup for different conditions. For thirsting (diabetic), coronary heart disease, angina pectoris and chronic diarrhea, Ge Gen (30 g) and rice (50 g) are cooked untill it turns sticky. For hypertension, Sha Shen (*Radix adenophorae*, 30 g) and Mai Dong (*Radix ophiopogon*, 30 g) are added to make soup. For child cold with fever, headache and vomiting, take 15 g of Ge Gen, 50 g of rice and 6 g of ginger to make soup.

## PHARMACOLOGICAL ACTIONS OF GE GEN RELATED TO ITS USE IN TRADITIONAL CHINESE MEDICINE

### Antipyretic action

As an exterior-relieving herb, the main function of Ge Gen is to allay fever and headache. Early research found that a 20 per cent water decoction or 20 per cent alcoholic extract of Ge Gen is able to lower the body temperature in rabbits with artificially induced fever. The alcoholic extract is more potent than the water decoction (Sun, 1956). The crude extract of the root of *P. lobata*, *P. omeiensis* and *P. thomsonii* and the main isoflavone constituent puerarin exhibited different degrees of antipyretic activities in rats with 2,4-dinitrophenol-induced fever (Zhou *et al.*, 1995). The antipyretic action of the herb

*Table 3.1* Antipyretic effects of Ge Gen (Lian, 1995)

| Treatment | Dose (mg/kg) | Normal body temperature | Body temperature change after treatment | | | | | | | |
|---|---|---|---|---|---|---|---|---|---|---|
| | | | 0.5 h | 1 h | 2 h | 3 h | 4 h | 5 h | 6 h | 8 h |
| Saline | 5 ml | 37.4±0.17 | +0.8 | +1.2 | +0.8 | +0.4 | +0.3 | +0.2 | +0.1 | 0 |
| Aspirin | 200 | 37.2±0.13 | −0.4 | −0.5 | −0.7 | −0.7 | −0.5 | −0.4 | −0.2 | −0.3 |
| P. lobata | 1250 | 37.6±0.11 | −0.2 | −0.5 | −0.7 | −1.5 | −1.0 | −0.6 | −0.6 | −0.1 |
| P. thomsonii | 2500 | 37.9±0.13 | +0.1 | +0.1 | +0.3 | +0.3 | +0.1 | 0 | 0 | −0.4 |

has been tested in rats and was compared with aspirin. Intraperitoneal administration of 1.25 g/kg (1/10 of the $LD_{50}$ dose) of alcoholic extract of the root of *P. lobata* produced an antipyretic effect in rats with fever induced by 2,4-dinitrophenol. The effect was more potent than that of 200 mg/kg of aspirin and lasted for 7–8 h. In contrast, the antipyretic effect of the root of *P. thomsonii* was significantly lower even at a much higher dose, 2.50 g/kg (Table 3.1).

To elucidate the mechanism by which Ge Gen relieves fever, Zeng *et al.* tested the anti-endotoxin activities of the aqueous extract of the root of several *Pueraria* plants in China. Among the species tested, *P. alopeculoides* has the highest anti-endotoxin activity, followed by *P. lobata, P. edulis, P. thomsonii, P. phaseoloides*, and *P. montana*. *P. peduncuralis* has the lowest anti-endotoxin effect. Among the roots of *P. lobata* collected from different habitats, those from Shandong, Hebei, Zhejiang and Sichuan exhibited higher anti-endotoxin activities than those from Liaoning, Shaanxi and Jiangxi, whereas the roots from Anhui, Guizhou and Yunan showed relatively weak anti-endotoxin activities (Zeng *et al.*, 1997a).

### Anti-diabetic effect

Ge Gen has been used for the treatment of diabetic symptoms nearly 2000 years ago as it is indicated in Shen Nong Ben Cao Jing for "thirsting and wasting symptoms" which often refers to diabetic symptoms in modern medicine. Today, Ge Gen is often prescribed together with other herbs to treat diabetes. Early researches found that the aqueous extract of Ge Gen had a mild hypoglycemic effect in rabbits (Luo, 1957). Puerarin, the main flavonoid constituent of Ge Gen, reduced blood glucose level at the oral dose of 500 mg/kg. Co-administration with aspirin increased the hypoglycemic effect of puerarin (Shen and Xie, 1985).

Recently, Zeng re-examined the anti-diabetic effect of Ge Gen extracts in alloxan- and streptozotocin (STZ)-induced diabetic rats. Oral administration of Ge Gen extracts (PT1 and PT2) for 6 consecutive days significantly decreased the blood glucose in alloxan-induced diabetic rats (Table 3.2).

In STZ-induced diabetic rats, daily oral dose of 1 and 4 g/kg of the Ge Gen extract (PT3) for 32 days also reduced the blood suger levels to normal (Table 3.3).

Futhermore, it was found that in STZ-induced diabetic rats, the same treatment with extract PT3 increased serum insulin and C-peptide levels (Table 3.4). This suggests that Ge Gen extracts may have a stimulatory effect on the excretion of serum insulin and C-peptide in diabetic animals (Zeng, 1999).

*Table 3.2* Hypoglycemic effect of Ge Gen in alloxan-
        induced diabetic rats (Zeng, 1999)

| Treatment | Blood glucose content (mM/L) | |
| --- | --- | --- |
| | Pretreatment | Posttreatment |
| Control | 20.47±6.17 | 18.60±5.47 |
| PT1 | 18.65±2.56 | 3.78±1.18* |
| PT2 | 18.78±4.31 | 3.30±1.82* |

Note

* $p < 0.01$ in comparison both with pretreatment level of the
  same group and with posttreatment level of the control group.

*Table 3.3* Hypoglycemic effect of Ge Gen in STZ-induced
        diabetic rats (Zeng, 1999)

| Treatment | Blood glucose content (mM/L) | |
| --- | --- | --- |
| | Pretreatment | Posttreatment |
| Normal | 3.60±0.59 | 2.99±0.48 |
| Control | 16.68±3.41 | 16.04±4.52 |
| PT1 | 16.56±2.11 | 4.53±3.93*[§] |
| PT2 | 15.42±1.15 | 3.25±1.95*[§] |

Notes

* $p < 0.01$ in comparison with pretreatment level of the same group.
§ $p < 0.05$ in comparison with posttreatment level of the control group.

*Table 3.4* Effect of Ge Gen on insulin and C-peptide in STZ-induced diabetic rats

| Treatment | Serum insulin level (μu/ml) | Serum C-peptide level (pM/ml) |
| --- | --- | --- |
| Normal | 0.29±0.22* | 4.05±1.16* |
| Control | 0.15±0.05 | 2.65±0.45 |
| PT3 (4 g/kg) | 0.24±0.04* | 7.28±2.31* |
| PT2 (1 g/kg) | 0.38±0.14* | 5.27±0.93* |

Note

* $p < 0.05$ in comparison with the control.

## QUALITY EVALUATION OF MEDICINAL GE GEN

It is generally accepted that the isoflavons, particularly puerarin, daidzein and daidzin, are the main active principles of Ge Gen and the quality of Ge Gen is measured in terms of the content of these isoflavones (Hayakawa *et al.*, 1984; Zhang and Yang, 1984; Lian *et al.*, 1995; Zhou *et al.*, 1995; Zeng *et al.*, 1997b).

### *Pueraria* species, habitats and isoflavone contents

The isoflavone contents of Ge Gen vary dramatically in samples from different *Pueraria* species and in samples from the same species but grow in different habitats. It is generally believed that the root of *P. lobata* has the highest isoflavone contents. This early belief was later confirmed by two comprehensive studies on this herb (Tables 3.5, 3.6 and 3.7) (Lian *et al.*, 1995; Zeng, 1999). The total isoflavone content in the root of *P. lobata* reaches 5.02–14.54 per cent while that of *P. thomsonii* contains only 0.60–3.67 per cent of

*Table 3.5* Isoflavone contents of *Pueraria* plants (Lian *et al.*, 1995)

| Species | Total flavonoids | Isoflavonoids | | |
|---|---|---|---|---|
| | | Puerarin | Daidzin | Daidzein |
| *P. lobata* | >5 | 1.03–6.44 | 0.028–0.87 | 0.024–0.22 |
| *P. thomsonii* | | | | |
| wild | 1–4 | 0.23–1.60 | 0.0003–0.33 | 0.033–0.14 |
| cultivated | <2 | 0.019–0.89 | 0–0.16 | 0.0076–0.16 |
| commercial | 0.55–1.40 | 0.28–0.76 | 0.03–0.14 | 0.012–0.027 |
| *P. omeiensis* | 1.27–1.85 | 0.21–0.48 | 0.11–0.37 | 0.036–0.13 |
| *P. montana* | 0.53–1.00 | 0–0.014 | 0.0077–0.06 | 0.0096–0.028 |
| *P. phaseoloides* | 1.58 | 0.0025–0.023 | 0.030–0.057 | 0.020–0.057 |
| *P. peduncularis* | 0.68–0.94 | 0.0019–0.027 | 0.0017–0.052 | 0.014–0.032 |

*Table 3.6* Total flavonoids and isoflavonoids contents of *Pueraria* species (Lian *et al.*, 1995)

| Species | Habitat and harvest time | Total flavonoids | Puerarin | Daidzin | Daidzein |
|---|---|---|---|---|---|
| *P. lobata* | Zhengan-1, Guizhou (June, 1987) | 5.02 | 3.47 | 0.35 | 0.044 |
| | Huairou, Beijing (May, 1978) | 5.50 | 2.16 | 0.71 | 0.95 |
| | Luoyang, Henan (1987) | 5.72 | 2.01 | 0.39 | 0.095 |
| | Nanchuan-1, Sichuan (1986) | 5.81 | 3.53 | 0.028 | 0.092 |
| | Hunan (1987) | 6.20 | 3.00 | 0.77 | 0.067 |
| | Zhengan-2, Guizhou (June, 1987) | 6.57 | 2.94 | 0.077 | 0.099 |
| | Quannan, Jiangxi | 6.80 | 3.00 | 0.45 | 0.084 |
| | Tonghua, Jilin (1976) | 6.90 | 2.78 | 0.127 | 0.12 |
| | Sichuan (1987) | 6.90 | 3.19 | 0.65 | 0.13 |
| | Xinyang, Henan (1987) | 7.26 | 2.90 | 0.38 | 0.11 |
| | Chonqing, Sichuan (August, 1987) | 7.33 | 2.92 | 0.60 | 0.22 |
| | Bose, Guangxi (September, 1983) | 7.57 | 2.88 | 0.38 | 0.076 |
| | Huairou, Beijing | 8.36 | 4.69 | 0.77 | 0.077 |
| | Guiyang, Guizhou (1989) | 8.60 | 2.97 | 0.46 | 0.11 |
| | Nanchuan-2, Sichuan (1986) | 8.84 | 3.36 | 0.21 | 0.18 |
| | Pinglu, Shaanxi | 10.00 | 3.69 | 0.36 | 0.12 |
| | Tonghua Jilin (August, 1978) | 10.50 | 4.84 | 0.87 | 0.11 |
| | Xinlong, Hebei (1987) | 10.60 | 5.40 | 0.53 | 0.024 |
| | Taibaishan, Shaanxi (June, 1986) | 14.54 | 6.44 | 0.86 | 0.16 |
| *P. thomsonii* | Daxin, Guangxi (September, 1988) | 0.60 | 0.25 | 0.018 | 0.0095 |
| | Zhengan, Guizhou (June, 1987) | 1.00 | 0.23 | 0.0003 | 0.036 |
| | Pingnan-1, Guangxi (1987) | 1.13 | 0.48 | 0.120 | 0.020 |
| | Pingnan-2, Guabgxi (1987) | 1.14 | 0.31 | 0.039 | 0.016 |
| | Jinfo, Sichuan (1987) | 1.21 | 0.48 | 0.130 | 0.033 |
| | Daxin, Guangxi (October, 1989) | 1.38 | 0.32 | 0.013 | 0.015 |
| | Daxin, Guangxi (June, 1987) | 1.68 | 0.35 | 0.023 | 0.077 |
| | Sichuan (1987) | 2.29 | 0.39 | 0.270 | 0.062 |
| | Pingnan-3, Guangxi (1987) | 2.94 | 0.80 | 0.040 | 0.047 |
| | Zixing, Hunan | 3.56 | 1.58 | 0.110 | 0.074 |
| | Nanchuan, Sichuan (August, 1987) | 3.67 | 1.46 | 0.330 | 0.140 |

*Table 3.6* (Continued)

| Species | Habitat and harvest time | Total flavonoids | Puerarin | Daidzin | Daidzein |
|---|---|---|---|---|---|
| P. omeiensis | Bailongdong, Sichuan | 1.55 | 0.43 | 0.150 | 0.510 |
| | Zunyi, Guizhou (1987) | 1.85 | 0.48 | 0.370 | 0.130 |
| P. montata | Quanxian, Guangxi | 0.53 | – | 0.040 | 0.028 |
| | Nandan, Guangxi | 0.67 | 0.01 | 0.0077 | 0.0096 |
| | Guangxi | 1.00 | 0.014 | 0.0001 | 0.027 |
| P. phaseoloides | Yulin, Guangxi | 1.57 | 0.0025 | 0.029 | 0.020 |
| | Cangwu, Guangxi | 1.58 | 0.023 | 0.057 | 0.057 |
| P. peduncularis | Jinfoshan, Sichuan (1987) | 0.68 | 0.027 | 0.052 | 0.032 |
| | Miyi, Sichuan (1973) | 0.71 | 0.0084 | 0.014 | 0.017 |
| | Fenghuangsgan, Guizhou (1987) | 0.75 | 0.0019 | 0.028 | 0.014 |
| | Nanchuan, Sichuan (1987) | 0.94 | 0.0099 | 0.0017 | 0.018 |

*Table 3.7* Total flavonoids and isoflavonoids contents of *Pueraria* species (Zeng, 1999)

| Species | Habitat and harvest time | Total flavonoids | Puerarin | Daidzin | Daidzein |
|---|---|---|---|---|---|
| P. lobata | Jinzai, Anhui (September, 1994) | 5.44 | 2.50 | 0.41 | 0.05 |
| | Lingan, Zhejiang (May, 1995) | 6.75 | 3.23 | 0.53 | 0.13 |
| | Taian, Shandong (September, 1994) | 8.51 | 4.56 | 0.09 | 0.02 |
| | Nanchang, Jiangxi (August, 1994) | 9.74 | 2.19 | 1.59 | 0.09 |
| | Cenggong, Guizhou (October, 1994) | 10.21 | 4.38 | 0.71 | 0.13 |
| | Xi'an, Shaanxi (October, 1994) | 10.80 | 4.04 | 1.90 | 0.18 |
| | Jingxing, Hebei (August, 1994) | 12.82 | 4.54 | 0.89 | 0.14 |
| | Jixian, Tianjin (October, 1994) | 16.44 | 8.03 | 1.82 | 0.17 |
| | Emei, Sichuan (October, 1994) | 20.18 | 7.28 | 4.05 | 0.12 |
| | Lüshun, Liaoning (August, 1994) | 21.68 | 11.50 | 3.95 | 0.09 |
| P. thomsonii | Lijiang, Yunnan (November, 1994) | 2.54 | 0.84 | 0.21 | 0.03 |
| | Fengxin, Jiangxi (August, 1994) | 2.79 | 1.24 | 0.30 | 0.03 |
| | Dali, Yunnan (November, 1994) | 3.32 | 1.51 | 0.45 | 0.05 |
| | Jinghong, Yunnan (January, 1995) | 4.11 | 2.77 | 0.27 | 0.02 |

isoflavone. Puerarin is the major component of the isoflavones, accounting for 34.5–69.1 per cent of the total isoflavones in the root of *P. lobata* and 17.03–44.38 per cent of the isoflavones found in the root of *P. thomsonii*. Daidzin and daidzein are the minor constituents of the isoflavone fraction, accounting for 0.48–12.91 per cent and 0.23–17.27 per cent of the total isoflavones in the root of *P. lobata* and 0.03–11.79 per cent and 1.09–4.58 per cent in the root of *P. thomsonii*, respectively (Lian *et al.*, 1995).

From geographical point of view, *P. lobata* from the north and north-eastern parts of China and Sichuan province often have higher content of isoflavones than those from the southern provinces (Tables 3.6 and 3.7).

## Harvest time and isoflavone contents

In the root of *P. lobata*, isoflavone contents could vary as much as five-fold within a year, from 3.58 per cent in June to 15.94 per cent in January in samples collected in Nanchang, Jiangxi on the 7th day of the month from February 1995 to January 1996. The total isoflavones including puerarin, daidzin and daidzein are highest in January,

October and November (Table 3.8). These results suggest that the root of *P. lobata* is best collected in January, October and November (Zeng, 1999).

In the root of *P. thomsonii* cultivated in Pingnan, Guangxi, the total isoflavone contents and the major isoflavone constituent puerarin are also highest in January and February (Table 3.9). It is of interest to note that it has been a long standing tradition for the locals to collect the root in late January and February (Lian *et al.*, 1995).

*Table 3.8* Contents of flavonoids in the root of *P. lobata* (Zeng, 1999)

| Harvest time | Total flavonoids | Puerarin | Daidzin | Daidzein |
|---|---|---|---|---|
| February 1995 | 5.71 | 2.99 | 0.46 | 0.02 |
| March 1995 | 5.26 | 2.82 | 1.10 | 0.04 |
| April 1995 | 3.37 | 1.31 | 0.49 | 0.04 |
| May 1995 | 6.02 | 1.92 | 0.99 | 0.02 |
| June 1995 | 3.58 | 1.35 | 0.60 | 0.03 |
| July 1995 | 4.84 | 1.54 | 0.74 | 0.07 |
| August 1995 | 9.74 | 2.99 | 1.59 | 0.09 |
| September 1995 | 8.61 | 3.40 | 1.28 | 0.06 |
| October 1995 | 10.06 | 3.73 | 0.38 | 0.02 |
| November 1995 | 11.22 | 3.43 | 4.74 | 0.04 |
| December 1995 | 9.98 | 3.59 | 0.72 | 0.09 |
| January 1996 | 15.94 | 3.28 | 1.19 | 0.01 |

*Table 3.9* Contents of flavonoids in the root of *P. thomsonii* (Lian *et al.*, 1995)

| Harvest time | Total flavonoids | Puerarin | Daidzin | Daidzein |
|---|---|---|---|---|
| 19 September, 1988 | 0.068 | 0.019 | 0.00075 | 0.010 |
| 28 September, 1988 | 0.60 | 0.25 | 0.018 | 0.0095 |
| 30 October, 1988 | 1.45 | 0.37 | 0.017 | 0.017 |
| 4 December, 1988 | 0.94 | 0.43 | 0.042 | 0.013 |
| 29 December, 1988 | 0.58 | 0.21 | 0.026 | 0.012 |
| 30 January, 1989 | 1.45 | 0.69 | 0.043 | 0.011 |
| 28 February, 1989 | 1.45 | 0.51 | 0.10 | 0.024 |

*Table 3.10* Total flavonoids content of the stem of *Pueraria* plants (Zeng *et al.*, 1999)

| Species | Habitat | Harvest time | Total flavonoids |
|---|---|---|---|
| P. lobata | Nanchang, Jiangxi | August, 1994 | 1.13 |
| | Taian, Shandong | October, 1994 | 4.51 |
| | Xi'an, Shaanxi | November, 1994 | 7.93 |
| P. thomsonii | Dali, Yunnan | November, 1994 | 0.93 |
| | Quxian-1, Zhejiang | July, 1997 | 2.53 |
| | Quxian-2, Zhejiang | July, 1997 | 7.05 |
| P. montata | Guangzhou, Guangdong | July, 1997 | 1.02 |
| P. alopecuroides | Jinghong, Yunnan | August, 1994 | 0.36 |
| P. calycina | Yongsheng, Yunnan | November, 1994 | 2.01 |
| P. edulis | Weishan, Yunnan | November, 1994 | 1.02 |
| P. phasedoides | Guangzhou, Guangdong | July, 1994 | 1.63 |
| P. peduncularis | Weishan, Yunnan | November, 1994 | 1.03 |
| | Kunming, Yunnan | August, 1994 | 1.83 |

## The stem of *Pueraria*

The stem of *Pueraria* plants is cast during the collection of the root. Recent studies have shown that the stem of *Pueraria* plants also contains fairly high content of isoflavones. In the stem of *P. thomsonii*, the isoflavone contents are even higher than those found in the root (Table 3.10). The composition of the isoflavone constituents in the stem are similar to those found in the root. It is therefore suggested that the stem of kudzu should also be considered as a source of *Pueraria* isoflavones (Zeng *et al.*, 1999).

## REFERENCES

Bensky, D. and Barolet, R. (1990) *Chinese Herbal Medicine: Formulas & Strategies*. Eastland Press, Seattle, pp. 51–52.

Bensky, D. and Gamble, A. (1986) *Chinese Herbal Medicine: Materia Medica*. Eastland Press, Seattle, pp. 66–67.

Deng, W.L. (1990) *Pharmacology and Applications of Chinese Herbal Formulas*. Chongqing Press, Chongqing, pp. 15–25, 171–176.

Jiangsu College of New Medicine (1977) *Encyclopedia of Traditional Chinese Medicinal Substances*. Shanghai People's Publishers, Shanghai, pp. 2307–2310.

Hayakawa, J., Noda, N. and Yamada, S. (1984) Studies on physical and chemical quality evaluation of crude drugs preparations. I. Analysis of *Pueraria radix* and species *Pueraria*. *Yakugaku Zasshi*, 104, 50–56.

Lian, W.T., Feng, R.Z., Chen, B.Z., Zhou, Y.P., Su, X.L., Zhong, Y. *et al.* (1995) Studies on Ge Gen, *Radix puerariae*. In Z.C. Lou and B. Qin (eds), *Species Systematization and Quality Evaluation of Commonly Used Chinese Herbs*, North China Edition, Vol. I, Beijing Medical University and China Union Medical University Press, Beijing, pp. 379–420.

Lian, W.T., Feng, R.Z. and Wang, H. (1989) Ge Gen. In Institute of Materia Medica, Chinese Academy of Medical Sciences (ed.), *Zhong Yao Zhi*, Vol. 1, People's health Publishers, Beijing, pp. 563–568.

Luo, H.W. (1957) Effect of several Chinese herbs used for thirsting and wasting syndrome on rabbit blood glucose level. *J. Nanjing Coll. Pharm.*, 2, 61.

Maesen, van der L.J.C. (1985) Revision of the genus *Pueraria* DC. *Agric. Univ. Wageningen Pap.*, 85, 37–62.

Pharmacopoeia Commission of the Ministry of Health of PRC (1995) *Pharmacopoeia of the People's Republic of China*. Chemical Industries Press, Beijing, pp. 610–611.

Shen, Z.F. and Xie, M.Z. (1985) Hypoglycemic action of puerarin and aspirin complex. *Acta Pharm. Sin.*, 20, 863–865.

Sun, S.X. (1956) Pharmacological studies of several antipyretic Chinese herbs. *Chin. Med. J.*, 42, 964.

Wang, J.M. (1985) *Chinese Herbal Pharmacology*. Shanghai Science & Technology Publishers, Shanghai, pp. 30–31.

Wang, Z.Y. (1992) *Encyclopedia of Chinese Medical Diets*. Dalian Press, Dalian, p. 117, 300, 432, 433, 443, 521, 628.

Zeng, M. (1999) Studies on resources utilization and quality evaluation of *Pueraria* root and other plants of *Pueraira* DC. in China. Ph.D. Thesis, Second Military Medical University, Shanghai, China.

Zeng, M., Zhang, H.M., Zheng, S.Q., Liu, K., Liu, S.N. and Zheng, X.H. (1997a) Comparison of anti-endotoxin activities of Ge Gen and related plants. *China J. Chin. Mater. Med.*, 22, 178–179.

Zeng, M., Zhang, H.M., Zheng, S.Q. and Zheng, X.H. (1997b) Ontogenic chemical changes of the active constituents in Ge Gen (*Pueraria lobata*). *Acad. J. Second Mil. Med. Univ.*, 18, 150–152.

Zeng, M., Zhang, H.M., Zheng, S.Q., Shao, F., Tao, C.Y. and Su, Z.W. (1999) Chemical analysis of the stems of *Pueraria* plants. *China J. Chin. Mater. Med.*, 24, 136–137, 149.

Zeng, M., Zhang, H.M., Zheng, S.Q., Su, Z.W. and Qian, Z.M. (2000) Herbological investigation on traditional Chinese medicine Ge Gen. *J. Chin. Med. Mater.*, 23, 46–48.

Zhang, Y.Z. and Yang, F. (1984) HPLC determination of isoflavonoids in Ge Gen and its tablet preparations. *Chin. J. Pharm. Anal.*, 4, 67.

Zhou, Y.P., Su, X.L., Cheng, B., Jiang, J. and Chen, H. (1995) Comparative study on pharmacological effects of various species of *Pueraria. China J. Chin. Mater. Med.*, 20, 619–621.

Zhu, Y.P. (1998) *Chinese Materia Medica: Chemistry, Pharmacology and Applications.* Harwood Academic Publishers, Amsterdam, pp. 11–15, 92–97.

# 4 *Pueraria tuberosa* DC: contraceptive efficacy and toxicological profile

*Sangeeta Shukla and R. Mathur*

## INTRODUCTION

In the past three decades, there has been a marked increase in the use of contraception worldwide. There is also a growing awareness of the urgent needs to improve the quality of family planning services and develop a wider range of methods for family planning that will meet the different and changing needs of individuals and couples throughout their reproductive lives. The challenge facing those working in the field is to respond adequately to these needs by paying due attention to quality of care in the service providing sector and developing new, safe, effective, acceptable and affordable methods for fertility control in the laboratories. The explosion of scientific advances in the past decades has generated an enormous amount of new knowledge about the physiology of human reproduction and information, critical to the development of new and improved methods of fertility regulation.

In India, medicinal plants have played a very prominent role in the practice of traditional medicine and our herbal pharmacopoeias include hundreds of plants. Many of our present medicines are derived directly or indirectly from plants. A number of novel lead compounds originated from medicinal plants have been developed into conventional Western medicines. Research on medicinal plants has made particularly rewarding progress in the discovery of anticancer, antimalarial, hepatoprotective and other therapeutic agents. In addition to drugs purified from plants, there is also an enormous market for herbal medicines in their crude forms. The executive board of World Health Organization (WHO) has passed a resolution, calling on countries to promote the role of traditional medicine practitioners in the health care systems of developing countries. World Health Organization's task force for fertility regulation has also completed a computer search for all information on indigenous plants which have been traditionally used in different parts of the world as contraceptives. Emphasis has been placed on their virtues like high efficacy, reversibility, no severe hazards, low cost, and acceptability in the light of the religious, ethical and cultural background.

Many indigenous plants are capable of preventing conception when administered orally. Research on plants with anti-fertility activity has been exhaustively reviewed (Kirtikar and Basu, 1935; Chopra and Chopra, 1955; Dhar *et al.*, 1968, 1973, 1974; Prakash and Mathur 1976; Chaudhary and Haq, 1980; Farnsworth *et al.*, 1981; Kamboj and Dhawan, 1982; Prakash, 1984; Satyavati, 1984; Dhall and Dogra, 1988; Al-Hamood and Al-Bayati, 1995; Alkofahi *et al.*, 1996; Elbetieha *et al.*, 1996). This chapter will be devoted mainly to the contraceptive principles found in the medicinal plant *Pueraria tuberosa*.

*Pueraria* is a plant widely used in traditional Indian Medicine. The tuber of *Pueraria* is sweet in taste and used in indigenous system of Indian medicine as tonic, aphrodisiac, antirheumatic, diuretic and galactogogue (Kirtikar and Basu, 1935). It is an important constituent of Ayurvedic medicines including Chywanprash, a popular tonic (Ayurvedic Formulary of India, 1978). In the traditional Indian medicine, the root of *P. tuberosa* (*Radix puerariae*) has been mentioned for its antispasmodic activity. Certain Sadhus of Terai area of Uttar Pradesh (India) have been reported to consume these tubers for increasing the general body resistance. Furthermore, it is active against angina pectoris, hypertension, deafness, optic nerve atrophy or retinitis. The root is also used as demulcent and refrigerent in fevers. Peeled and bruised into a cataplasm, it is used to reduce swelling of the joints. We have worked on this plant for about two decades and have been provided opportunities to investigate in depth the contraceptive efficacy of this medicinal plant and assess the manner and extent by which it can be used for birth control.

*P. tuberosa* is distributed widely in India, from the Western Himalaya to Sikkim, up to 4000 ft in Kumaon. It is found in the lower hills of the Punjab, Mount Abu, the hilly tracts of Bengal and in most of Southern India. Morphologically, *P. tuberosa* is a large deciduous climber or a twiner with shrubby stem and tuberous roots. The leaves are three-foliate with subeoriaceous leaflets and long petioles about 10–15 cm. Flowers are in lax (sometimes panicled), leafless racemes, 15–30 cm long with 6–8 mm long calyx. The corolla is bluish. The pods are 5.0–7.5 cm long, membranous, flat and constricted between the seeds. The seeds are in 3–6 numbers in each pod.

## CONTRACEPTIVE AND ABORTIFICIENT EFFECTS OF *P. TUBEROSA*

### Post-coital contraceptive efficacy of extracts of *P. tuberosa* in adult animals

The tubers of *P. tuberosa* DC have been used for birth control by the nomads in Jammu and Kashmir for years (Sharma, 1979). Animal studies have also shown that various extracts of *P. tuberosa* posses significant *post coital* contraceptive effect in rats, mice, hamsters and guinea pigs. Chandhoke *et al*. (1980) reported the absence of ova in the fallopian tubes of animals treated with an alcoholic extract of *P. tuberosa* DC (125 mg/kg p.o.). This result indicates that a or more active principle(s) of *P. tuberosa* is able to suppress ovulation. Further, the alcoholic extract of *P. tuberosa* given orally (125 mg/kg) to pregnant rats from Day 1 to Day 7 of pregnancy exerted 100 per cent antizygotic activity. While studying the hormonal profile of *P. tuberosa* DC, Gupta *et al*. (1980) revealed that the alcoholic extract possesses 100 per cent post coital antifertility activity in rats, guinea pigs and hamsters. According to Regional Research Laboratory (1981), when pregnant rats were administered from Day 1 to Day 7 with an alcoholic extract of *P. tuberosa*, 100 per cent anti-implantation activity was evident at 250 mg/kg p.c. dose. Moreover, when the extract was given on Day 6 and Day 7 of pregnancy, the rats failed to deliver the pups. In this situation, it appears that the extracts acted as an early abortificient agent. Jani *et al*. (1981) have reviewed the plant *P. tuberosa* DC. They have reported pharmacognosy, pharmacology and the clinical trials of this plant. Mathur *et al*. (1983) reported that the crude powder of *P. tuberosa*, at a dose of 500 mg/rat, induced resorption of foetuses in early stages of pregnancy. This result suggests that implantation of embryos occurred normally but as pregnancy advanced, some unspecified factors created disturbances in the uterus which caused the death of some foetuses. Prakash (1984) evaluated the contraceptive efficacy of a number of medicinal plant extracts and reported a significant antifertility activity in the crude powder, methanolic and ethanolic extracts of *P. tuberosa* DC in rats. Further, Prakash *et al.* (1985) showed that the crude powder, ethanolic, butanolic, petroleum ether and benzene extracts of *P. tuberosa* interfered with early pregnancy in rats, mice and hamsters when administered at doses ranging from 500 to 150 mg/kg. Chloroform and methanolic extracts also exhibited significant anti-implantation activity at 200 mg/kg dose. Hexane extract did not show any significant activity. Aqueous extract was also ineffective in preventing implantation in mice and rats. Prakash *et al.* (1985, 1986) and Shukla *et al.* (1989b) have reported significant anti-implantation activity in the ethanolic and butanolic extracts of *P. tuberosa* DC. Among the extracts tested, the butanolic extract was found to be the most effective at low doses (~150 mg/kg) administered according to the 7-day schedule.

## Minimum effective contraceptive dose of butanolic extract of *P. tuberosa* in adult rats and hamsters

Two doses of butanolic extract of *P. tuberosa* were found effective in preventing nidation successfully. Administration of a butanolic extract of *P. tuberosa* on Day 1 of gestation prevented pregnancy in 66.6 per cent of the rats. However, when the extract was administered on Day 1 and Day 2 p.c., it inhibited implantation in all animals. Furthermore, the extract was tested for anti-zygotic and blastocidal activities in rats and was administered at different durations, i.e. D1–D3, D2–D3 and D3–D5. In all cases, 100 per cent anti-implantation effect was observed.

At a dose of 100 mg/kg, the butanolic extract suppresses implantation in golden hamsters by 75 per cent and at higher doses, 150 and 200 mg/kg, it inhibited implantation completely. Correlation coefficient between the doses administered and anti-implantation activities suggested a strong positive correlation between the two ($r = 0.96$) (Shukla, 1993). At the various doses tested, the butanolic extract did not cause foetal resorption at later days of pregnancy.

## Anti-implantation activity of different fractions of *P. tuberosa*

Three known compounds namely tuberosin, puerarin and diadzein isolated from the chloroform and butanolic extracts of *P. tuberosa* showed significant anti-implantation activity in rats and hamsters (Gupta *et al.*, 1990). Puerarin Fr. was found to be most effective (100 per cent) whereas other compounds such as anhydrotuberosin, hydroxytuberosin and saponinglycoside mixture (from postpuerarin Fr.) were not active at all. Shukla (2000) reported that the *n*-butanol and pre-puerarin fractions prevented pregnancy (100 per cent) in female rats when administered orally on Day 1 to Day 8 *post coitum*. Puerarin fraction also showed significant activity. Pure compounds diadzein and tuberosin exhibited 75 per cent and 50 per cent anti-implantation activity.

In pre-coital studies, animals were pretreated with the butanolic extract at three different doses: 50 mg/kg, 100 mg/kg and 150 mg/kg. These doses were given orally for 2, 4 and 8 days, respectively. These animals were then left with males of proven fertility.

Rats with positive sperm were selected and observed for implantation on Day 10 p.c. No significant anti-implantation effect was observed in rats received the lowest dose of extract (50 mg/kg) for 2 and 4 days. However, when the extract was given orally for 8 consecutive days, anti-implantation activity was observed in two rats. At 100 mg/kg dose, significant activity was also observed after 8 days of treatment. At $ED_{100}$, i.e. 150 mg/kg dose, 50 per cent inhibition was observed after two doses and 100 per cent anti-implantation effect was resulted after four consecutive doses.

## Hormonal profile

Studies on the biological profile indicate that crude powder, ethanolic, petroleum ether and benzene extracts of *P. tuberosa* possess significant estrogenic property. In addition to this activity, the crude powder also showed progestational and mild antiprogestational activities whereas, ethanolic and butanolic extracts showed only estrogenic activity (Prakash *et al.*, 1985). Oral administration of diethylstilbestrol (DES) at different doses increased the uterine wet weight in immature rats, the dose response was indicated by a sigmoidal curve. Similar type of sigmoidal response was observed when butanolic extract of *P. tuberosa* was administered at various doses. When administered conjointly with DES, the extract induced a dose-dependent reduction in uterine wet weight. Maximum reduction was observed at the highest dose administered (600 mg/kg). Open vagina was observed at lower doses whereas at higher doses vagina remained closed. In adult rats undergoing experimentally induced delayed implantation, DES, at a dose of 2 mg/kg, successfully induced implantation of the blastocysts in all rats. Furthermore, 150 and 200 mg/kg dose of butanolic extract of *P. tuberosa* also induced implantation in all rats. As the dose of extract increased from 300 mg to 600 mg/kg, a successive decrease in the number of rats with implantation. In the traumatization method for the assessment of progestational activity in adult rats, standard progesterone induced deciduoma in all rats at a dose of 3 mg/day. Administration of butanolic extract at doses ranging from 150 to 500 mg/kg could not induce any significant decidual response. However, 600 mg/kg dose induced deciduoma in 83.3 per cent of the rats showing thereby a significant progestational effect. The administration of different doses of butanolic extract of *P. tuberosa* (150–600 mg/kg) conjointly with 3 mg progesterone per day did not inhibit the formation of deciduoma as induced by progesterone *per se* clearly indicating antiprogestational activity (Mathur *et al.*, 1987).

## Effect of butanolic extracts of *P. tuberosa* on serum and pituitary hormone levels

A series of experiments were conducted to evaluate whether the butanolic extract of *P. tuberosa* acts on the hypothalamo-pituitary axis or enters the feed back action as a result of its effect on gonads. Animals were bilaterally ovariectomized and kept at rest for 8 days. Extract was orally administered for a period of 10 days. Ovariectomy resulted in an increase in serum and pituitary levels of LH and FSH in control animals. The extract successfully prevented the post castration rise of these gonadotropins, hence confirming its estrogen like activity. It was deemed worthwhile to explore the effect on the levels of steroids in rat serum. This experiment was also performed in spayed rats.

Ovariectomized rats after a post-operative rest of 15 days were divided into various groups. Group I served as controls, group II received estradiol dipropionate for 7 days, group III received extract and group IV received conjoint administration of both EDP and extract. Serum estrogen level was increased in both the EDP and extract treated animals and these levels were further increased in rats receiving EDP/extract combination treatment. Progesterone level was significantly decreased after EDP and conjoint treatment with EDP and extract inference confirmed estrogenic activity at this dose as extract did not counter the effect of EDP.

In the next experiment the mode of antifertility action relates to its evaluation of its estrogenecity and antiestrogenecity with time using estrogen sensitive uterine and liver glycogen as biochemical parameters in rats during delayed implantation. A comparison of the sequence of action of estradiol-17-$\beta$ and extract revealed a dose dependent response. Extract treatment indicated a linear rise in the wet weight and glycogen content up to 5 days, almost a parallel response to DES reaching a peak at 18 h and was comparable to controls, at later durations of 48 h, clearly indicating that the effect of single administration of extract remained for 5 days. These findings are further substantiated with the results of post-coital efficacy in which 100 per cent antinidatory effect was seen after 2 days of feeding of the extract (Shukla, 2000).

## *P. TUBEROSA* ON PRE-NATAL DEVELOPMENT IN RATS

As an important facet of contraceptive development, it is essential to evaluate its effects on fetal development. It is also pertinent to note that contraceptives may not produce congenital malformations but show other side effects. Accordingly, studies were conducted to evaluate the influence of the butanolic extract of *P. tuberosa* on prenatal development in the foetus of rats. Adult female rats were selected on Day 1 of pregnancy. The mated females were caged individually and divided into various groups. The animals received the extract on gestation Day 8 through Day 15 (period of organogenesis). All animals were delivered by caesarian section on Day 20. The number of live and still born foetuses were noted for each female. Results demonstrate that both the litter size and weight were, comparable to those of the controls. The body weight gain of mothers and offsprings also were normal and comparable to those of the controls. No abnormalities were detected in the offsprings during post natal growth up to 6 weeks. The animals did not show any overt toxicological effects immediately after drug administration and thereafter. The offsprings were also normal.

## EFFECT OF *P. TUBEROSA* ON ESTROUS CYCLE

Mathur *et al.* (1984) have reported the effects of various extracts of *P. tuberosa* DC on the oestrous cycle of adult rats. Petroleum ether and benzene extracts prolonged the cornified stage which persisted till the last day of treatment whereas, crude powder and ethanolic extract induced cornification but after the 13th day, the cornified stage changed into diestrous stage which continued till the last day of treatment. Butanolic extract, at a dose that is contraceptive, also induced the cornified stage which persisted for four consecutive cycles (Shukla *et al.*, 1987a).

## BIOCHEMICAL AND HISTOLOGICAL EFFECT OF *P. TUBEROSA* IN CONTROL AND OVARIECTOMIZED RATS

Shukla *et al.* (1987b, 1989a) have reported physiological and biochemical changes in female genital tract of ovariectomized and control rats treated with butanolic extract of *P. tuberosa*. Administration of the butanolic extract of *P. tuberosa* provoked a successive and significant increase in the wet weight and protein contents of the ovary, uterus, cervix and vagina. Glycogen content of the ovary, uterus and cervix increased significantly at all time intervals measured (from 6 to 24 days), but decreased in the vaginal samples. The level of total cholesterol did not change appreciably in the organs examined in the initial days of treatment, i.e. 6 and 12 days, but after prolonged treatments, i.e 18 and 24 days, cholesterol levels increased significantly. The extract provoked sharp elevation in acid and alkaline phosphatase activity in all organs at consecutive time intervals. Further Shukla *et al.* (1987c) have also observed the role of butanolic extract of *P. tuberosa* DC in the movement of glycogen in reproductive organs of cyclic and ovariectomized rats.

Following administration of the extract for 6 days, ovarian histoarchitecture showed normal matured follicles, along with follicles in different stages of development. However, in the peripheral region connective tissue septa were observed. After 12 days of treatment, ovum in some follicles had undergone atrophy and appeared deformed and granulated. Some freshly formed corpora lutea were also visible along with maturing Graafian follicles. Although, at later stages normal follicles were observed, atretic follicles had increased considerably in which younger stages were also involved. Newly formed corpora lutea were also observed as indicated by the large central cavity. In some medium sized follicles, the follicular cells were normal although the ovum showed shrinkage. Stroma was loose and exhibited prominent vascularity. The histoarchitecture of the uterus after 6 days of treatment showed increased height of luminal epithelium with basally placed nuclei and edematous stroma. After administration of extract for 12 days, uterus showed stimulated histoarchitecture and endometrial folds were increased. Uterine glands were enlarged, hypertrophied and showed evidence of darkly stained secretory material. At later stages there was a significant increase in the luminal epithelium. Leucocytic infiltration was also observed, and the uterine glands had increased. Cervical histoarchitecture showed stimulation. Stroma was loose and richly vascular. There was significant cellular alteration after 6 days of treatment. After12 days, the epithelium was markedly keratinized with epithelial cells desquamating in the lumen. As treatment prolonged, the extract appeared to cause metaplastic changes in the epithelium with marked keratinization; plicae palmatae and uterine glands were also increased significantly. Stroma were loose and exhibited a rich vascularity. Vaginal histological structures after 6 days of treatment provoked no significant alteration in the histoarchitecture. Administration of the extract for 12 days caused an increase in the cellular layers by multiplication showing significant stratification. Rugae had increased with desquamating superficial cells. Pyknotic nuclei were also seen. However, after a further interval of treatment distinct stimulation in all these structures was evident and cornification was markedly increased. Rugae also showed an increase in number, and the stroma was loose and fibrotic (shukla *et al.* 1989a).

Biochemical and histological alterations have been observed in the uterus, cervix and vagina of ovariectomized rats given butanolic extracts of *P. tuberosa* in the presence and absence of estradiol dipropionate (ED) and progesterone (P). Ovariectomy resulted in the decrease of glycogen contents, protein concentrations, activity of acid and alkaline phosphatases and total cholesterol in the uterus, cervix and vagina. However, the level of esterified cholesterol remained unaffected. EDP or progesterone tend to restore this loss, EDP being more active. Progesterone antagonized the effects EDP on the above parameters. Administration of the butanolic extract of *P. tuberosa per se* increased glycogen contents, protein concentration, and acid and alkaline phosphatase activity, and total cholesterol level in all these organs. However, in vagina, the glycogen contents decreased significantly. In the combined treatment of the butanolic extract with steroids, EDP acted synergistically whereas progesterone antagonized the action of extract. In the histo-logical studies, the butanolic extract of *P. tuberosa* stimulated uterine structures, caused metaplastic changes in the cervix epithelium with marked cornification, and increased the number of vaginal epithelial cells by multiplication along with metaplastic changes. Its combined treatment with EDP *per se* induced changes. Biochemical findings, supported by the histological observations, further confirm the estrogenic mode of action of the butanolic extract of *P. tuberosa* in adult ovariectomized rats (Shukla *et al.*, 1987b).

## MODE OF CONTRACEPTIVE ACTION OF *P. TUBEROSA*

Shukla *et al.* (2000) conducted experiments to elucidate the transitional cellular and biochemical changes and to highlight the role of factors/enzymes responsible for anti-implantation effect. Rats were selected at Day 1 of pregnancy and were divided into ten groups of five rats each. The extract was administered from Day 1 to Day 5 of pregnancy and later necropsy was performed 24 h after the last treatment. Ovaries and uterus were examined for changes during the period of implantation.

The plasma levels of estradiol-17-$\beta$ in control animals began to rise from Day 1 of pregnancy, showed a peak at Day 3 and the level was maintained upto Day 4 and then it fell at the day of implantation. This estrogen surge observed on Day 3 and Day 4 was missing in the treated rats. The progesterone level increased steadily between Day 1 and Day 2 p.c. and rose further showing significant values. In contrast, a fall in the level of plasma progesterone had been recorded in the extract treated group. Thus, the prerequisite factor for the attachment of blastocyst and preparation of endometrium was lacking (Psychoyos, 1970; Ahmed, 1971).

## *P. TUBEROSA* ON THE BIOCHEMICAL AND HISTOLOGICAL CHANGES DURING OVUM-IMPLANTATION IN RATS

Biochemical changes during pre-implantation period revealed a successive and significant increase in wet weight, protein and glycogen contents of both organs of the control animals, except the glycogen content in the ovary. Administration of the butanolic extract of *P. tuberosa* also provoked a successive increase from Day 1 to Day 5, when com-pared with their respective controls, however, the values were statistically significant at Day 4 and Day 5 only. As glycogen is located in circular muscles in rats, it may be involved in rhythmic uterine contractions. Therefore, increased uterine glycogen level observed in this study may be involved in providing energy for uterine contractions and thus facilitate the expulsion of fertilized eggs.

The same study also showed a successive and significant increase in the activity of acid and alkaline phosphatases in the ovary and uterus of control animals. In animals treated with a butanolic extract of *P. tuberosa*, no significant change in these enzyme activities were observed initially. However, 4 and 5 days after treatment, the activity of acid phosphatase was significantly depleted in the uterus. No significant changes were observed in the activity of alkaline phosphatase of the ovary. However, in the uterus, a sharp increase in enzyme activity was observed from Day 2 to Day 5 p.c. Acid phosphatase within the luminal epithelial cells causes capillary permeability, initiates decidualization, adhesion and attachment of blastocyst and destruction of the epithelial cells which facilitates trophoblast invasion during implantation. Thus, low acid phosphatase in the uterus of treated rats did not favor the preparation of the uterus for implantation. On the other hand, alkaline phosphatase is associated with induction of decidual cell reaction and enhancement of cellular permeability. Extract induced elevation of alkaline phosphatase may result in unfavorable conditions for the formation of implantation chamber. This could be achieved by altering the permeability and transport of nutrients by the uterine cells. This effect is also augmented by a significant decrease in ATPase activity in the uterus from Day 3 onwards. On the contrary, the activity of glucose-6-phosphatase was increased considerably from Day 1 to Day 5 in the uterus when compared with their respective controls (Shukla, 1996).

Histologically, uterus of control pregnant rats at Day 1 and Day 2 of pregnancy revealed stimulation. At Day 3 of pregnancy, endometrium was increased, stromal cells showed darkly stained nuclei. In the *P. tuberosa* treated animals, uterine lumen was lined with tall columnar epithelial cells, and many of LE and GE cells demonstrated vacoulated degeneration. The height of the uterine luminal epithelium was considerably increased, glands were enlarged. At Day 4 of control the endometrium was on the way to becoming a progestational type and the lumen was reduced and shifted towards the mesometrial side. The endometrium of extract treated rats showed numerous foldings of luminal epithelial lining. Glands were large and empty. At Day 5 of pregnancy, control uterus exhibited typical progestogenic endometrium with highly developed uterine glands. The stroma, having reached a maximum edematous stage, and seems to contain more blood vessels than earlier, obliterated the lumen to compress the blastocyst within the epithelial surface towards the antimesometrial side. Stroma was hypertrophied to decidual cells and apposition sites were clearly visible. The uterine glands were tortous and the lumen was slit like. The epithelial nuclei were found to have moved towards the basal membrane. In the treated rats significant stimulation like estrogenic endometrium was seen, the epithelial lining being composed of tall columnar type of cells. Uterine glands were large and empty, no decidualization was seen.

Electron microscopy has shown that the surface of the uterine luminal epithelium is changed characteristically with extract at the onset of implantation. It clearly revealed that the signs of blastocyst activation were present in the form of abundant glycogen like particles and numerous polysomes in the trophoblast cytoplasm but the attachment reaction was not discernible. The stromal cells were spindle shaped. Microvilli were shorter, had blunt apical ends and were irregular in distribution. Apposing luminal cell surfaces showed no tendency of intimate contact. The stroma displayed no signs of decidualization. Glands with wide lumen were seen. In addition, presence of significant amount of uterine secretion acts as an insulator inhibiting the close contact between the blastocyst and the epithelium. Thus, the luminal epithelium may not transduce the stimulus of the blastocyst for the attachment and the message which triggers decidualization of the underlying stroma never reached (Shukla, 2000).

## CLINICAL STUDIES ON *P. TUBEROSA*

Preliminary clinical trials have been conducted in about 250 cases, at various primary health centers of Jammu for its contraceptive, abortifacient and emmenagogue activities. The capsules containing the alcoholic extract of *P. tuberosa* and lactose in equal quantities was supplied. Results showed that the capsules were highly effective against dysmenorrhea, dysfunctional uterine bleeding and menopausal syndrome. Patients resistant to routine hormonal therapy, responded remarkably to these capsules. Out of the 153 cases studied, 128 showed marked improvement in dysfunctional uterine bleeding (DUB), 13 moderate and 12 poor or nil. The study shows that the effective in DUB case resistant to routine treatment might be due to type of estrogen or progestational or due to ideal ratio of the two naturally occurring hormonal activities in the same plant (Jani *et al.*, 1981).

## EFFECT OF *P. TUBEROSA* ON MALE REPRODUCTION

At doses of 500 mg/kg and 150 mg/kg, respectively, the crude powder and a butanolic extract of *P. tuberosa* exhibited 100 per cent antifertility effects in male rats. The crude powder of the tuber administered at a dose of 500 mg/kg body weight for 60 consecutive days, elicited antifertility effect from Day 30 till Day 60. Antifertility activity of the crude powder was 50 per cent from Day 30 onward, 70 per cent from Day 45 onward, and 100 per cent on Day 60. In rats receiving the butanolic extract at a dose of 150 mg/kg for 10 consecutive days, 100 per cent antifertility effect was developed after the 12th day till Day 60. These plant preparations were administered to castrated immature rats to perform the androgen bioassay. 500 mg/kg body weight of crude powder of tubers of *Pueraria tuberosa* inhibited the testosterone induced gain in weight of seminal vesicles, ventral prostate and levator antimuscle, whereas the 150 mg/kg body weight of butanolic extract significantly inhibited the testosterone induced increase in weight of seminal vesicles, ventral prostate and levator antimuscle of castrated immature rats, suggesting antiandrogenic activity. Both plant preparations significantly decreased the testicular glycogen content. Fructose content also declined in seminal vesicles, prostate and coagulating glands. Acid and alkaline phosphatase activity of testes, seminal vesicles and prostate was enhanced remarkably. Of the two, the butanolic extract has been found to be more efficacious than the crude powder for producing antifertility effect with a comparatively lower dose and within short time frame. The increase in glucose-6-phosphatase activity in testis, seminal vesicles, prostate and coagulating glands is suggestive of an overall disturbance in carbohydrate metabolism. In liver, the activities of acid and alkaline phosphatases and the level of glucose-6-phosphatase did not show any significant alteration at any dose and duration of treatment, suggesting that these plant preparations do not possess any side effect on this vital organ. The study on spermatological studies revealed that crude powder and butanolic extract does affect the seminal quality, which is mainly concerned with testicular size, sperm count and motility, which of course are related with fertilization. Administration of above mentioned plant preparations showed a marked decline in sperm motility and sperm count. At the same time deformities in sperm morphology increased. Thus *P. tuberosa* affects the complete spermatological aspect of experimental rats by exerting negative influence on fertilizing capacity of spermatozoa which is in conformity with results of antifertility screening. The effect of crude powder and butanolic extract of *P. tuberosa* on the anatomy and histology of testes, seminal vesicle, prostate, liver, kidney and adrenal gland. Both preparations decreased the weights of testis and accessory reproductive organs. Histologically, crude powder, at 500 mg/kg body weight dose produced atrophic changes in the testis. There was exfoliation of germ cells in the lumen. Varying degrees of conspicuous nuclear degeneration in spermatocytes with cytoplasmic granulation were seen on Day 30 and Day 45. On Day 15 and onwards, centrally placed tubules showed shrinkage. Anastamosis of tubules was also observed. Most of the tubules had an atrophic look and almost all tubules showed arrest of spermatogenesis. On day 45 tubules showed high degree of deformation. Complete atrophy of germ cells in most of the tubules was observed (Singh, 1997).

## TOXICOLOGICAL PROFILE

Acute toxicity of *P. tuberosa* has been evaluated by administering the extract once at doses ranging from 400, 800 and 1600 mg/kg. Necropsy was performed on 10th day. Results reveal mild to moderate changes in liver and kidney. Significant alterations were observed only at 1600 mg/kg dose. Evaluation of organ weights in various groups showed normal values. Leucocyte counts and haemoglobin values were within normal limits. The blood biochemistry consisting of blood sugar, total proteins and serum transaminase (GOT and GPT) showed normal values within the limits of variation at low doses. However, increase in GOT and GPT was observed at the highest dose. Biochemical analysis revealed no significant change in protein and glycogen contents in liver and kidney except that at 1600 mg/kg dose, liver glycogen content was reduced significantly. No significant change was observed in acid and alkaline phosphatases activities in rats received 400 and 800 mg/kg doses. However, at a dose of 1600 mg/kg, liver acid phosphatase activity was elevated. On the contrary, liver and kidney alkaline phosphatase activity was decreased. No significant alterations in the activity of adenosine triphosphatase and glucose-6-phosphatase was observed in rats given low doses (400 and 800 mg/kg). However, at 1600 mg/kg, the activities of these liver enyzmes were significantly elevated.

Calculated $LD_{50}$ of a butanolic extract is 3.16 g/kg. When expressed in terms of body surface area ($g/m^2$), it comes to be 15.80, suggesting high margin of safety.

Histopathology showed no severe lesions with single administration at 400 and 800 mg/kg doses. At 1600 mg/kg dose hepatic cells were deformed in some places, canaliculi showed irregular diameter. Nuclei were also affected. Cytoplasmic granulation along with increase in the number of Kupffer cells were observed. Renal tubules in the periphery had undergone hypertrophy and showed exfoliation. Dysplasia was observed in the tubules and nucleus and cytoplasm ratio was disturbed.

An ultrastructural examination of hepatocytes at 400 mg/kg dose showed normal hepatocytes with sharply defined boundaries. The mitochondria were spherical and better formed and were closely packed near the nucleus and in association with endoplasmic reticulum. The outer and the inner membranes were intact with almost normal crests. Nuclear membrane was intact with uniform distribution of chromatin material. With 800 mg/kg dose mild ultrastructural changes were observed. At 1600 mg/kg marked dilatation and degranulation was observed in the endoploasmic reticulum, and swelling of the mitochondria was seen. Nucleus was deformed. Electron microscopical examination of kidney at low dose showed normal proximal convulated tubules. Plasma membrane exhibited clear and deep basal infoldings. The podocytes, foot processes were clear with uniform glomerular basement membrane. Medullary region was normal. At 1600 mg/kg dose proximal convulated tubules revealed scanty and short microvilli. Deep infoldings of the plasma membrane showed degenerated mitochondria; lysosomes were less in number.

In subchronic treatment (Shukla, 1995), extract was fed daily at $ED_{100}$ (150 mg/kg) for short and for a period of 30, 45 and 60 days. Haemoglobin of all the animals showed a slight fall towards the end of the experimental period, but it was not significant. There was no significant change in the leucocyte counts, however, serum urea, creatinine and bilirubin values registered high value at 45 days regimen. Serum transaminase activity was high at the terminal time. Blood sugar and total serum proteins at all periods showed slight variation in values which were within the normal range. Tissue biochemistry showed insignificant alterations at 30 and 45 days regimen, however, after 60 days of feeding there was significant fall in the glycogen content of liver. On the contrary, elevation was observed in the activity of acid phosphatase and glucose-6-phosphatase. Adenosine triphosphatase and succinic dehydrogenase demonstrated descending trend. Initial durations presented no significant change in hepatic lipid peroxidation and glutathione level, however at 60 days LPO showed elevation whereas glutathione level decreased. Control values registered no significant change.

In chronic exposure, butanolic extract was administered orally at a dose of 75 mg/kg for 3, 4 and 5 months. Results demonstrate no significant mortality rate in different groups including the controls. During the course of 20 weeks treatment the body weights of the rats showed a steady gain in weight, leucocyte counts and haemoglobin values were within normal limits of variation. The blood biochemistry including blood sugar and serum transaminase (GOT and GPT) showed elevation only at the terminal time. The total serum proteins were also within normal range. Tissue biochemistry performed at all the periods till the termination of the experiment revealed mild to moderate changes in liver and kidney. Linear and successive fall in the glycogen content and activity of alkaline phosphatase was observed in liver which was significant after 20 weeks of treatment. Similar pattern was seen in kidney. Total activity of protein content and acid phosphatase increased linearly at successive time intervals as revealed by analysis of variance.

## CONCLUSION

*P. tuberosa* is a potent antifertility plant and appears to be safe according to preliminary toxicological studies.

## REFERENCES

Ahmed, N. (1971) Maintenance of pregnancy in rats with progesterone and estrone: Effect on the corpora lutea. *Biol. Reprod.*, 4, 106.

Al-Hamood and Al-Bayati, Z.A.F. (1995) Effects of *Trigonella foenum* Graceum, *Nerium oleander* and *Ricinus communis* on reproduction in mice. *Iraqi J. Sci.*, 36, 425.

Alkofahi, A., Al-Hamood, M.H. and Elbetieha, A. (1996) Antifertility evaluation of some medicinal plants in female rats. *Arch. of STD/HIV*, 10, 189.

Ayurvedic Formulary of India (1978) *Part I. 1st Edition.* Government of India, Ministry of Health and Family Planning, Department of Health.

Chandhoke, N., Gupta, S., Daftari, P., Dhar, S.K. and Atal, C.K. (1980) Effect of *Pueraria tuberosa* on the reproductive performance in rats. *Indian J. Pharmacol.*, 12, 57.

Chaudhary, R.R. and Haq, M. (1980) Review of plants screened for antifertility activity II. *Bull. Med. Ethno. Bot. Res.*, 3, 420.

Chopra, R.N. and Chopra, I.C. (1955) *A Review of Work on Indian Medicinal Plants.* I.C.A.R., New Delhi.

Dhall, K. and Dogra, M. (1988) Phase I and II clinical trials with *Vicoa indica* (Banjauri), a herbal medicine, as an antifertility agent. *Contracept.*, 37(1), 75.

Dhar, M.L., Dhar, M.M., Dhawan, B.N., Mehrotra, B.N., Srimal, R.C. and Tandon, J.S. (1973) Screening of Indian plants for biological activity, Part 4. *Ind. J. Exp. Biol.*, 11, 43.

Dhar, M.L., Dhar, M.M., Dhawan, B.N. and Ray, C. (1968) Screening of Indian plants for biological activity, Part I. *Ind. J. Exp. Biol.*, 6, 232.

Dhar, M.L., Dhawan, B.N., Prasad, C.R., Rastogi, R.P., Singh, K.K. and Tandon, J.S. (1974) Screening of Indian plants for biological activity, Part 5. *Ind. J. Exp. Biol.*, 12, 512.

Elbetieha, A., Al-Hamood, M.H. and Alkofahi, A. (1996) Anti-implantation potentials of some medicinal plants in female rats. *Arch. of STD/HIV*, 10, 181.

Farnsworth, N.R., Bingel, A.S., Soejarto, D.D., Wijesekera, R.O.B. and Perea-Sasiain, J. (1981) Prospects for higher plants as a source of useful fertility regulating agents for human use. In C.C. Fen, D. Griffin and A. Woolman (eds), *Recent Advances in Fertility Regulation*, Atar S.A., Geneva, p. 332.

Gupta, S., Chandhoke, N., Daftari, P. and Atal, C.K. (1980) *Hormonal Profile of Pueraria tuberosa. Souvenir of XIII Annual Conference of IPS*, Regional Research Laboratory, Ab.No.E-14.

Gupta, D.N., Lakshmi, V., Mehrotra, B.N., Kapil, R.S. and Kamboj, V.P. (1990) Post-coital contraceptive efficacy and hormonal profile of *Pueraria tuberosa. Indian Drugs*, 27, 1

Jani, Y.K., Patel, M.R. and Patal, R.B. (1981) *Pueraria tuberosa*: An overview. *Indian Drugs*, 93.

Kamboj, V.P. and Dhawan, B.N. (1982) Research on plants for fertility regulation in Indian. *J. Ethnopharmacol.*, 6, 191.

Kirtikar, K.R. and Basu, B.D. (1935) *Indian Medicinal Plants.* Lalit Mohan Basu, Allahabad, Vol.1, pp. 792.

Mathur, R., Saxena, V. and Prakash, A.O. (1983) Antifertility screening of *Pueraria tuberosa* DC. *IRCS Med. Sci.*, 11, 522.

Mathur, R., Saxena, V. and Prakash, A.O. (1984) Effect of *Pueraria tubersosa* on the oestrous cycle of adult rats. *Acta Eur. Fertil.*, 15, 393.

Mathur, R., Shukla, S., Mathur, A. and Prakash, A.O. (1987) Hormonal/antihormonal profile of butanolic extract of *Pueraria tuberosa* DC. *Asian J. Exp. Sci.*, 4, 21.

Prakash, A.O. (1984) Biological evaluation of some medicinal plant extracts for contraceptive efficacy in females. *Contracept. Deliv. Syst.*, 5, No.9.

Prakash, A.O. and Mathur, R. (1976) Screening of Indian plants for antifertility activity. *Indian J. Exp. Biol.*, 14, 623.

Prakash, A.O., Saxena, V., Shukla, S. and Mathur, R. (1985) Contraceptive potency of *Pueraria tuberosa* DC and its hormonal status. *Acta Eur. Fertil.*, 16, 59.

Prakash, A.O., Shukla, S., Gupta, A. and Mathur, R. (1986) Effect of embelin – an oral contraceptive of plant origin on some biochemical constituents of the rat uterus in the presence and absence of estradiol dipropionate and progesterone. *Comp. Physiol. Ecol.*, 11, 4.

Psychoyos, A. (1970) Hormonal requirements for egg-implantation. In G. Raspe (ed.), *Advances in the Biosciences 4*, Pergamon, Vieweg, p. 275.

Regional Research Laboratory (1981) *News Letter*, 8(2), 6.

Satyavati, G.V. (1984) Indian plants and plant products with antifertility effect. *Ancient Sci. Life*, 4, 193.

Sharma, K.C. (1979) Vidarikand – an antifertility plant. Published in Avkash, 2nd July by Javed Ansari, No. 486, 23.

Shukla, S. (1993) Contraceptive potency of butanolic extract. *Indian Drugs*, 30, 510.

Shukla, S. (1995) Toxicological studies of *Pueraria tuberosa*, a Potent Antifertility Plant. Inter. *J. Pharmacognosy*, 33(4), pp. 324–329.

Shukla, S. (1996) Post-coital contraceptive action of *Pueraria tuberosa* DC. extract in rats. *Phytother. Res.*, 10, 95.

Shukla, S. (2000) Contraceptive efficacy of *Pueraria tuberosa*. D.Sc thesis, Jiwaji University.

Shukla, S., Mathur, R. and Prakash, A.O. (1989a) Butanolic extract of *Pueraria tuberosa* DC: Physiological response in the genital tract of cyclic female rats. *Phytother. Res.*, 1, 1.

Shukla, S., Mathur, R. and Prakash, A.O. (1989b) Cogency of extracts of *Pueraria tuberosa* DC to interrupt pregnancy in rats. *Comp. Physiol. Ecol.*, 14, 163.

Shukla, S., Mathur, R. and Prakash, A.O. (1987a) Butanolic extract of *Pueraria tuberosa* DC. and its effect on oestrous cycle in adult rats. *Indian J. Pharmacol.*, 19, 49.

Shukla, S., Mathur, R. and Prakash, A.O. (1987b) Physiology and biochemistry of female genital tract of ovariectomized rats treated with butanolic extract of *Pueraria tuberosa*. *Fitoterapia*, 58, 9.

Shukla, S., Mathur, R. and Prakash, A.O. (1987c) Role of butanolic extract of *Pueraria tuberosa* DC in the movement of glycogen in reproductive organs of cyclic and ovariectomized rats. *Med. Sci. Res.*, 15, 545.

Singh, B. (1997) Ph. D. Thesis Auther. Antifertility effect of *Pueraria tuberosa* with reference to male reproductive.

# 5 Chemical constituents of *Pueraria* plants: identification and methods of analysis

*Haojing Rong, Denis De Keukeleire, and Luc De Cooman*

## INTRODUCTION

As widely used material in oriental herbal medicines, *Pueraria* plants have been extensively studied for their chemical content. It appears that the plants are rich sources of polyphenols and polyphenolic glycosides. Isoflavones and their glycosides are principal bioactive constituents. The wide scope of the nature of secondary metabolites in *Pueraria* plants is reflected by the presence of complex triterpene saponins as well as of various volatile flavor components. Therefore, a review on the isolation, identification, and analysis of pertinent constituents should be of interest to researchers studying the properties and bioactivities of *Pueraria* plants.

## POLYPHENOLS AND POLYPHENOLIC GLYCOSIDES

The bioactivities of *Pueraria* plants can to a great extent be attributed to the presence of isoflavones and their glycosides. Kinjo *et al.* (1987), Ohshima *et al.* (1988), and Hirakura *et al.* (1997) separated a number of isoflavones by silica column chromatography including the well-known daidzein (1, Scheme 5.1), genistein (2), formononetin (3), and biochanin A (4), while these aglycones prevailed as well as glycosides having varying sugar moieties at different positions. Puerarin (5), the most abundant component, was first identified in *Pueraria lobata*, while daidzin (6) was originally detected in soybeans (Walz, 1931) and formononetin (3) is widely distributed in leguminous plants (Price and Fenwick, 1985).

Various procedures have been applied for extraction of isoflavones from *Pueraria* plants. Ohshima *et al.* (1988) obtained isoflavone aglycones, daidzein (1), formononetin (3) together with puerarol (7, Scheme 5.2), a coumestan derivative, in an acetone extract. Extraction of the residue with methanol followed by distribution using aqueous butan-1-ol furnished glycosidic compounds that were fractionated on Sephadex LH-20 with methanol as the eluent. Hirakura *et al.* (1997) isolated and identified six polyphenolic glycosides, including kudzubutenolide A (8), 3′-hydroxypuerarin-4′-0-D-glucoside (9), and 3′-methoxydaidzin (10), in addition to three daidzein glycosides from a methanol extract of *Pueraria lobata* followed by chromatography on a DIAION HP-20 column, eluted with water, 50 per cent methanol in water, and methanol successively. After defattening of *Pueraria lobata* with petroleum ether and hexane, Rong *et al.* (1998) used exhaustive extraction with methanol–water 3:1 (v/v) to deliver an extract containing essentially all low-molecular weight polyphenolic material.

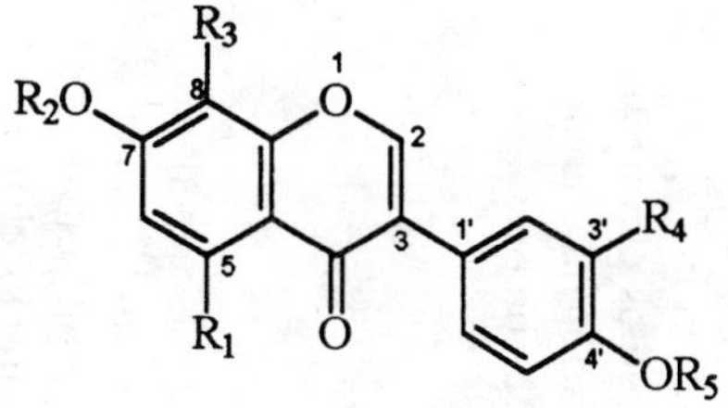

| | $R_1$ | $R_2$ | $R_3$ | $R_4$ | $R_5$ |
|---|---|---|---|---|---|
| Daidzein (1) | H | H | H | H | H |
| Genistein (2) | OH | H | H | H | H |
| Formononetin (3) | H | H | H | H | Me |
| Biochanin A (4) | OH | H | H | H | Me |
| Puerarin (5) | H | H | Glc | H | H |
| Daidzin (6) | H | Glc | H | H | H |
| 3'-Hydroxypuerarin-4'-O-D-glucoside (9) | H | H | Glc | OH | Glc |
| 3'-Methoxydaidzin (10) | H | Glc | H | OMe | H |
| 6''-O-D-Xylosylpuerarin (11) | H | H | $Glc^6$-Xyl | H | H |
| 3'-Methoxydaidzein (12) | H | H | H | OMe | H |
| 6''-O-Malonyl ester of daidzin (25) | H | 6''-O-Malonyl-Glc | H | H | H |
| 6''-O-Malonyl ester of genistin (26) | OH | 6''-O-Malonyl-Glc | H | H | H |
| 6''-O-Malonyl ester of puerarin (27) | H | H | 6''-O-Malonyl-Glc | H | H |
| Genistin (35) | OH | Glc | H | H | H |
| Puerarin-6''-monoacetate (36) | H | H | Glc-6''-OAc | H | H |
| Puerarin-4'-O-D-glucoside (61) | H | H | Glc | H | Glc |
| 3'-Hydroxypuerarin (62) | H | H | Glc | OH | H |
| 3'-Hydroxypuerarin-4'-O-deoxyhexoside (63) | H | H | Glc | OH | Deoxyhexosyl |
| 3'-Methoxypuerarin (64) | H | H | Glc | OMe | H |
| 3'-Methoxy-6''-O-D-xylosylpuerarin (65) | H | H | $Glc^6$-Xyl | OMe | H |
| 4'-Glucosyldaidzin (66) | H | Glc | H | H | Glc |
| 3'-Hydroxy-7-O-glucosylformononetin (67) | H | Glc | H | OH | Me |
| Daidzein-7-O-methyl ether (68) | H | Me | H | H | H |
| 3'-Methoxydaidzein-7-O-methyl ether (69) | H | Me | H | OMe | H |

Scheme 5.1  Structures of isoflavones and isoflavonoid glycosides from *Pueraria lobata*.

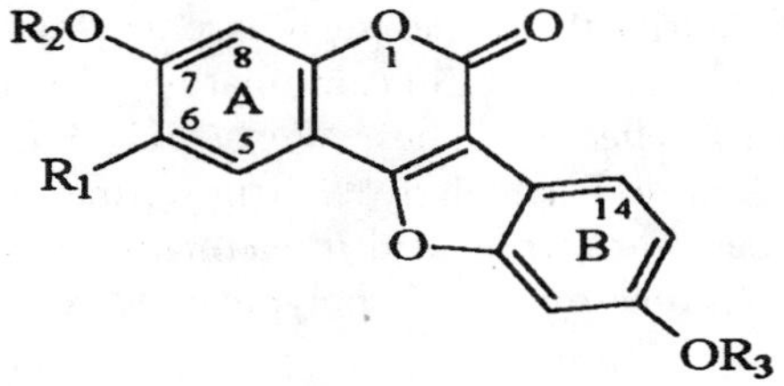

|  | $R_1$ | $R_2$ | $R_3$ |
|---|---|---|---|
| Puerarol (7) | 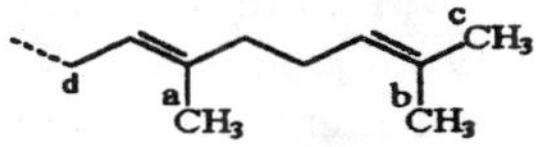 (geranyl chain) | H | H |
| Psoralidin dimethyl ether (18) | H | Me | Me |
| Puerarol dimethyl ether (19) | 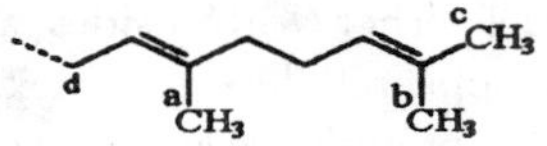 (geranyl chain) | Me | Me |

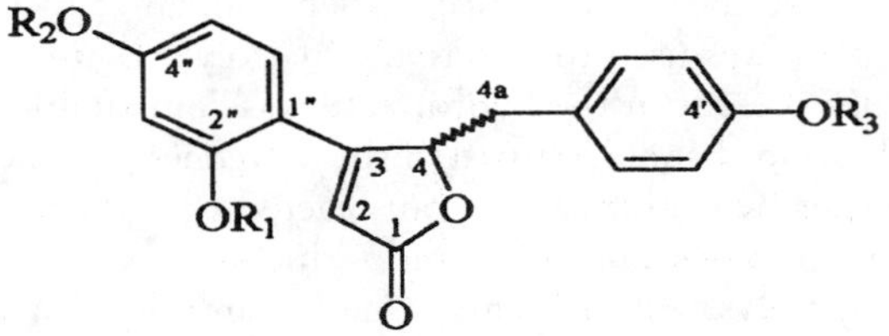

|  | $R_1$ | $R_2$ | $R_3$ |
|---|---|---|---|
| Kudzubutenolide A (8) | Glc | H | H |
| (±)-Puerol B 2-*O*-glucopyranoside (20a, 20b) | Glc | Me | H |
| Pueroside A (21) | Glc$^6$-Rha | H | H |
| Pueroside B (22) | Glc | Me | Glc |
| Puerol B (23) | H | Me | H |
| Aglycone of pueroside A (24) | H | H | H |

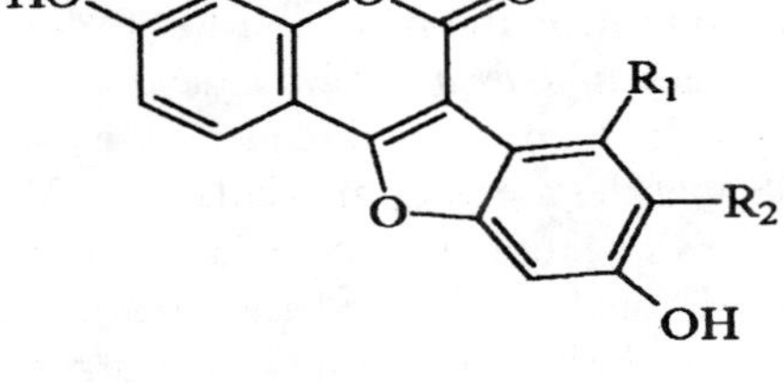

|  | $R_1$ | $R_2$ |
|---|---|---|
| Coumestrol (13) | H | H |
| Mirificoumestan (14) | -CH$_2$CH=C(CH$_3$)$_2$ | OMe |
| Mirificoumestan hydrate (15) | -CH$_2$CH$_2$C(CH$_3$)$_2$OH | OMe |
| Mirificoumestan glycol (16) | -CH$_2$CH(OH)C(CH$_3$)$_2$OH | OMe |

*Scheme 5.2* Structures of puerarol and derivates, but-2-enolides, and coumestans.

Although special HPLC stationary phases such as phenyl-1252-*N* can be used to separate phenolic mixtures, octadecylsilica in combination with gradient elution using polar solvents proved to be as effective. Thus, Wen *et al.* (1993) separated isoflavone glycosides and quantified puerarin. Particularly efficient separations were achieved using a gradient composed of 1 per cent formic acid in water and 5 per cent acetonitrile in methanol. A representative chromatogram of a polyphenolic extract of *Pueraria lobata* is shown in Figure 5.1.

Ultra Violet (UV) spectral data of early and late eluting peaks indicate that peaks a–h are most likely glycosides derived from aglycones eluting at longer retention times. An UV-spectrum with an absorption maximum in the wavelength region of 250–265 nm and a shoulder around 300 nm is suggestive of an isoflavone structure and the presence of a glycosyl residue does not alter these features significantly. Both acidic and enzymatic ($\beta$-glucosidase) hydrolysis of the glycosides were effected to differentiate between aglycones, *O*-$\beta$-glucosides, other *O*-glycosides, and *C*-glycosides. After acidic hydrolysis, the glycoside fraction appeared less complex, as peaks a, f, g, and h were absent, suggesting that these compounds are *O*-glycosides. Moreover, the intensities of peaks in the less polar region were more pronounced when compared to the analysis prior to hydrolysis. $\beta$-Glucosidase was used to selectively hydrolyze *O*-$\beta$-glucosides. The HPLC profile of the aglycones present in the hydrolysate was comparable to that obtained on acidic hydrolysis, indicating that at least part of the aglycones prevailed as *O*-$\beta$-glucosides. Peak f resisted hydrolysis by $\beta$-glucosidase, but reacted on acidic treatment, suggesting an *O*-glycosidic nature, however not involving $\beta$-glucose.

Differences in the hydrolysate profiles give clues regarding the nature of the glycoside bond, i.e. an *O*- or a *C*-glycoside, but the position of sugar moieties remains uncertain. HPLC-MS/MS has proved to be a powerful tool to obtain detailed structural information including distinction between flavonoid *O*- and *C*-glycosides (Markham, 1982), and establishment of glycosyl sequences (Sägesser and Deinzer, 1996). The mass spectrum of an *O*-glycoside is generally characterized by the presence of an abundant fragment ion by atmospheric pressure chemical ionization (APCI) resulting from (terminal) glycosyl cleavage ($Y_0^+$ for the aglycone fragment of a monosaccharide or $Y_1^+$ in case of a disaccharide). In contrast, the aglycone moiety of *C*-glycosides is not normally produced (Domon and Costello, 1988). Thus, puerarin (5) and daidzin (6) can be readily distinguished (Scheme 5.3). Interestingly, a striking difference has been noted between 6-*C*- and 8-*C*-glycosides, the $[MH-150]^+$-fragment ion being most prominent for a 6-*C*-glycoside, while the prevalent fragment ion for an 8-*C*-glycoside is $[MH-120]^+$. It appears that all glycosides in *Pueraria lobata* have an 8-*C*-linkage. Tandem mass spectral analysis gave information on glycosyl sequences, e.g. 6″-*O*-D-xylosylpuerarin (11) exhibited in the first MS analyzer a pseudomolecular ion $[MH]^+$ at m/z 549 and a fragment ion at m/z 417 corresponding to loss of a pentosyl residue ($Y_1^+$, *O*-glycoside). Furthermore, the second MS analyzer showed the presence of a $[Y_1-120]^+$-ion at m/z 297 as the predominant peak thereby providing evidence for $Y_1$ being an 8-*C*-glycoside. Thus, the structure was confirmed as 6″-*O*-D-xylosylpuerarin.

The mass spectral data of the isoflavone glycosides and aglycones of *Pueraria lobata* are summarized in Table 5.1. All major glycosides are derived from daidzein and most are 8-*C*-glycosides with the exception of daidzin (6) and 3′-methoxydaidzin (10). In addition to the respective molecular ions, the aglycones showed characteristic fragment ions arising from retro Diels-Alder (RDA) type fission of the $\gamma$-pyrone ring. Hence, valuable information could be obtained on the A- and B-ring substitution patterns (Ma *et al.*,

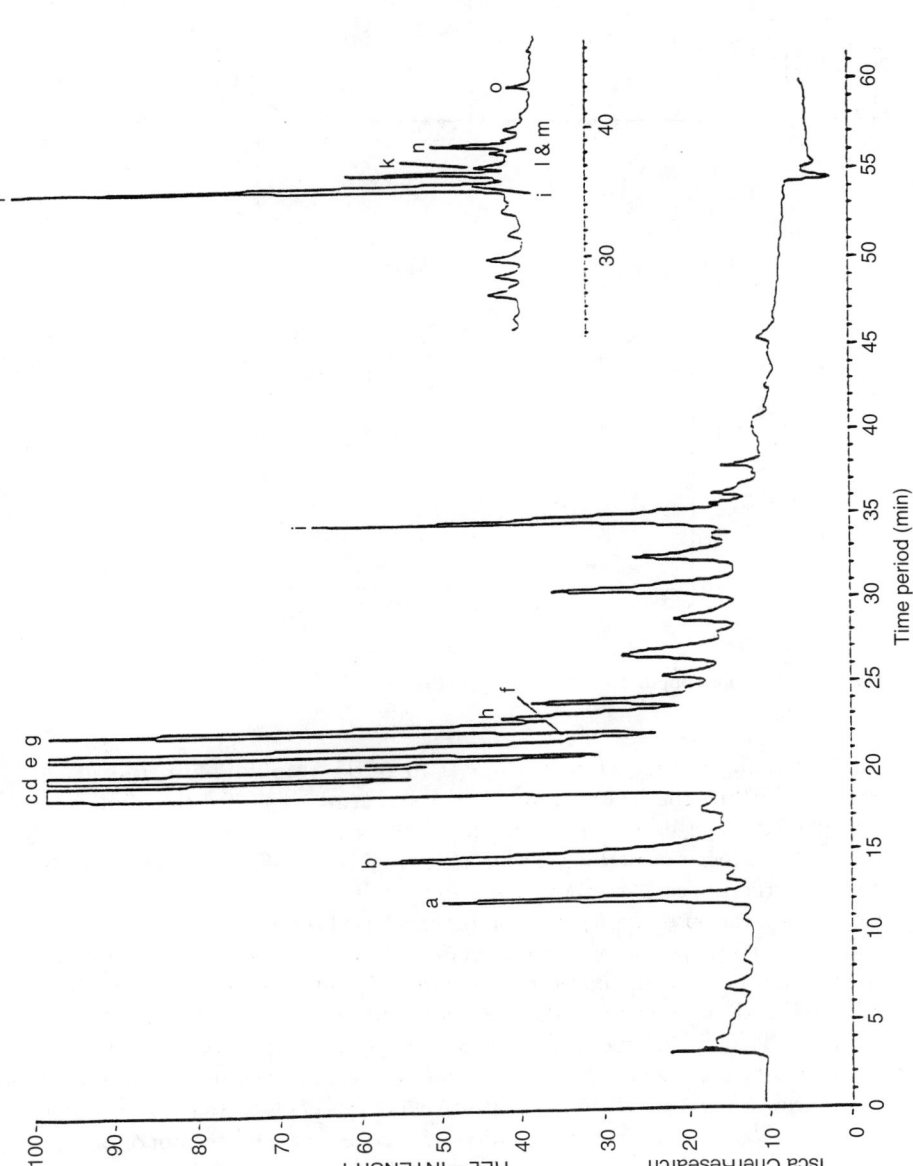

*Figure 5.1* HPLC-UV chromatogram of a polyphenolic extract of the root of *Pueraria lobata* and of the aglycone part (inset) of the acid hydrolysate (peak labels correspond to compound numbering). Chromatographic conditions: LiChrospher RP-18 (5 μm, 250 mm × 4 mm); mobile phase: 1 per cent aqueous formic acid (solvent A) and 5 per cent acetonitrile in methanol (solvent B), linear gradient from 15 to 95 per cent B in A in 50 min, detected at 280 nm (Rong *et al.*, 1998). Peak assignment, see Table 5.1.

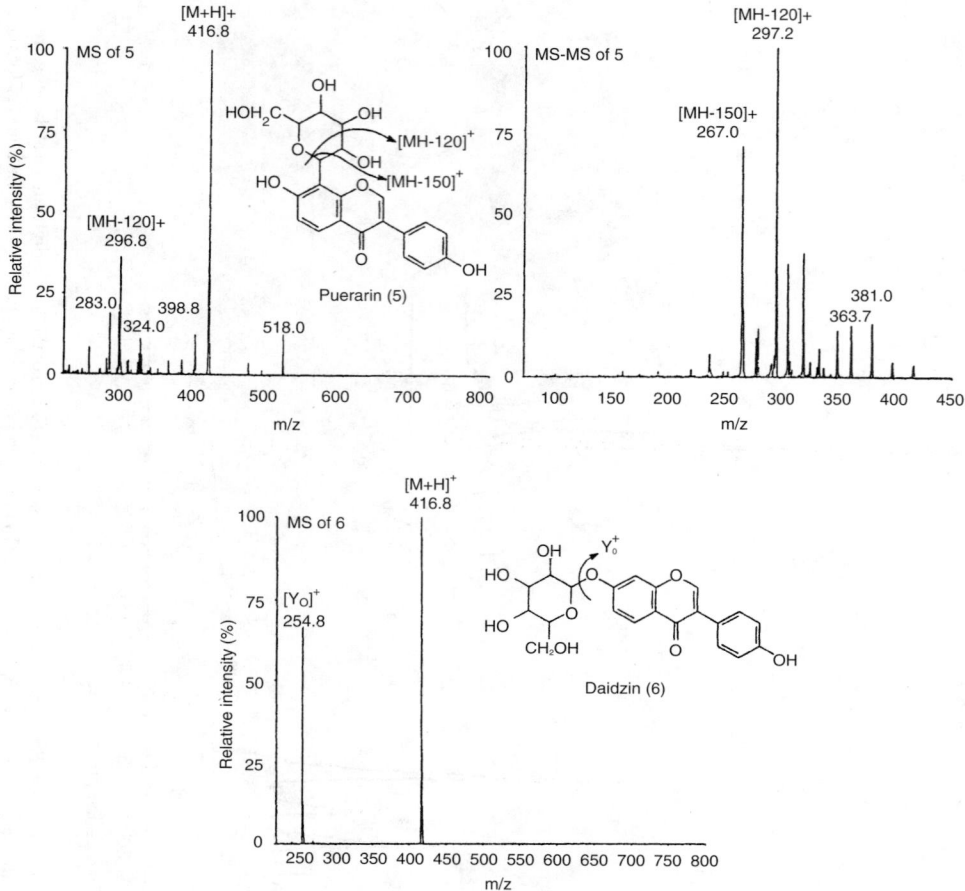

*Scheme 5.3* Mass fragmentations of puerarin (5) and daidzin (6).

1997). The pseudomolecular ion of peak j was at m/z 285, implying a methoxylated daidzein derivative. While the RDA fragment [1,3]A was at m/z 137 (Scheme 5.4) as in daidzein, [1,3]B yielded m/z 149. Comparison with the fragment ion at m/z 119 derived from daidzein indicates that a methoxy group is located on the B-ring leading to the assignment of the structure as 3′-methoxydaidzein (12).

Besides isoflavones, various polyphenolic compounds have been isolated and identified. The most important coumestan derivative is puerarol (7), while, additionally, coumestrol (13), mirificoumestan (14), mirificoumestan hydrate (15), and mirificoumestan glycol (16) have been isolated along with isoflavones from the roots of *Pueraria mirifica*, an oriental traditional Thai herbal medicine for gynecological symptoms (Ingham *et al.*, 1988). Puetuberosanol (17, Scheme 5.5), an epoxychalcanol, was derived from the fresh tubers of *Pueraria tuberosa* (Khan *et al.*, 1996). Puerarol (7) was first isolated from an acetone extract of *Pueraria* roots accompanied by daidzein (1) and formononetin (3) and its structure was characterized by Ohshima *et al.* (1988). The UV-spectrum of

*Table 5.1* Mass spectral data of the isoflavones of *Pueraria lobata* (Rong *et al.*, 1998)

| Peak | Compound | Single MS | | Daughter ions of [MH]⁺ ions on CAD | | |
|------|----------|-----------|---|---|---|---|
| | | [MH]⁺ | $Y_1^+$ | $Y_0^+$ | [MH-120]⁺, % | [MH-150]⁺, % |
| **Glycosides** | | | | | | |
| a | Puerarin-4'-O-D-glucoside (61) | 579 | 417 | | 297 (100) | 267 (27) |
| b | 3'-Hydroxypuerarin (62) | 433 | | | 313 (100) | 283 (66) |
| | 3'-Hydroxypuerarin-4'-O-deoxyhexoside (63) | 579 | 433 | | 313 (100) | 283 (66) |
| c | Puerarin (5) | 417 | | | 297 (100) | 267 (71) |
| d | 3'-Methoxypuerarin (64) | 447 | | | 327 (100) | 297 (75) |
| e | 6"-O-D-Xylosylpuerarin (11) | 549 | 417 | | 297 (100) | 267 (27) |
| f | 3'-Methoxy-6"-O-D-xylosylpuerarin (65) | 579 | 447 | | 327 (100) | 297 (79) |
| g | Daidzin (6) | 417 | | 255 | | |
| h | 3'-Methoxydaidzin (10) | 447 | | 285 | | |
| | | [MH]⁺ | $^{1,3}A^+$ | $^{1,3}B^+$ | | [MH-15]⁺ |
| **Aglycones** | | | | | | |
| i | Daidzein (1) | 255 | 137 | 119 | | |
| j | 3'-Methoxydaidzein (12) | 285 | 137 | 149 | | |
| k | Genistein (2) | 271 | 153 | 119 | | |
| l | Daidzein-7-O-methyl ether (68) | 269 | 151 | | | 254 |
| m | 3'-Methoxydaidzein-7-O-methyl ether (69) | 299 | | | | 284 |
| n | Formononetin (3) | 269 | 137 | 133 | | 254 |
| o | Biochanin A (4) | 285 | 153 | 133 | | 270 |

puerarol is superimposable with that of psoralidin dimethyl ether (18) (Khastgir *et al.*, 1961), suggesting that it is a coumestan derivative. The bathochromic shifts of the UV maxima induced by addition of sodium acetate indicate the presence of a free hydroxyl at C(7). The presence of a geranyl group was revealed by the ¹H NMR signals at $\delta=1.55$, 1.66, and 1.80 (3 $CH_3$), 2.15 (2 allylic $CH_2$), 3.47 (benzylic $CH_2$), 5.21, and 5.68 (2 vinylic H), while an 1,2,4-trisubstituted aromatic ring system is deduced from the signals at $\delta=8.09$ (d, $J=9.0$ Hz), 7.41 (d, $J=2.0$ Hz), and 7.26 (dd, $J_1=2.0$ Hz, $J_2=9.0$ Hz). With reference to the ¹H NMR data of coumestans, the signal at $\delta=8.09$ was assigned to the proton at C(14), which is shifted by the diamagnetic effect exerted by the neighboring carbonyl. The singlet signals at $\delta=7.10$ (s, 1 H) and 7.83 (s, 1 H) are assigned to the protons at C(5) and C(8), respectively. Therefore, the geranyl group is located at C(6). The *trans*-configuration of the geranyl side chain was proved by a NOE experiment using puerarol dimethyl ether (19) in pyridine-$d_5$. Irradiation of the benzylic methylene (d) at $\delta=3.47$ in compound 19 resulted in collapse of a triplet vinylic proton signal at $\delta=5.68$ to a singlet. Thus, this vinyl proton was assigned to the double bond next to the benzylic proton. Irradiation of the allylic methyl protons (a) ($\delta=1.55$) gave an 11 per cent NOE on the benzylic proton signal (d), indicating that protons (a) and (d) are located on the same face thus forming a *trans*-configuration of the double bond. Similarly, methyl (c) ($\delta=1.66$) was irradiated to result in an 18 per cent NOE at the

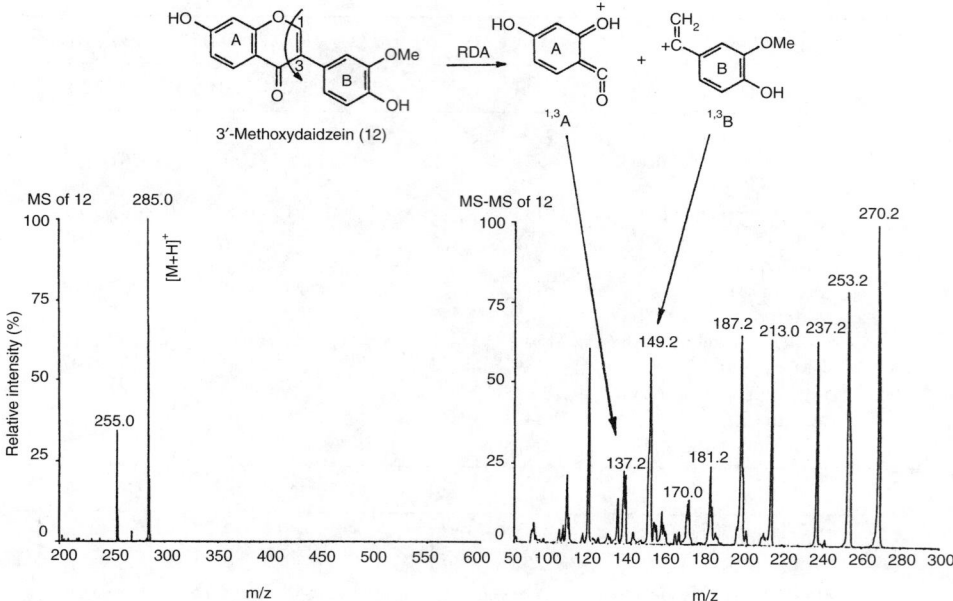

*Scheme 5.4* Retro Diels-Alder cleavage of 3′-methoxydaidzein (12).

vinylic proton at $\delta = 5.21$. Therefore, the signal (c) is assigned to the *cis*-methyl group. Consequently, puerarol was identified as 6-geranyl-7,4′-dihydroxycoumestan.

But-2-enolides, another typical class of compounds present in *Pueraria* plants, include (±)-puerol B 2-O-glucopyranoside (**20a, 20b**, Scheme 5.2), pueroside A (**21**), pueroside B (**22**), and kudzubutenolide A (**8**) (Nohara *et al.*, 1993). These compounds have been obtained by column chromatography on silica, MCI gel CHP 20 P and Sephadex LH-20 from a methanolic extract of the roots of *Pueraria lobata*. During structure verification of compound **20a** using X-ray analysis, it became apparent that the previously proposed structures of pueroside A (**21**) and pueroside B (**22**) (Kinjo *et al.*, 1985) ought to be revised to related but-2-enolides. Compound **20a** was subjected to enzymatic hydrolysis (β-glucosidase) and the aglycone exhibited [13]C-NMR signals similar to the aglycone residue of pueroside B, namely (+)-puerol B (**23**). However, the specific rotation was opposite and an inconsistency was observed in the H-C long range COSY. An X-ray analysis was obtained of the aglycone brosyl (*p*-bromobenzene sulfonyl) derivative, obtained by acidic hydrolysis of the fraction including pueroside B and subsequent esterification with *p*-bromobenzene sulphonyl chloride. It was concluded that the previously proposed structure for puerol B should be revised to a but-2-enolide. Moreover, the X-ray crystallographic result also indicated that the crystal of aglycone puerol B derived from acid hydrolysis was a racemate, while the aglycone obtained from enzymatic hydrolysis was (−)-puerol B. Subsequently, (+)-puerol B was identified from enzymatic hydrolysis of (+)-puerol B 2-O-glucopyranoside (**20b**), while the absolute configuration was confirmed by the specific rotation and the opposite CD-curve of (−)-puerol B. Since glycosylation shifts of −0.9, +1.5, +1.7 and +2.5 ppm were observed in [13]C-NMR at C(2″), C(1″), C(5″), and C(2), respectively, between the glucoside and the aglycone,

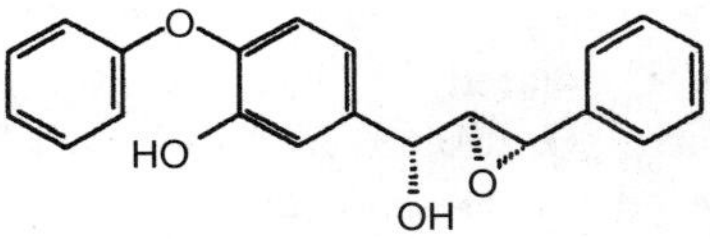

Puetuberosanol (17)

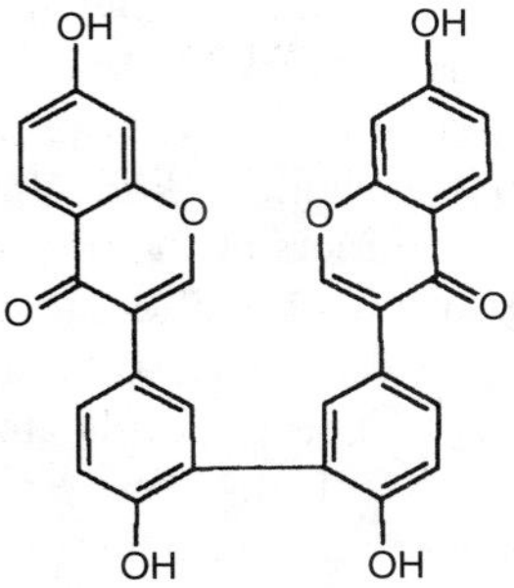

Kudzuisoflavone A(28)

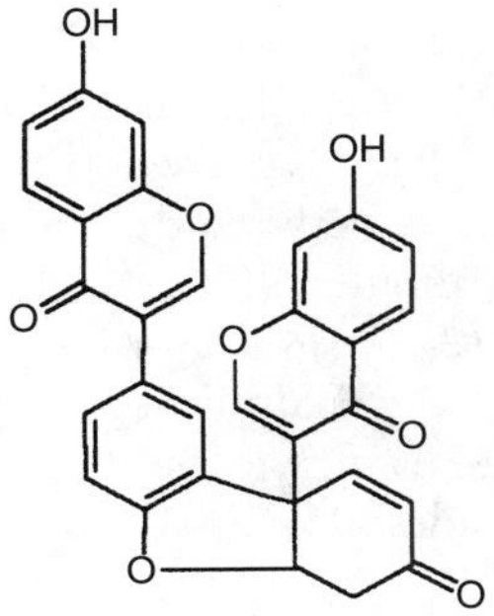

Kudzuisoflavone B (29)

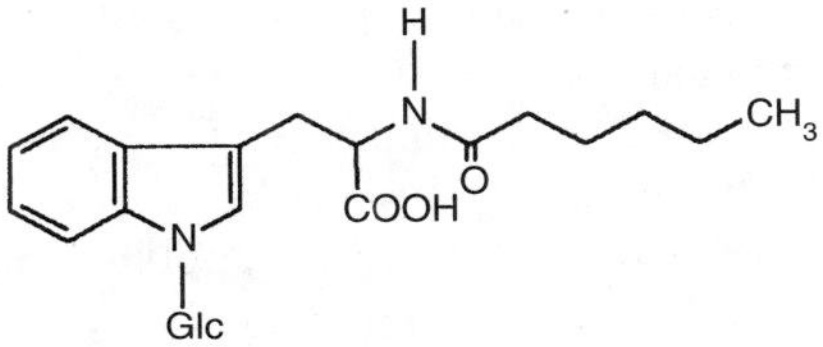

PF-P (30)

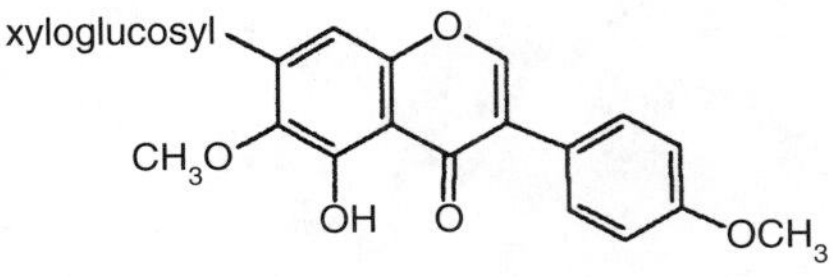

Glc

xyloglucosyl

CH₃O

OH

O

OCH₃

Kakkalide (31)

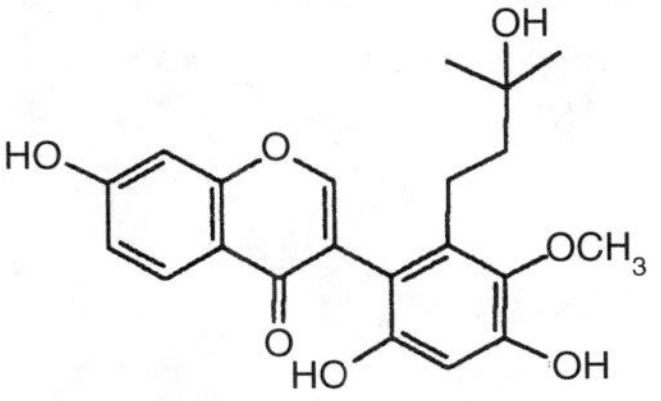

HO

OH

OCH₃

HO

OH

Kwakhurin hydrate (37)

R₄

OH

R₃O

OR₁

R₂

O

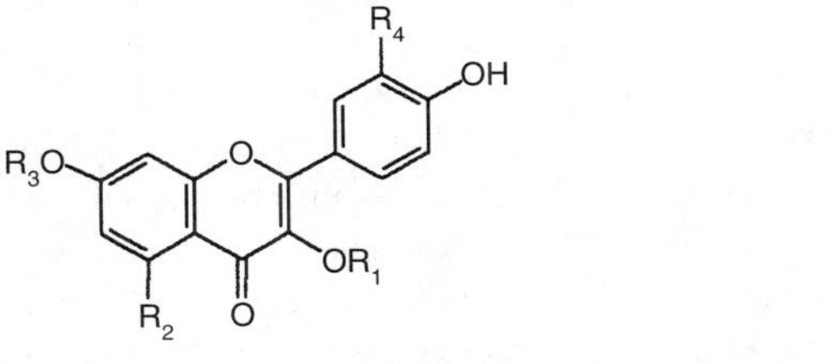

| | R₁ | R₂ | R₃ | R₄ |
|---|---|---|---|---|
| Rutin (32) | rhamnoglucosyl | OH | H | OH |
| Robinin (33) | rhamnogalactosyl | OH | rhamnosyl | H |
| Nicotiflorin (34) | rutinosyl | H | H | H |

*Scheme 5.5* Minor constituents from the *Pueraria* plant.

the glucosyl linkage was located at C(2″), not at C(4″). Consequently, the structure of **20a** was elucidated as 3-(2-*O*-glucopyranosyl-4-methoxyphenyl)-4-(4-hydroxybenzyl) but-2-en-4-olide or (−)-puerol B 2-*O*-glucopyranoside.

Another but-2-enolide, kudzubutenolide A (**8**), was isolated by Hirakura *et al.* (1997) from dried roots of *Pueraria lobata*. The UV spectrum exhibited high similarity to the spectra of but-2-enolides. The IR spectrum showed the presence of a hydroxyl (*br* $3300\,cm^{-1}$), a carbonyl ($1692\,cm^{-1}$), a double bond ($1604\,cm^{-1}$), and an aromatic ring ($1516\,cm^{-1}$). The molecular formula was determined to be $C_{23}H_{24}O_{10}$ by HR-FAB MS (High Resolution-Fast Atom Bombardment Mass Spectroscopy). The $^1H$ and $^{13}C$-NMR data of this compound were similar to those of 3-(2,4-dihydroxyphenyl)-4-(4-hydroxybenzyl)but-2-en-4-olide (**24**), the aglycone of pueroside A previously identified by Nohara *et al.* (1993). Enzymatic hydrolysis (*β*-glucosidase) of kudzubutenolide A (**8**) gave glucose and compound **24**, suggesting that it was glucosylated **24**. The location of the glucose residue was confirmed to be at C(2″) from a cross peak between the anomeric proton and C(2″) in a HMBC experiment. The structure of kudzubutenolide A, including the absolute configuration at C(4), was determined as (4*R*)-3-(2-*β*-D-glucopyranosyloxy-4-hydroxyphenyl)-4-(4-hydroxybenzyl)but-2-en-4-olide by X-ray analysis.

Furthermore, other minor polyphenols were found in the stems, flowers, and cultured cells of *P. lobata*, e.g. 6″-*O*-malonyl esters of isoflavone glycosides (**25**, **26**, **27**) from the fresh root and stems as well as cultured cells (Park *et al.*, 1992); isoflavone dimers, kudzuisoflavone A (**28**) and B (**29**) from yeast extract-treated cell suspension cultures (Hakamatsuka *et al.*, 1992); a tryptophan derivative, PF-P (**30**), from the flowers (Kinjo *et al.*, 1988b); kakkalide (**31**), rutin (**32**), robinin (**33**), and nicotiflorin (**34**) from the leaves (Kinjo *et al.*, 1988a). Isoflavones such as genistin (**35**), puerarin-6″-mono-acetate (**36**), and kwakhurin hydrate (**37**) were obtained from the roots of *Pueraria mirifica* (Ingham *et al.*, 1989).

## TRITERPENOIDS

Oleanene-type triterpene glycosides have been found in the roots of *Pueraria lobata* (Arao *et al.*, 1995, 1997), namely soyasapogenol A (**38**, Scheme 5.6), soyasapogenol B (**39**), kudzusapogenol A (**40**), and kudzusapogenol C (**41**) glycosides. The glycosides having soyasapogenol A as aglycone were identified as kudzusaponins SA$_1$ (**42**), SA$_2$ (**43**), SA$_3$ (**44**), and SA$_4$ (**45**), and soyasaponin A$_3$ (**46**). These triterpenes are featured by having two methyl groups at C(20) and hydroxyl groups at C(21) and C(22), the latter one being glycosylated. Kudzusapogenol A (**40**) is the aglycone of kudzusaponins A$_1$ (**47**), A$_2$ (**48**), A$_3$ (**49**), A$_4$ (**50**), and A$_5$ (**51**). The only difference between these compounds and the soyasapogenol A derivatives is the presence of a hydroxymethyl group at C(20). Kudzusaponin SB$_1$ (**52**) and soyasaponin I (**53**) have soyasapogenol B (**39**) as aglycone, while kudzusaponin C$_1$ (**54**) is the glycoside of kudzusapogenol C (**41**). These aglycones are structurally related to soyasapogenol A (**38**) except for the absence of a hydroxy group at C(21) in soyasapogenol B (**39**) and at C(22) in kudzusapogenol C (**41**). The triterpene glycosides could be extracted from fresh roots of *Pueraria lobata* using methanol, followed by partition with butan-1-ol and water or ethyl acetate and 40 per cent methanol in water. Other triterpenoidal saponins, a sophoradiol derivative (**55**), and kaikasaponin III (**56**) were isolated from the flowers of *Pueraria lobata* (Kinjo *et al.*, 1988a). Kudzusaponin B$_1$ (**57**), subproside V (**58**), acetylkaikasaponin III (**59**), and acetylsoyasaponin I (**60**)

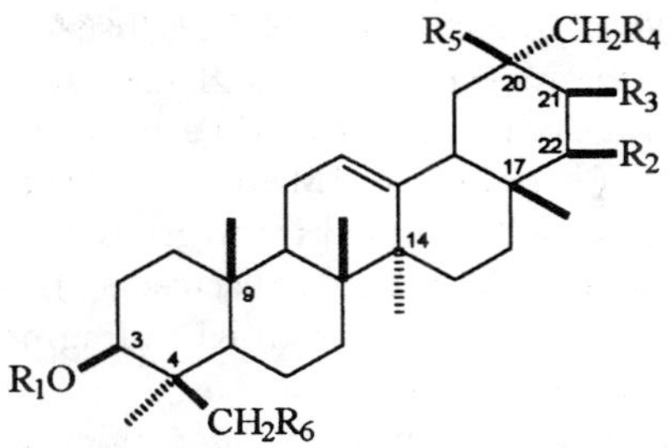

| | $R_1^*$ | $R_2$ | $R_3$ | $R_4$ | $R_5$ | $R_6$ |
|---|---|---|---|---|---|---|
| Soyasapogenol A (38) | H | OH | OH | H | $CH_3$ | OH |
| Soyasapogenol B (39) | H | OH | H | H | $CH_3$ | OH |
| Kudzusapogenol A (40) | H | OH | OH | OH | $CH_3$ | OH |
| Kudzusapogenol C (41) | H | H | OH | H | $CH_3$ | OH |
| Kudzusaponin $SA_1$ (42) | Glc A²-gal | OH | OH | H | $CH_3$ | OH |
| Kudzusaponin $SA_2$ (43) | Glc A²-gal | O-ara | OH | H | $CH_3$ | OH |
| Kudzusaponin $SA_3$ (44) | Glc A²-gal²-rha | O-ara | OH | H | $CH_3$ | OH |
| Kudzusaponin $SA_4$ (45) | Glc A²-glc A | O-ara | OH | H | $CH_3$ | OH |
| Soyasaponin $A_3$ (46) | Glc A²-gal²-rha | OH | OH | H | $CH_3$ | OH |
| Kudzusaponin $A_1$ (47) | Glc A²-ara²-rha | O-xyl | OH | OH | $CH_3$ | OH |
| Kudzusaponin $A_2$ (48) | Glc A²-gal | OH | OH | OH | $CH_3$ | OH |
| Kudzusaponin $A_3$ (49) | Glc A²-gal²-rha | OH | OH | OH | $CH_3$ | OH |
| Kudzusaponin $A_4$ (50) | Glc A²-glc | OH | OH | OH | $CH_3$ | OH |
| Kudzusaponin $A_5$ (51) | Glc A²-glc²-rha | OH | OH | OH | $CH_3$ | OH |
| Kudzusaponin $SB_1$ (52) | Glc A²-gal²-rha | O-ara | H | H | $CH_3$ | OH |
| Soyasaponin I (53) | Glc A²-gal²-rha | OH | H | H | $CH_3$ | OH |
| Kudzusaponin $C_1$ (54) | Glc A²-gal²-rha | H | OH | H | $CH_3$ | OH |
| Sophoradiol derivative (55) | Glc A²-ara²-rha | OH | H | H | $CH_3$ | H |
| Kaikasaponin III (56) | Glc A²-gal²-rha | OH | H | H | $CH_3$ | H |
| Kudzusaponin $B_1$ (57) | Glc A²-gal²-rha | OH | OH | H | COOH | OH |
| Subproside V (58) | Glc A²-gal²-rha | OH | OH | O-glc | $CH_3$ | OH |
| Acetylkaikasaponin III (59) | Glc A²-gal²-rha | O–$COCH_3$ | H | H | $CH_3$ | H |
| Acetylsoyasaponin I (60) | Glc A²-gal²-rha | O–$COCH_3$ | H | H | $CH_3$ | OH |

Abbreviations: glc A: glucuronic acid; gal: galactose; rha: rhamnose; ara: arabinose

*Scheme* 5.6  Structures of triterpenes and glycosides from *Pueraria*.

were isolated and identified from *Pueraria thomsonii*, another *Pueraria* species used as a perspiration, antipyretic, and antispasmodic agent (Arao *et al.*, 1996).

Arao *et al.* (1997) elucidated the structure of the oleanene-type triterpene glycosides by [13]C-NMR and tandem MS techniques in conjunction with acidic hydrolysis, furnishing sapogenols and sugar residues. The sapogenols were preliminarily identified with reference material on TLC and further confirmed with [13]C-NMR data. Glycosylated carbons were shifted downfield in comparison with the sapogenols, thereby providing information on the location of the sugar residues. The nature of the sugars was determined according to the procedure developed by Hara *et al.* (1987).

The application of MS/MS to the structural determination of oleanene-type triterpene glycosides proved to be a very convenient alternative to conventional degradation methods, such as selective cleavage of the glucuronide linkage in oligoglycosides (Kitagawa *et al.*, 1980) and enzymatic methods (Sakaki *et al.*, 1988), as usually only a limited amount of material was available. A bisdesmoside, having a glycosylated carboxy group, was characterized and selective cleavage of this ester could be achieved (Isobe *et al.*, 1992). In the MS/MS of kudzusaponin SA$_3$ (44), [M-H]$^-$, [M+H]$^+$, and [M+Na]$^+$ ions were observed as parent ions, while daughter ions derived from glycosidic cleavages such as [M-ara (arabinose)]$^+$, [M-rha (rhamnose)]$^+$, [M-rha-gal (galactose)]$^+$, [M-rha-gal-ara]$^+$, [M-rha-gal-glc A (glucuronic acid)]$^+$, and ions derived from RDA fragmentation, were obtained in the second mass analyzer (Arao *et al.*, 1995).

It was, furthermore, noted that saponins of *Radix puerariae* are characterized by the presence of a great variety of sugar residues connected to C(3). Six sugar chains have been obtained from the crude drug, which is a rare phenomenon in leguminous saponins. For example, despite the isolation of 23 saponins from *Abrus cantoniensis*, only two sugars (β-fabatriose and β-abritetraose) are present. Furthermore, all the *Pueraria* saponins have glucuronic acid as the first sugar moiety, while triglycosides contain rhamnose as terminal sugar, the only difference being the nature of the second sugar. The complexity as well as the structural similarities of the saponins occuring in *Pueraria lobata* rendered isolations difficult (Arao *et al.*, 1997).

## VOLATILE CONSTITUENTS

*Pueraria lobata* exhibits a slightly sweet note and a mild, fruity-winey odor, although a medicinal flavor is noted also. Volatile constituents, identified by GC, GC-MS, IR,

*Table 5.2* Volatile components in roots of *Pueraria lobata* (Miyazawa and Kamioka, 1988)

| Compound | % (w/w) |
|---|---|
| Acetone | 1.0 |
| Methyl propyl ketone | 1.2 |
| Acetyl carbinol | 4.5 |
| 2-Propanoyl acetate | 0.3 |
| 2-Methoxyethyl acetate | 4.8 |
| Octanal | 2.2 |
| Butanoic acid | 4.1 |
| Methyl caprylate | 0.6 |
| Acetonyl acetone | 1.2 |
| Methyl caprate | 0.7 |
| Furfuryl alcohol | 2.5 |
| Methyl laurate | 0.5 |
| Phenol | 2.3 |
| Methyl myristate | 2.8 |
| *p*-Cresol | 1.5 |
| *m*-Cresol | 1.4 |
| Methyl pentadecanoate | 0.8 |
| Dimethyl azelate | 0.7 |
| *p*-Ethylphenol | 0.4 |
| Methyl palmitate | 42.2 |
| Paeonol | 0.3 |
| Methyl margarate | 1.0 |
| Methyl stearate | 5.2 |

and ${}^{1}$H-NMR, are listed in Table 5.2 (Miyazawa *et al.*, 1988). Methyl palmitate was recognized as the main constituent and dimethyl azelate was associated to the fruity-winey odor. The contents of methyl and dimethyl esters of dibasic acids in the volatile oil were 53.8 per cent and 0.7 per cent, respectively. A mixture prepared from acetyl carbinol (42 per cent), paeonol (26.5 per cent), furfuryl alcohol (23.5 per cent), dimethyl azelate (6.5 per cent), and dimethyl suberate (0.5 per cent) resembled very much the aroma characteristics of the root of *Pueraria lobata*.

## REFERENCES

Arao, T., Kinjo, J., Nohara, T. and Isobe, R. (1995) Oleanene-type triterpene glycosides from *Puerariae Radix*. II. Isolation of saponins and the application of tandem mass spectrometry to their structure determination. *Chem. Pharm. Bull.*, 43, 1176–1179.

Arao, T., Idzu, T., Kinjo, J., Nohara, T. and Isobe, R. (1996) Oleanene-type triterpene glycosides from *Puerariae Radix*. III. Three new saponins from *Pueraria thomsonii*. *Chem. Pharm. Bull.*, 44, 1970–1972.

Arao, T., Kinjo, J., Nohara, T. and Isobe, R. (1997) Oleanene-type triterpene glycosides from *Puerariae Radix*. IV. Six new saponins from *Pueraria lobata*. *Chem. Pharm. Bull.*, 45, 362–366.

Domon, B. and Costello, C.E. (1988) A systematic nomenclature for carbohydrate fragmentations on FAB-MS MS spectra of glycoconjugates. *Glycoconjugate J.*, 5, 397–409.

Hakamatsuka, T., Shinkai, K., Noguchi, H. and Ebizuka, S.U. (1992) Isoflavone dimers from yeast extract-treated cell suspension cultures of *Pueraria lobata*. *Z. Naturforsch.*, 47C, 177–182.

Hara, S., Okabe, H. and Mihashi, K. (1987) Gas–liquid chromatographic separation of aldose enantiomers as trimethylsilyl ethers of methyl 2-(polyhydroxyalkyl)-thiazolidine-4(R)-carboxylates. *Chem. Pharm. Bull.*, 35, 501–506.

Hirakura, K., Morita, M., Nakajima, K., Sugama, K., Takagi, K., Nijitsu, K., Ikeya, Y., Maruno, M. and Okada, M. (1997) Phenolic glucosides from the root of *Pueraria lobata*. *Phytochemistry*, 46, 921–928.

Ingham, J.L., Tahara, S. and Dziedzic, S.Z. (1988) Coumestans from the roots of *Pueraria mirifica*. *Z. Naturforsch.*, 43C, 5–10.

Ingham, J.L., Tahara, S. and Dziedzic, S.Z. (1989) Minor isoflavones from the roots of *Pueraria mirifica*. *Z. Naturforsch.*, 44C, 724–726.

Isobe, R., Higuchi, R. and Komori, T. (1992) Negative-ion fast-atom-bombardment mass-spectrometry of native gangliosides using a high-polar matrix system. *112th Ann. Meet. Pharm. Soc. Jpn., Fukuoka, Japan, Book of Abstracts*, p. 224.

Khan, R., Agrawal, P.K. and Kapil, R.S. (1996) Puetuberosanol, an epoxychalcanol from *Pueraria tuberosa*. *Phytochemistry*, 42, 243–244.

Khastgir, H.N., Duttagupta, P.C. and Sengupta, P. (1961) Structure of psoralidin. *Tetrahedron*, 14, 275–283.

Kinjo, J., Takeshita, T., Abe, Y., Terada, N., Yamashita, H., Yamasaki, M., Takeuchi, K., Murakami, K., Tomimatsu, T. and Nohara, T. (1988a) Studies on the constituents of *Pueraria lobata*. IV. Chemical constituents in the flowers and the leaves. *Chem. Pharm. Bull.*, 36, 1174–1179.

Kinjo, J., Furusawa, J., Baba, J., Takeshita, T., Yamasaki, M. and Nohara, T. (1987) Studies on the constituents of *Pueraria lobata*. III. Isoflavonoids and related compounds in the roots and the voluble stems. *Chem. Pharm. Bull.*, 35, 4846–4850.

Kinjo, J., Furusawa, J. and Nohara, T. (1985) Two novel aromatic glycosides, pueroside-A and -B, from *Puerariae Radix*. *Tetrahedron Lett.*, 26, 6101–6102.

Kinjo, J., Takeshita, T. and Nohara, T. (1988b) Studies on the constituents of *Pueraria lobata*. V. A tryptophan derivative from *Puerariae Flos*. *Chem. Pharm. Bull.*, 36, 4171–4173.

Kitagawa, I., Kamigauchi, T., Ohmori, H. and Yoshikawa, M. (1980) Saponins and sapogenols XXIX. Selective cleavage of the glucuronide linkage in oligoglycosides by anodic oxidation. *Chem. Pharm. Bull.*, **28**, 3078–3086.

Ma, Y.L., Li, Q.M., van den Heuvel, H. and Claeys, M. (1997) Characterization of flavone and flavonol aglycones by collision-induced dissociation tandem mass spectrometry. *Rapid Commun. Mass Spectrom.*, **11**, 1357–1364.

Markham, K.R. (1982) *Techniques of Flavonoid Identification*. Academic Press, New York, USA, pp. 90–93.

Miyazawa, M. and Kameoka, H. (1988) Volatile flavor components of *Puerariae Radix* (*Pueraria lobata* Ohwi). *Agric. Biol. Chem.*, **52**, 1053–1055.

Nohara, T., Kinjo, J., Furusawa, J., Sakai, Y., Inoue, M., Shirataki, Y., Ishibashi, Y., Yokoe, I. and Komatsu, M. (1993) But-2-enolides from *Pueraria lobata* and revised structures of puerosides A, B and sophoroside A. *Phytochemistry*, **33**, 1207–1210.

Ohshima, Y., Okuyama, T., Takahashi, K., Takizawa, T. and Shibata, S. (1988) Isolation and high performance liquid chromatography (HPLC) of isoflavonoids from the *Pueraria* root. *Planta Med.*, **54**, 250–254.

Park, H., Hakamatsuka, T., Noguchi, H., Sankawa, U. and Ebizuka, Y. (1992) Isoflavone glucosides exist as their 6″-O-malonyl esters in *Pueraria lobata* and its cell suspension cultures. *Chem. Pharm. Bull.*, **40**, 1978–1980.

Price, K.R. and Fenwick, G.R. (1985) Naturally occurring estrogens in foods – A review. *Food Addit. Contam.*, **2**, 73–106.

Rong, H., Stevens, J.F., Deinzer, M.L., De Cooman, L. and De Keukeleire, D. (1998) Identification of isoflavones in the roots of *Pueraria lobata*. *Planta Med.*, **64**, 620–627.

Sägesser, M. and Deinzer, M.L. (1996) HPLC-ion spray tandem mass spectrometry of flavonol glycosides in hops. *J. Am. Soc. Brew. Chem.* **54**, 129–134.

Sakaki, Y., Morita, T., Kuramoto, T., Mizutani, K., Ikeda, R. and Tanaka, O. (1988) Substrate-specificity of glycyrrhizinic acid hydrolase. *Agric. Biol. Chem.*, **52**, 207–210.

Walz, E. (1931) Isoflavon- und Saponin-glucosides in *Soja hispida*. *Justus Liebigs Ann. Chem.*, **498**, 118–155.

Wen, K.C., Huang, C.Y. and Lu, F.L. (1993) Determination of baicalin and puerarin in traditional Chinese medicinal preparations by high performance liquid chromatography. *J. Chromatogr.*, **631**, 241–250.

# 6 Chemical components and pharmacology of the rejuvenating plant *Pueraria mirifica*

*John L. Ingham, Satoshi Tahara, and Gerald S. Pope*

## INTRODUCTION

The woody perennial climber *Pueraria mirifica* (Figure 6.1) was first described by Kashemsanta *et al.* (1952) following the detailed re-examination of a Thai medicinal plant initially thought to be identical with *Butea superba*, another climbing legume widely found across south-east Asia. Interest in the plant was especially strong in Thailand, notably around the northern hill town of Chiengmai, where it was known locally as kwao keur (or sometimes kwao khua, kwao kua or hua kwao) and where, for many years, its globular or pear-shaped tuberous roots[1] had been collected in large quantities by local tribesmen for processing, either in Chiengmai or Bangkok, into various traditional Thai medicines (Figures 6.2 and 6.3).

*Pueraria mirifica* is mainly found growing in the deciduous forests of Chiengmai Province where it thrives at altitudes between 300 and 800 m on the steep slopes of mountain gulleys characterized by slightly acidic soils formed from quartz-containing rocks which have been weathered to a sandy or gravelly texture by the summer monsoon rains (Pendleton and Suvatabandhu, 1952). After being sliced and sun-dried, the powdered tubers are typically mixed with honey to give peppercorn-size pills which, when taken once-a-day at night-time for 3–6 months, are reputed to possess a range of remarkable properties, including the ability to induce a form of menstruation in elderly women, and cause the regrowth of hair in bald men (Kerr, 1932; Wanandorn, 1933). The early literature also suggests that kwao keur has tubers of different colors (white, red and black), and that preparations made from the white tubers have the weakest rejuvenating action, whilst those made from the black tubers are the strongest (Kerr, 1932). Tubers of kwao keur have a sweet taste because of their sucrose content of up to 3 per cent on a dry weight basis (Jones and Pope, 1961).

The use of kwao keur as a possible rejuvenator appears to have first been drawn to the attention of the wider scientific community in 1932 by Kerr, although its botanical identity was only resolved in the early 1950's by Kashemsanta and Suvatabandhu who, with the help of Kew Director H.K. Airy Shaw, realized that the plant was a new species of *Pueraria*, giving it the specific name *mirifica* (Latin, *mirificus* = amazing or wonderful) in recognition of its reputed miraculous powers (Kashemsanta *et al.*, 1952).

---

1 The tubers of *P. mirifica* occur singly or in strings along the roots (Figures 6.2 and 6.3). In size they can be up to 20 cm in length and 14 cm in width (average 12 cm by 9 cm), and weigh 2 kg or more (Tirawat and Smitasuwana, 1949).

*Figure 6.1 Pueraria mirifica*: Plant approx. 8–10 weeks old grown from seed at Reading. Note the characteristic right-turning, climbing habit. This plant, and several others, have now been donated to the Chelsea Physic Garden, London.

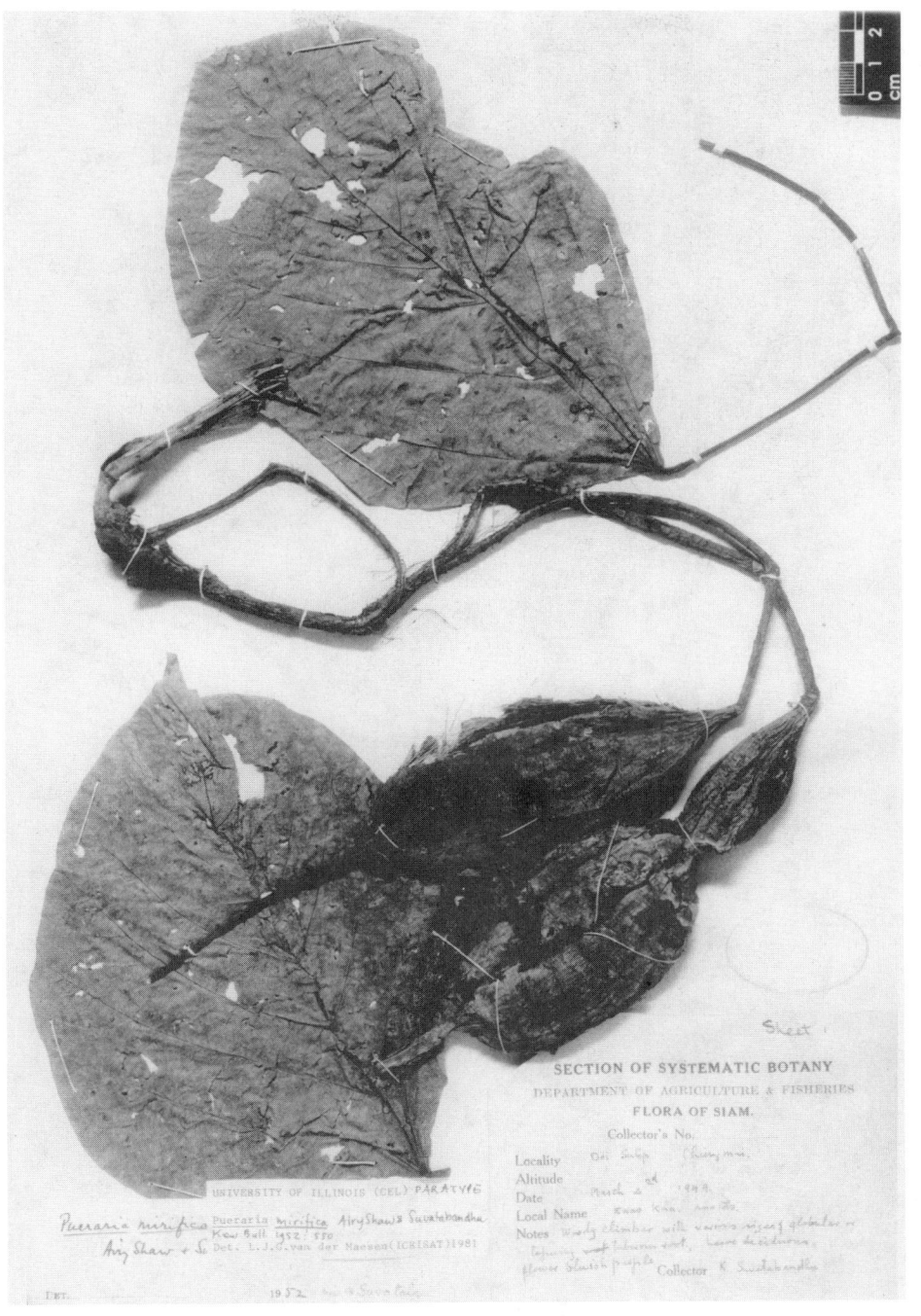

*Figure 6.2 Pueraria mirifica*: Herbarium sheet showing leaves and three tubers (two linked) of varying sizes. Collected in 1949 on Doi Sutep mountain near Chiengmai (courtesy of Kew Herbarium).

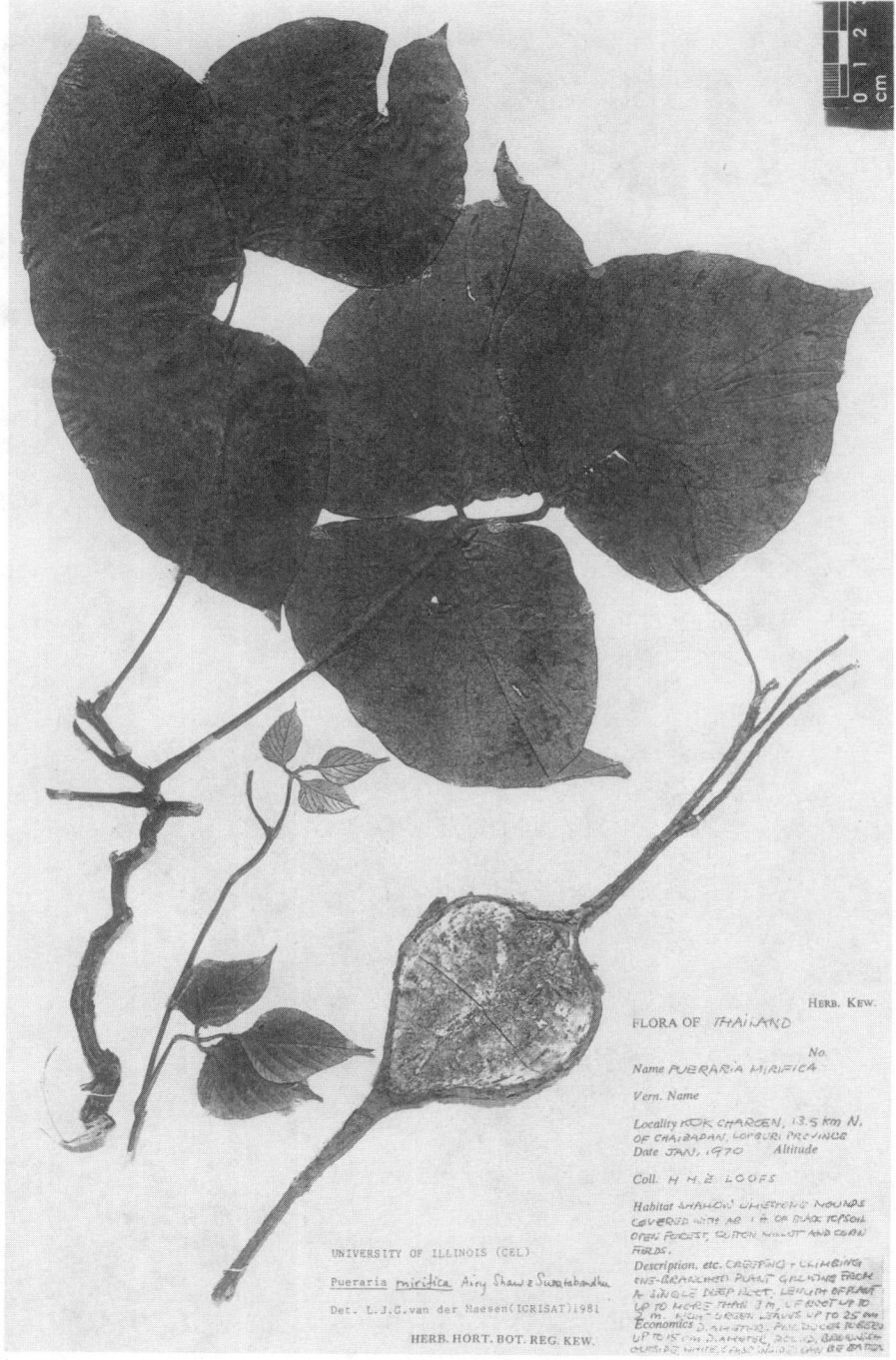

*Figure 6.3 Pueraria mirifica*: Herbarium sheet of a comparatively young plant with leaves of various ages and a single tuber sectioned to show the internal structure (brownish outside; white, crisp, inside). Collected in 1970 on a limestone hill near Chaibadan (upper Chao Phraya delta, N.E. of Bangkok) (courtesy of Kew Herbarium).

Both *P. mirifica* and *Butea superba* (the species with which *P. mirifica* was at first confused) grow in the same general localities around Chiengmai, and appear very similar in the non-flowering (vegetative) state. The two species can, however, be distinguished by the absence of tuberous roots in *B. superba*, and by the flowers which are small and purple-blue in *P. mirifica* (but which are only rarely produced except on old plants) and large, red to deep orange in color, in *B. superba*.

Kashemsanta *et al.* (1952) also mentioned the close similarity between *P. mirifica* and another woody climber, *P. candollei* which likewise grows around Chiengmai as well as elsewhere in Thailand, and in Burma, Bangladesh and parts of India (van der Maesen, 1985, 1994). Although Kashemsanta *et al.* (1952) kept these species as separate entities, in a more recent study (Niyomdham, 1992) they have been combined to give *P. candollei* var. *candollei* (formerly *P. candollei*) and *P. candollei* var. *mirifica* (formerly *P. mirifica*). Both varieties are deciduous and many intermediate forms are said to occur (Niyomdham, 1992). The only major difference between these two varieties appears to lie in the relative size of the corolla and calyx (smaller in var. *mirifica* when compared with var. *candollei*), and the length of the inflorescence (shorter in var. *mirifica*) (Niyomdham, 1992). It is assumed, although not yet confirmed, that var. *candollei* will, like var. *mirifica*, also be found to have tuberous roots.

Some of the rejuvenating effects reported by Kerr (1932) and Wanandorn (1933) for preparations containing powdered *P. mirifica* root are consistent with the presence of one or more oestrogenic compounds, and this was confirmed by a series of simple tests carried out some years later by Vatna (1939). Powdered kwao keur was first extracted with 95 per cent ethanol over a period of several days. After evaporation, the oily residue was suspended in olive oil and amounts equivalent to 20 g of dried powder were injected subcutaneously in three doses over a 24 h period into mature, spayed mice. Within 72 h of the first injection, all the treated mice were found to be in full oestrus whereas mice injected only with olive oil were unaffected. Injections equivalent to a total of only 2 mg of dried powder were sufficient to induce oestrus, a clear indication of the very high oestrogenic potency of the root preparations.

Probably the best documented legume phytoestrogens are the simple isoflavones such as daidzein (7,4′-dihydroxyisoflavone, 1) and genistein (5,7,4′-trihydroxyisoflavone, 2) which occur in soybeans (*Glycine max*), and in soy-based foods, mainly as the corresponding 7-O-glucosides (daidzin 3 and genistin 4, respectively), or as the 6″-O-acetyl or 6″-O-malonyl derivatives of these latter compounds (Figure 6.4). After ingestion, all these compounds may eventually be converted by the action of intestinal microflora to molecules of significantly greater oestrogenic potency. Another type of isoflavonoid phytoestrogen is exemplified by the coumestan derivative coumestrol (3,9-dihydroxy-coumestan, 5) which possesses considerably stronger activity than either daidzein or genistein (Bickoff *et al.*, 1969).

The isoflavones, of which upwards of 900–1000 examples have now been reported, are the most numerous and hence most commonly encountered of the many different types of isoflavonoid, the vast majority of which are only associated with species belonging to the subfamily Papilionoideae of the Leguminosae (Ingham, 1983). In some legumes, isoflavones with 2′-hydroxylation may be produced, a feature which allows these compounds to undergo cyclization to give comparatively rare isoflavonoids such as the pterocarpans, pterocarpenes and coumestans. Although isoflavones may occur in plants as free aglycones, it is not uncommon to find them combined with one or more sugars (notably glucose) as either O- or C-glycosides in which the sugar residue is connected to one or other of the aromatic rings via an ether link (as in O-glucosides such as daidzin

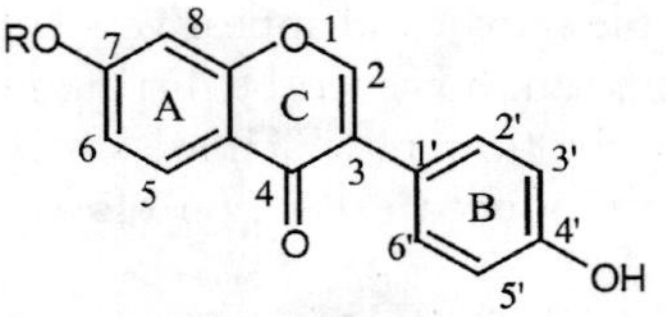

1 : R=H : daidzein
3 : R=Glucose : daidzin

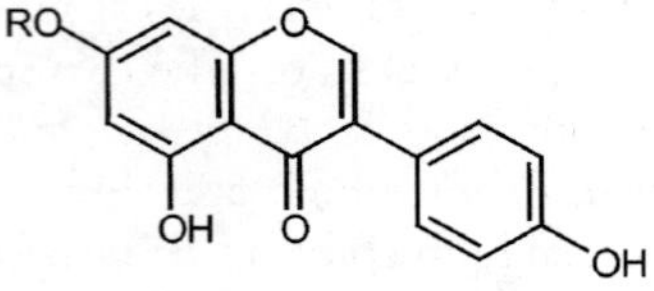

2 : R=H : genistein
4 : R=Glucose : genistin

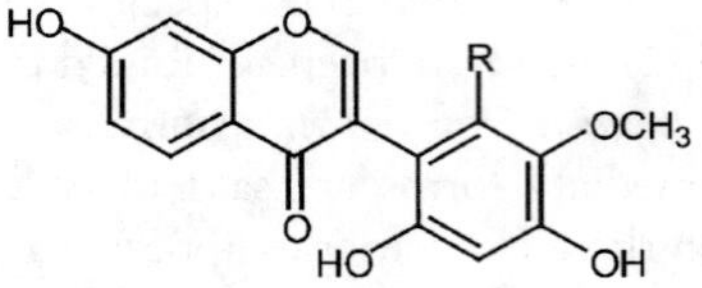

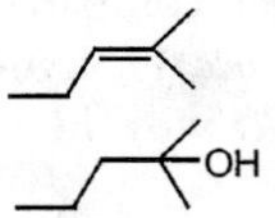

6 : R=    kwakhurin

7 : R=    kwakhurin hydrate

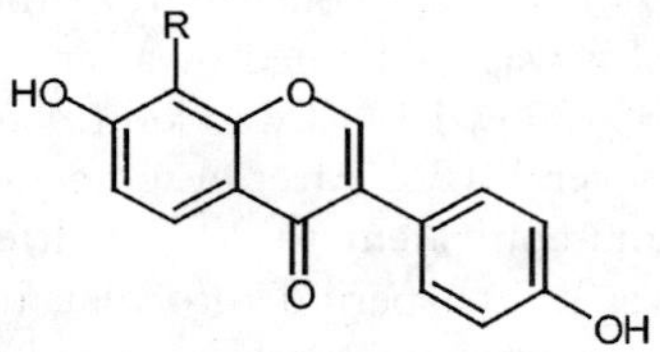

8 : R=Glucose : puerarin
9 : R=Glucose(6)CH$_2$OC=OCH$_3$ : puerarin-6″-monoacetate
10 : R=Glucose(6)-(1)Apiose : mirificin

*Figure 6.4* Isoflavones of *P. mirifica* roots.

3 and genistin 4) or, much less frequently, by a direct C–C bond (as in *C*-glycosides). These sugar units may in turn be substituted at the 6″ position with other sugars, or with acyl (e.g. acetyl) groups.

As yet, no studies appear to have been carried out on the chemical constituents of the aerial parts of *P. mirifica* which possess little or no oestrogenic activity (Jones and Pope, 1961). For obvious reasons, attention has principally been focused on the components of the tuberous roots from which a variety of known and previously unrecognized isoflavonoid compounds have recently been isolated and identified. As discussed in Part 1 of this review, some of these compounds may contribute, in a limited way, towards the oestrogenic activity of kwao keur. Of much greater interest, however, is the discovery that the tuberous roots contain two highly potent phytoestrogens, miroestrol (17) and deoxymiroestrol (18) which have a carbon skeleton found nowhere else in the Plant Kingdom. These compounds, which almost certainly are together responsible for most of the oestrogenic and rejuvenating properties of kwao keur, are dealt with in detail in Part 2 of this review.

## ISOFLAVONOIDS IN THE TUBEROUS ROOTS OF *P. MIRIFICA*

Detailed studies on the isoflavonoid constituents of *P. mirifica* (Figure 6.4 and 6.5) were first carried out between 1985 and 1990 in the Department of Food Science at Reading University where the dried tuberous roots were powdered and then exhaustively extracted with warm 95 per cent aqueous methanol over a 72 h period. These extracts were initially chromatographed on silica gel thin-layer plates in a non-acidic solvent system (chloroform–methanol, 20:1) to give seven principal bands (with $R_f$ values ranging from 0.52–0.05), all of which fluoresced blue or pale blue under long wavelength (354 nm) UV light (Ingham *et al.*, 1986a). From these bands, nine isoflavones (four aglycones, two O-glycosides and three C-glycosides) and four coumestans were eventually isolated and identified by chemical and spectroscopic methods, or by direct comparison with authentic samples (Ingham *et al.*, 1986a,b, 1988, 1989; Tahara *et al.*, 1987). In more recent work, Corey and Wu (1993) described the isolation of a compound with a structure superficially similar to that of a 3-phenylcoumarin, a relatively uncommon type of isoflavonoid, whilst Chansakaow *et al.* (2000a) obtained a new pterocarpene from the ethyl acetate extracts of defatted *P. mirifica* tubers. Apart from many of the compounds revealed by the earlier Reading investigation, Chansakaow *et al.* (2000a) also found that both the ethyl acetate and ethanolic root extracts contained a pterocarpan derivative not previously found in *P. mirifica* but known to occur in other *Pueraria* species.

### Isoflavone aglycones

Only two simple (non-prenylated) aglycones, daidzein (7,4′-dihydroxyisoflavone, 1) and genistein (5,7,4′-trihydroxyisoflavone, 2) have so far been obtained from the tuberous roots (Ingham *et al.*, 1986a). Daidzein and genistein occur in soybean (*Glycine max*), a close botanical relative of *Pueraria* species, and are widely found elsewhere in the Leguminosae including other members of the genus *Pueraria* (Ingham, 1983). Both compounds are of interest mainly because they are weak oestrogens and as such presumably contribute, if only in a limited way, to the overall oestrogenic activity of medicinal root preparations.

In addition to daidzein and genistein, *P. mirifica* roots also contain a third isoflavone aglycone which has been isolated by column chromatography in quantities sufficient to permit its identification as 7,2′,4′-trihydroxy-5′-methoxy-6′-[3,3-dimethylallyl]isoflavone (6). This compound has been named kwakhurin after one of the local Thai names (kwao khua) for *P. mirifica* (Tahara *et al.*, 1987). The B-ring substitution pattern of kwakhurin is very rare amongst the naturally occurring isoflavones, and no comparable compounds have been found in any other *Pueraria* species. Apart from kwakhurin, the tuberous roots of *P. mirifica* also contain very small quantities of the related isoflavone kwakhurin hydrate (7), in which the 6′-prenyl attachment of kwakhurin is end-hydroxylated (Ingham *et al.*, 1989).

Kwakhurin has recently been found to have oestrogenic potency comparable with that of daidzein (but much less than that of either genistein, or the coumestan, coumestrol) as determined by its relative ability to promote growth of human breast cancer cells being cultured in the presence of the oestrogen antagonist toremifene (Chansakaow *et al.*, 2000a).

## Isoflavone-*O*-glycosides

The only isoflavone *O*-glycosides known to occur in *P. mirifica* are daidzin (daidzein-7-*O*-glucoside, 3) and genistin (genistein-7-*O*-glucoside , 4) (Ingham *et al.*, 1986a, 1989). Daidzin and genistin occur in other *Pueraria* species (Ingham,1983), and are well recognized pro-estrogens in *Glycine max* (soybean) and some fodder legumes (clovers), being converted to daidzein and genistein respectively, and then to more highly oestrogenic compounds, by the action of intestinal microflora.

## Isoflavone-*C*-glycosides

Three isoflavone-*C*-glycosides have been isolated from *P. mirifica*, the principal compound being puerarin (daidzein-8-*C*-glucoside, 8) (Ingham *et al.*, 1986a). The corresponding genistein derivative (genistein-8-*C*-glucoside) occurs in *P. lobata* (Kinjo *et al.*, 1987) but has not been found in *P. mirifica* although its presence in root extracts, even if only in trace amounts, would not be unexpected.

In addition to puerarin and its acylated derivative puerarin-6″-monoacetate (9) (Ingham *et al.*, 1989), *P. mirifica* roots have also yielded mirificin (10), a compound which gives puerarin and the rare sugar apiose after treatment with boiling 2*N*-hydrochloric acid (Ingham *et al.*, 1986b). Mirificin has been identified as puerarin-6″-*O*-apiofuranoside, with the apiose residue being linked 1 → 6 to the glucose unit, the first time that such an arrangement has been recognized in a natural product. Kinjo *et al.* (1987) have also obtained mirificin (which they refer to as daidzein-8-*C*-apiosyl(1 → 6)glucoside) and the corresponding genistein derivative from *P. lobata*.

The names puerarin and mirificin were earlier given to two crystalline components of unknown identity isolated from the ether extracts of *P. mirifica* roots (Nilanidhi *et al.*, 1957). Both compounds were said to be oestrogenic, with puerarin being the most active of the two, but few physico-chemical details were reported, and no further work on either compound appears to have been undertaken. It seems unlikely, however, that the puerarin and mirificin described by Nilanidhi *et al.* (1957) are identical with the isoflavone *C*-glucosides to which these names have also been given, in particular because of melting point differences. Thus, puerarin and mirificin had melting points of 90–91° and 151–152° respectively, much lower than those of puerarin (8; about 190°) and mirificin (10; 188–190°).

## Pterocarpans, coumestans, and other isoflavonoids

In addition to the known *Pueraria* pterocarpan tuberosin (11) (Ingham, 1983), Chansakaow *et al.* (2000a) also isolated a new pterocarpene, puemiricarpene (12) from *P. mirifica* (Figure 6.5). In view of their structural similarity, this latter compound is probably derived biosynthetically from the isoflavone kwakhurin (6) via an intermediate such as kwakhurin pterocarpan which has previously been synthetically prepared (Tahara *et al.*, 1987).

Apart from the well documented weak oestrogen, coumestrol (3,9-dihydroxycoumestan, 5), a series of three unique coumestans named mirificoumestan (13), mirificoumestan hydrate (14) and mirificoumestan glycol (15) have been identified as minor components of *P. mirifica* roots (Ingham *et al.*, 1986a, 1988). Mirificoumestan is the coumestan analogue of the pterocarpene puemiricarpene (12) and is likely to be derived from it (Figure 6.5).

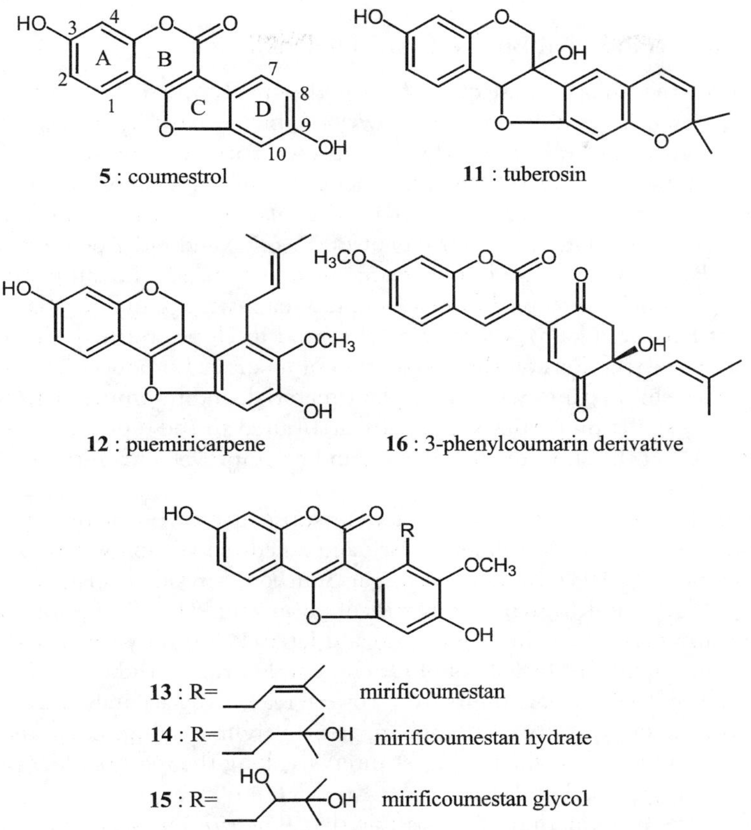

**5** : coumestrol

**11** : tuberosin

**12** : puemiricarpene

**16** : 3-phenylcoumarin derivative

**13** : R=    mirificoumestan

**14** : R=    mirificoumestan hydrate

**15** : R=    mirificoumestan glycol

*Figure 6.5* Coumestan, pterocarpan, pterocarpene and 3-phenylcoumarin derivatives of *P. mirifica* roots. Note that pterocarpan/pterocarpene derivatives are ring numbered as shown for coumestrol. The ring numbering of 3-phenylcoumarins follows the system used for isoflavones.

Finally, during their work on the total synthesis of miroestrol, Corey and Wu (1993) isolated about 1 mg of compound **16** from an ether extract of 500 g of dried *P. mirifica* root. Although nothing directly comparable has ever previously been isolated from a plant source, the structure of **16** suggests that it is a greatly modified 3-phenylcóumarin (a rare type of isoflavonoid), examples of which occur in some genera (*Pachyrrhizus* and *Neorautanenia*) related to *Pueraria* (Ingham, 1983).

## Relative concentrations of isoflavonoids in *Pueraria mirifica*

Of the isoflavonoids for which data are available, only daidzin (460 mg/kg dried root; 0.046 per cent) is reasonably abundant as a root constituent. With the exception of puerarin (64 mg/kg), mirificin (30 mg/kg) and puemiricarpene (18 mg/kg), the remaining compounds are present at very low levels (daidzin, 13 mg/kg; kwakhurin 7 mg/kg), with tuberosin, genistein and coumestrol occurring at a concentration of about 1 mg/kg, or less (Chansakaow *et al.*, 2000a).

## MIROESTROL, DEOXYMIROESTROL, AND RELATED COMPOUNDS

### Use of *P. mirifica* in traditional Burmese and Thai medicine

The first detailed description on the use of *Paukse*, a traditional Burmese drug principally derived from the tuberous roots of kwao keur (*P. mirifica*), but perhaps also incorporating root material from other *Pueraria* or non-*Pueraria* species, appears to be an account of uncertain date, but approximately 7000 words in length, which was published in their own language by two Burmese authors, U Nasada and U Nandiya. Their account, which is now probably lost, incorporates a shorter (about 600 words) and older description of the preparation of *Paukse*, and its use in the treatment of various specific ailments and for general ill-health, which was copied from a palm leaf manuscript found in a damaged Burmese temple at Pagan (Pukam), a former royal capital in Upper Burma. This palm leaf manuscript gives instructions on the preparation of medicinal products from kwao keur, including its use in mixtures with other plant material, and recommendations on dosage levels. Amongst the properties specifically attributed to these preparations are "rejuvenating effects" on the skin, teeth and eyes, and an improvement in male sexual activity.

The writings of U Nasada and U Nandiya were subsequently translated into the Thai character and language by Nai Plian Kitisri, and an edited version was published in pamphlet form in early 1931 by Luang Anusar Sunthon, a resident of Chiengmai who was convinced of the health-giving properties of kwao keur. This Thai publication, which was partly summarized by Bunnag (1938) and later (1955) fully translated into English by Mr S. Simmonds of the School of Oriental and African Studies (University of London; see also Cain, 1960)[2], has on its front cover a reasonably accurate drawing of kwao keur (*P. mirifica*) showing its climbing, apparently right-turning, habit and the characteristic tuberous swellings which occur at intervals along the roots of older plants (see Kerr, 1932, and Figures 6.1 and 6.2).

As mentioned in the Introduction, it is possible that *P. mirifica* may produce tubers of different colors (white, red and black) which differ in their degree of "rejuvenating potency," black tubers being the most active, and white the least (Kerr, 1932). In the Thai pamphlet produced by Luang Anusar Sunthon, it is said that daily doses of preparations made from the black tubers should be one-third, and those from the red tubers one-half, of those made from white tubers. It remains to be established, however, if the more potent red and black tubers definitely originate from *P. mirifica*, or are produced, predominantly or exclusively, by closely related *Pueraria* species that are either known (e.g. *P. candollei* which has a range extending into Burma; van der Maesen, 1985) or yet to be formally recognized. The possibility that the red and black tubers may be produced by species superficially similar to *P. mirifica* but belonging to genera other than *Pueraria* also cannot be firmly excluded at the present time.

According to U Nasada and U Nandiya, *white* kwao (keur) is a climbing plant found in Cambodia and Burma as well as in the vicinity of Chiengmai in Thailand. A single root may have up to three or four tubers which can eventually reach the size of a coconut.

---

2 One of the original pamphlets in Thai, and its English translation ("Treatise on the Drug from Tubers of the Kwao Vine") has been placed in the archives of Reading University Library together with a large collection of printed and written material (the property of the late Dr S. Bartlett) relating to the collection of, and work on, *P. mirifica* during the 1950's at Reading.

In contrast, *red* kwao (keur) often develops a more bushy habit and does not necessarily require other plants (trees) for support. Red kwao (keur) is said to form tubers which resemble "the trunk or tusk of an elephant," and thus are distinguishable from the coconut-like (globular) tubers of white kwao (keur). When cut, red kwao tubers typically exude a reddish sap. *Black Roman kwao* (keur) appears to be very similar to red kwao in its botanical habit, but the cut tubers afford a much darker (blackish) exudate.

## Early reports on the pharmacological properties of *P. mirifica*

Some of the vaguer properties attributed to preparations containing kwao keur include giving more restful sleep, and promoting appetite and general vigor, particularly in old and infirm people (Kerr, 1932). This may explain the use of kwao keur as a general tonic in northern Thailand. The medicines are also said to "rejuvenate" and "prolong life" (Kerr, 1932). More specifically, the tuber-based preparations are claimed to turn white or grey hair black, remove freckles, produce a youthful or improved complexion, cause regrowth of hair on bald heads, stimulate development of the breasts (in both men and women) and induce menstruation (or something resembling it) in women of 60 to 80 years of age (Kerr, 1932; Wanandorn, 1933). Kerr (1932) indicates that kwao keur preparations are normally forbidden to adults under 40 years of age, whilst Wanandorn (1933) suggests that abstinence from alcohol is necessary for the drug to exert its full beneficial effects.

Early evidence for the presence of oestrogenic material in *P. mirifica* tubers came from Wanandorn herself (1933) who took, on a daily basis and for a period of one month, a medicine consisting of kwao keur, honey and three myrobalan fruits (belleric, chebulic and emblic)[3]. She reported (Wanandorn, 1933) a gradual swelling and soreness of the breasts, with the symptoms disappearing slowly over several weeks when the medicine was discontinued.

Vatna (1939) later found that ethanol extracts of powdered kwao keur were capable, based on examination of vaginal smears, of inducing full oestrus in ovariectomized mice when given in doses equivalent to 2 mg of dried tuber, results that were subsequently confirmed by studies on rats (Sukhavachana, 1941). Aqueous extracts of the powdered tubers (made after initial extraction with ethanol) were lethal to mice at a dose equivalent to 60 mg dried tuber but, surprisingly, still gave the oestrogen response at a 15 mg equivalent dose (Vatna, 1939). These experiments not only established the oestrogenic potency of kwao keur but additionally provided some support for the anecdotal toxic effects of kwao keur on humans which have occasionally been mentioned (Wanandorn, 1933). The discovery (Vatna, 1939) that some oestrogen is extractable from kwao keur into water is in good agreement with the later findings of Jones and Pope (1961) who reported that about one-fifth of the total oestrogenic activity of powdered *P. mirifica* tubers partitioned into aqueous phases, although the nature of the compound (or compounds) concerned has never been established.

---

3 Myrobalan: trees of the genus *Terminalia* (family Combretaceae) found in south-east Asia and used for a variety of purposes. The main examples are *T. bellerica* (belleric myrobalan) and *T. chebula* (chebulic myrobalan). The drug used by Wanandorn (1933) contained these and emblic myrobalan the three together being known in traditional Thai medicine as tri-pala. *Author's note*: "Emblic (or embelic) myrobalan" may possibly be the dried fruit of a species of *Embelia* (family Myrsinaceae), rather than a *Terminalia* species, as these are also used medicinally in Asia.

In other studies, Pangsriwongse (1938) obtained small quantities of two glucosidic compounds from kwao keur, whilst Ketu-Sinh (1941) isolated a single crystalline glucoside ("butenin") from water extracts of the swollen roots. The latter compound exhibited various toxic effects (slow heart rate and decreased respiration) when tested on small animals, and was lethal to rabbits and dogs at a dose of 20–25 mg/kg body weight, but no information on its structure is available.

More recently, it has been found that the powdered tubers mixed with cooked rice (8 per cent by weight) or water (2 per cent w/v) can inhibit courtship and mating behavior in male pigeons, and decrease the number of eggs laid by female birds (Smitasiri, 1995). A significant antifertility effect was also observed when 86 female dogs showing signs of oestrus were each given 1 g of the powder in food on a daily basis for 30 days. Although most of the treated dogs mated during this period, none became pregnant, in contrast to the 81 untreated control animals (all of which mated) where there was a pregnancy rate of 85 per cent (Smitasiri *et al.*, 1990). Similar antifertility effects were seen when food mixed with the powdered tubers was fed to female bandicoots, mice (Y. Smitasiri, unpublished data) and rats. In the latter case, the powdered tubers were most effective at preventing pregnancy when 50 mg/day was given one to 3 days after insemination (Smitasiri *et al.*, 1986). Finally, when fed to immature female goats, the tuber powder caused an increase in body weight, promoted udder growth and an increase in teat length and diameter, and induced milk formation (Y. Smitasiri, unpublished data).

## Isolation and identification of miroestrol

Miroestrol was probably first obtained from kwao keur in a slightly impure state (mp. 260°) by Butenandt and Jacobi at the University of Danzig (see Butenandt, 1940). When tested by Schoeller *et al.* (1940) in the Berlin laboratories of Schering A.G., it was found that given subcutaneously, the compound had one-quarter the activity of $17\beta$-oestradiol, and double the activity of oestrone, in the Allen-Doisy rat vaginal cornification test, a standard assay which measures cellular changes associated with the onset of oestrus. In the same test, the compound had activity at least 55 times greater than either $17\beta$-oestradiol or oestrone when administered orally (Schoeller *et al.*, 1940), and was calculated to be present in kwao keur at a level equivalent to 150 mg oestrone/kg tuber. Apart from its effects in the Allen-Doisy vaginal cornification test, the compound also caused growth of the vagina, uterus and mammary glands of infantile, and ovariectomized adult, rats and rabbits, and induced oestrus in ovariectomized baboons (Schoeller *et al.*, 1940).

In later studies carried out at the National Institute for Research in Dairying in Reading, the dried tubers of *P. mirifica* were extracted on a semi-industrial scale to eventually yield several grams of an optically active ($[\alpha]_D + 301°$ in EtOH) compound which crystallized from dry methanol as colorless rectangular plates (mp. 268–270°) and which was named *miroestrol* (Pope *et al.*, 1958; Pope, 1959; Jones and Pope, 1961). Yields typically were 6–20 mg miroestrol/kg dry tuber (Cain, 1960; Jones and Pope, 1961). The potency of miroestrol was such that a methanol extract equivalent to 1 mg of powdered tuber gave a result in the mouse uterine weight assay comparable to that of 0.02–0.04 µg of $17\beta$-oestradiol (Kashemsanta *et al.*, 1957; Jones and Pope, 1961). Based on melting point similarities, and the results of oestrogen assays, it can be assumed that miroestrol is identical with the compound first studied by Butenandt (1940) and by Schoeller *et al.* (1940).

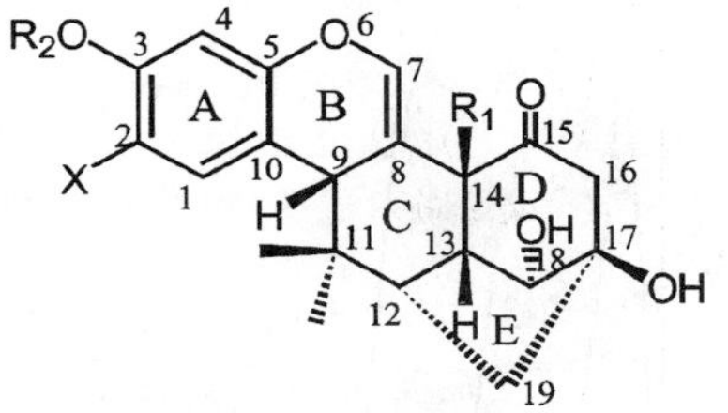

**17** : R$_1$=OH, R$_2$=X=H ; miroestrol
**18** : R$_1$=R$_2$=X=H ; deoxymiroestrol
**19** : R$_1$=OH, R$_2$=H, X=Br ; 2-bromomiroestrol
**25** : R$_1$=OH, R$_2$=CH$_3$, X=H ; miroestrol-3-methyl ether

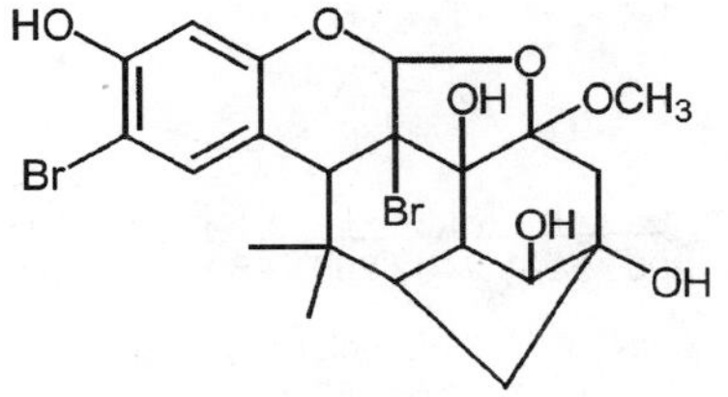

**26**

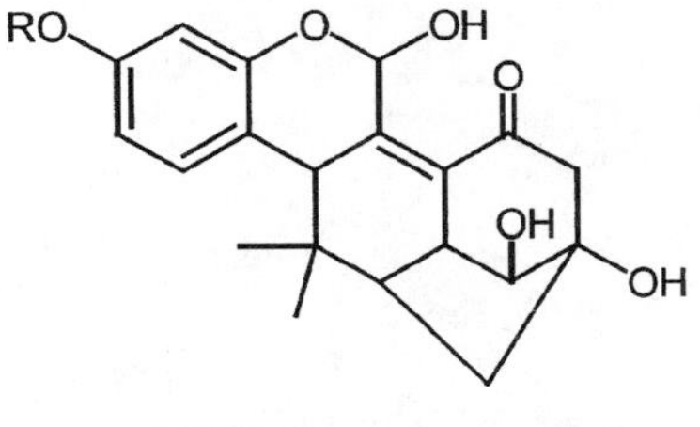

**27** : R=H ; isomiroestrol
**28** : R=CH$_3$ ; isomiroestrol-3-methyl ether

*Figure 6.6* Structures of miroestrol, deoxymiroestrol and related compounds. Note that the ring numbering/lettering system differs from that of isoflavones (Figure 6.4)

The identification of miroestrol as **17** (Figure 6.6)[4] was achieved by X-ray diffraction analysis using a crystal of monobromomiroestrol (2-bromomiroestrol, **19**) (Bounds and Pope, 1960; Taylor *et al.*, 1960). This structure has recently been confirmed by [1]H and [13]C NMR analysis (Corey and Wu, 1993; Chansakaow *et al.*, 2000b), and by enantio-selective total synthesis (Corey and Wu, 1993). The UV and mass spectra of miroestrol are shown in Figures 6.7 and 6.8. On silica gel TLC plates (layer thickness, 0.25 mm),

4 Miroestrol: C$_{20}$H$_{22}$0$_6$; *Chem. Abstr.* Index No. [2618-41-9]. Chemical Abstracts use the following systematic name: 2,12-methano-1*H*-benzo[*b*]naphtho[2,1-*d*]pyran-4(4a*H*)-one, 2,3,10b,11,12,12a-hexahydro-1,2,4a,8-tetrahydroxy-11, 11-dimethyl-[1*R*-(1α,2β,4aβ,10bβ,12α,12aβ)].

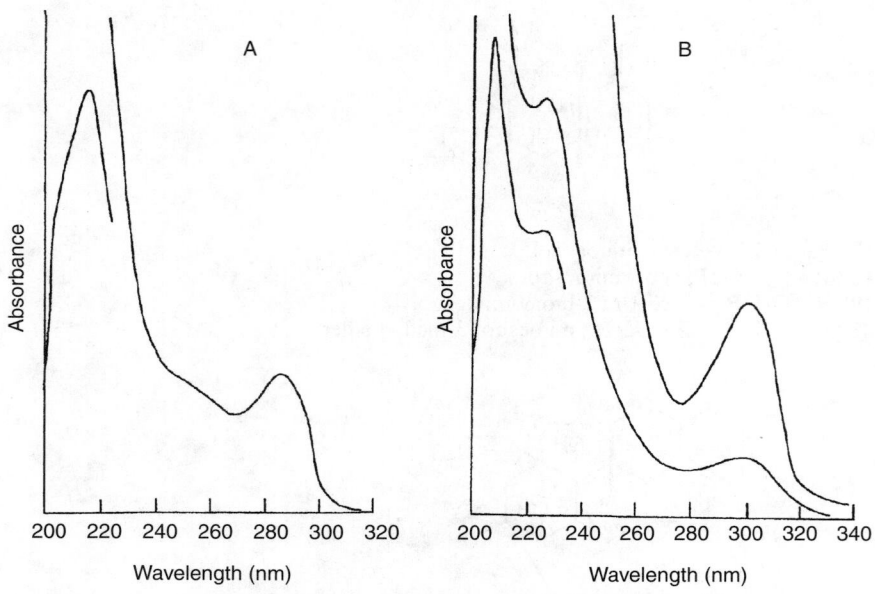

*Figure 6.7* UV spectrum of miroestrol determined in methanol (A) and methanol + aqueous sodium hydroxide (B).

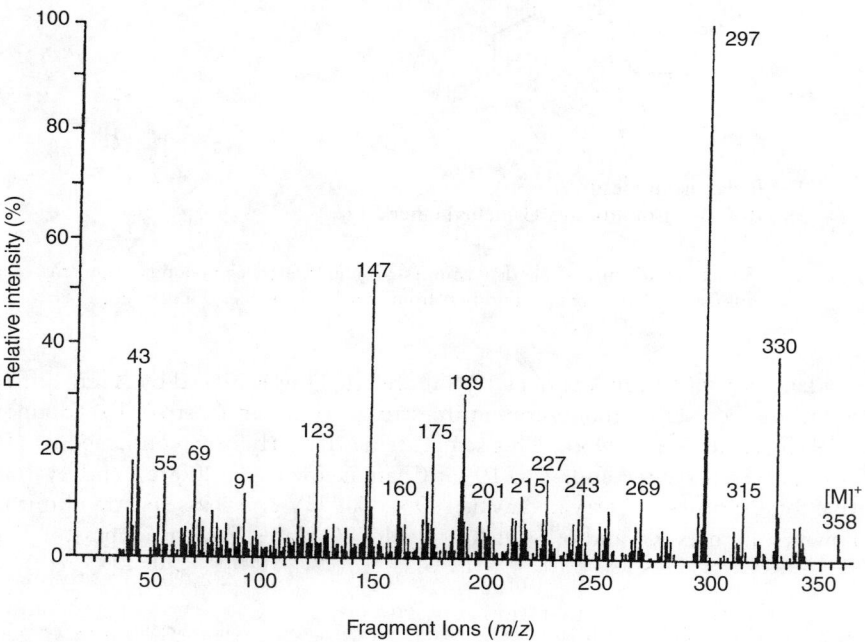

*Figure 6.8* Mass spectrum of miroestrol. Note the low intensity molecular ion at *m/z* 358.

miroestrol ($R_f$ 0.21) runs below the isoflavone daidzein (1, $R_f$ 0.33) in chloroform – methanol (10:1). Both compounds give an orange–brown color when sprayed with diazotized *p*-nitroaniline reagent. The strong oestrogenic activity of miroestrol can be assumed to reflect the non-planar nature of the molecule in which the distance between the 3-OH and the 18-OH (and the 3-OH and 17-OH) (Figure 6.6) is similar to that between the 3-OH and 17$\beta$-OH of 17$\beta$-oestradiol, thereby allowing miroestrol to effectively attach to oestrogen receptor sites (Jones and Pope, 1961).

## Isolation and identification of deoxymiroestrol

Apart from miroestrol, the tuberous roots of *P. mirifica* have recently been found to contain similar quantities (about 20 mg/kg dry tuber) of deoxymiroestrol 18 (mp. 213–216°; $[\alpha]_D + 217°$ in MeOH) (Chansakaow *et al.*, 2000b). The identification of deoxymiroestrol was based principally on a $^1$H NMR comparison with miroestrol, to which 18 can be oxidized *in vitro* (Chansakaow *et al.*, 2000b).

## Synthesis of miroestrol

Despite various attempts (see Miyano and Dorn, 1972a,b), it was not until 1993 that Corey and Wu eventually succeeded in preparing miroestrol. Two key intermediates, one a bicyclic vinylstannane (20) and the other an $\alpha$-bromo-$\alpha,\beta$-enone derivative (21) were first independently synthesized from 2-hydroxy-4-methoxybenzaldehyde (4-methoxy-salicylaldehyde) and 3-bromo-4-methoxyphenol, respectively. As shown in Figure 6.9, these silyl-protected intermediates were then coupled in the presence of a palladium catalyst to give a tricyclic keto-isoflavone (22), from which the miroestrol-type ring system of (24) was obtained in good yield in only two isomerization steps. Oxidation of 24 with selenium dioxide, and subsequent removal of the silyl protecting groups gave totally synthetic, optically active miroestrol which was identical by TLC, melting point and other physico-chemical characteristics ($^1$H and $^{13}$C NMR, MS, IR) with natural *P. mirifica*-derived material (Corey and Wu, 1993).

## Oestrogenic activity in the non-tuberous parts of *P. mirifica*, and in other *Pueraria* species

The *fibrous* roots of mature *P. mirifica* plants possess oestrogenic potency comparable with that of the tubers (Jones and Pope, 1961). Although no chemical studies have been carried out, it is reasonable to assume that much of this oestrogenic activity is due to miroestrol and related compounds such as deoxymiroestrol. The fibrous roots of a young (non-tuberous) plant grown at Reading had oestrogenic activity 20 times lower than that detected in the tubers of a mature plant growing at Kew Gardens, an indication that oestrogenic compounds accumulate in the whole root system as the plant ages and tubers are formed. Oestrogenic activity in the *leaves* and *stems* (including leaf petioles) of *P. mirifica* was minimal, being less than 1 per cent of that in the tuber (Jones and Pope, 1961).

The aerial parts (leaves + stems) and roots of tropical kudzu (*P. phaseoloides*), and the leaves, stems and roots of kudzu (*P. lobata*), also have very low oestrogenic activity, comparable with that found in the leaves and stems of *P. mirifica* (Jones and Pope, 1961).

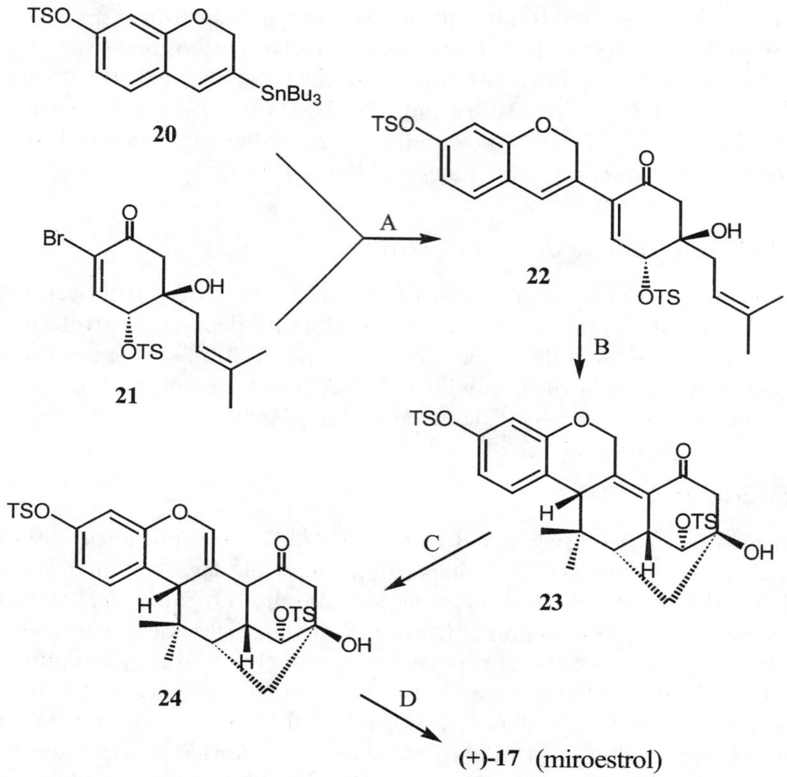

*Figure 6.9* Synthetic route to miroestrol (Corey and Wu, 1993). TS = tri-isopropylsilyl protecting group; SnBu$_3$: tri-*n*-butylstannyl residue; A = coupling reaction, B = cyclo-isomerization, C = isomerization, D = oxidation/desilylation. Note the close structural similarity between intermediates 23 and 24, and compounds 27 (isomiroestrol) and 18 (deoxymiroestrol), respectively. The similarity between 22 and 16 (Figure 6.4) is also notable from the viewpoint of the biogenesis of miroestrol.

## Pharmacological studies on miroestrol, deoxymiroestrol and related compounds

### Miroestrol

When given subcutaneously to ovariectomized rats, miroestrol had about one-third the activity of diethylstilboestrol, and about one-quarter the activity of *meso*-hexoestrol, in causing cornification of the vagina (Jones *et al.*, 1961). In the same test, miroestrol had a relative potency of 0.25 compared to 17$\beta$-oestradiol when given subcutaneously, and 0.8 compared to diethylstilboestrol when given orally to rats (Pope *et al.*, 1958; Jones *et al.*, 1961). In 20-day old ovariectomized mice, both miroestrol and oestrone (0.055 µg given subcutaneously on a daily basis for the next 21 days) caused an increase in uterine weight almost comparable with that of intact controls. Uterine responses to lower doses (0.01 µg daily) suggested that miroestrol had about twice the potency of oestrone (Benson *et al.*, 1961). In immature mice, miroestrol was more than three times as potent as diethylstilboestrol when given orally, and equal in activity to 17β-oestradiol when given subcutaneously, in six doses over a 3 day period as measured by the increase in uterine wet weight (Pope *et al.*, 1958; Jones and Pope, 1960; Jones *et al.*, 1961). However, using a different strain of immature mice, Terenius (1968) reported that miroestrol had only one-fifth the activity of 17β-oestradiol as determined by the increase in uterine dry weight, when the compounds were administered subcutaneously once daily for 3 days.

Using strips of mouse uterus, Terenius (1968) also found that when tested *in vitro*, miroestrol was much less active than 17β-oestradiol in competing for oestrogen receptor sites. Somewhat similar *in vitro* results were later obtained by Shutt and Cox (1972). These authors found that the binding affinity of miroestrol to receptors in protein (cytosol) preparations derived from sheep uteri was only 5 per cent (comparable with that of coumestrol) relative to 17β-oestradiol, a value much lower than was expected based on the *in vivo* studies of Jones and Pope (1960) which suggested that both compounds had similar oestrogenic potency.

Apart from its uterotrophic effects, miroestrol also has mammogenic activity (Benson *et al.*, 1961). Over a 30 day period, mammary growth in ovariectomized rats given 0.1 μg miroestrol daily was similar to that of untreated intact animals, and significantly greater than in ovariectomized of controls treated only with 70 per cent propylene glycol. Even at a miroestrol level of 0.01 μg/day, the mammary glands increased substantially in size relative to those ovariectomized, untreated, controls. Compared to 17β-oestradiol, the potency of miroestrol was about 0.70 (Benson *et al.*, 1961). Similar results were obtained using ovariectomized mice in which miroestrol (0.055 μg given daily) was found to be twice as potent as oestrone (Benson *et al.*, 1961). The mammary glands of castrated male rats also increased in size after daily administration of miroestrol (0.01 μg or 0.1 μg).

No marked differences were evident in the histology of the mammary glands, uteri, adrenal or thyroid glands of female rats receiving miroestrol or 17β-oestradiol. However, an increase in the weight of the thyroid gland was noted in those animals receiving 0.1 μg miroestrol daily. The weights of other glands in the female rat (pituitary, adrenal and thymus) were largely unaffected (Benson *et al.*, 1961). In adult male rats, miroestrol (96 μg given subcutaneously on a daily basis for 10 days) caused a significant decrease in the size of the testes, prostate gland and seminal vesicles, apparently by decreasing the output of gonadotrophins from the pituitary gland. Although, the adrenal glands became enlarged, the pituitary gland itself remained comparable in size with those of control rats (Jones *et al.*, 1961). The activity of miroestrol was about 0.3 compared with diethylstilboestrol, and 0.25 compared with *meso*-hexoestrol. Miroestrol was also found to interrupt the normal course of pregnancy, as measured principally by pre-implantation foetal loss, when 20–40 μg were given orally as a "morning after" treatment to inseminated female rats. Miroestrol had about one-half the activity of diethylstilboestrol, but was four times as effective as *meso*-hexoestrol (Jones *et al.*, 1961).

Finally, miroestrol given subcutaneously was found to reduce the weight gained by ovariectomized rats in comparison with ovariectomized but untreated animals. Thus, the body weights of rats receiving a daily miroestrol dose of 0.1 μg remained comparable with those of intact (non-ovariectomized) controls. When compared with 17β-oestradiol, miroestrol was only slightly less effective at reducing weight gain after ovariectomy (Benson *et al.*, 1961).

In non-mammalian studies, there was no evidence that miroestrol had auxin- or gibberellin-like activity. The compound had no growth promoting or cell elongating properties, and failed to induce parthenocarpic development in unfertilized tomato ovaries, or stimulate callus formation on the wounded stems of young tomato plants (Jones and Pope, 1961). Miroestrol (100 µg) had no detectable antifungal activity when tested against *Cladosporium herbarum* using the Homans and Fuchs (1970) TLC plate bioassay method, nor did it show any zoospore repelling action in tests using *Aphanomyces cochlioides*, although synthetic and natural oestrogens such as diethylstilboestrol, 17$\beta$-oestradiol and oestrone were all highly active (Islam and Tahara, 2001).

### Deoxymiroestrol and other compounds

Miroestrol and deoxymiroestrol (18, Figure 6.6) have been found to promote growth of human breast cancer cells in the presence of toremifene, an oestrogen antagonist (Chansakaow *et al.*, 2000b). Deoxymiroestrol was about 10 times more potent than miroestrol, and both compounds were more active than 17$\beta$-oestradiol at the highest concentrations tested. However, no data are available on the *in vivo* activity of deoxy-miroestrol, and the potency of this compound, relative to miroestrol, in whole animal systems remains unclear. Isomiroestrol (27) has also been isolated from *P. mirifica* (Chansakaow *et al.*, 2000b), but this compound had no detectable action on the growth of breast cancer cells.

In the mouse uterine weight assay, 2-bromomiroestrol (19), the dibromomethoxy deriv-ative (26), isomiroestrol (27) and isomiroestrol-3-methyl ether (28) (Figure 6.6) were all inactive when given subcutaneously at a level of 2 µg/mouse (Bounds and Pope, 1960; Jones and Pope, 1961). In contrast, miroestrol-3-methyl ether (25) had a potency of 0.59 relative to miroestrol when given subcutaneously, and comparable activity when given orally (Jones and Pope, 1961). Findlay *et al.* (1980) also synthesized a range of steroid-type compounds having some structural features in common with miroestrol. Several of these showed uterotrophic effects when given subcutaneously to immature rats but, with one exception, all were significantly less active than 17$\beta$-oestradiol (Mebe *et al.*, 1992).

### Human trials

After taking a kwao keur preparation daily for one month, Wanandorn (1933) described a swelling and soreness of the breasts which gradually disappeared when the medicine was discontinued. Sukhavachana (1941) gave an alcoholic extract of kwao keur in syrup to two patients and noted various oestrogen-related changes (e.g. an increase in the size of the uterus) but no ill-effects. In contrast, trials by Schoeller *et al.* (1940), using a crystalline compound probably identical with miroestrol, were abandoned when three women patients developed severe nausea after taking a 1 mg dose. Earlier tests on rats, rabbits, cats and monkeys had shown no adverse side effects even at very high dose levels (Schoeller *et al.*, 1940).

In clinical trials carried out at the Chelsea Hospital for Women in London (see Cain, 1960), nine patients with amenorrhoea, and one with artificial menopause, were treated orally with 1 mg or 5 mg of pure miroestrol on a daily basis for up to 14 days. Both dose levels caused significant vaginal cornification as well as effects such as an enlarge-ment of the breasts and an increase in the pigmentation of the nipples. In the patient with artificial menopause, the hot flushes diminished in frequency and severity. Oestrogen-mediated withdrawal bleeding was observed in some patients 7–18 days after the treat-ment ended. These clinical trials were eventually discontinued as, even at a dose rate of

1 mg daily, miroestrol was found to cause unacceptable side effects (malaise, headaches, nausea/vomiting) in the majority of patients.

## CONCLUSIONS

Although the tuberous roots of *P. mirifica* (white kwao keur) have probably been used for centuries in traditional Thai and Burmese medicine, doubts still remain over the exact botanical identity of red and black kwao. Both are mentioned along with white kwao in some early texts, but are claimed to have beneficial effects greater than, and somewhat different from, those of white kwao (Kerr, 1932). Red kwao, for example, is said not to cause breast enlargement, a fact which suggests that red and white kwao may be distinct botanical species. The term 'red kwao' seems currently to be used to describe *Butea superba*, from the roots of which a herbal preparation with reputed anti-impotency properties can be prepared. However, whether the red kwao of antiquity, with tubers described in the pamphlet by Luang Anusar Sunthon as "resembling in shape the tusk or trunk of an elephant" is identical with *B. superba*, or is a *Pueraria* species (either known or not yet formally described) related to *P. mirifica* is uncertain at the present time.

Jones and Pope (1961) obtained evidence to suggest that miroestrol occurs in both the tuberous and fibrous roots of *P. mirifica*. As interest in the medicinal properties of *P. mirifica* is rapidly increasing, it would be useful to determine if the plant can be cultivated, either on a farm scale, or in small village plots to provide local income and reduce the need for collecting tubers from the wild. Although no significant oestrogenic activity was detectable in *P. lobata* or *P. phaseoloides* (Jones and Pope, 1961), this early study should perhaps be extended to include the roots (tuberous and fibrous) of other *Pueraria* species, some of which might possibly have medicinal properties comparable with those of kwao keur. *Pueraria candollei*, a species almost identical with *P. mirifica*, would be an obvious first choice for investigation.

Further chemical work on *P. mirifica* itself is also required. Vatna (1939) found that some oestrogenic material could be extracted from the tubers with water, whilst Jones and Pope (1961) reported that about one-fifth of the total oestrogenic activity of the tubers partitioned into aqueous phases. Although, compounds such as the isoflavone glucoside daidzin (**3**) may contribute to this activity, the exact chemical identity of the hydrophilic material remains to be firmly established, as also does the nature of the toxic principle reported by both Vatna (1939) and Ketu-Sinh (1941).

Apart from miroestrol, the tuberous roots of *P. mirifica* are now also known to contain the related compound deoxymiroestrol (Chansakaow *et al.*, 2000b). At present, however, the relative importance of these compounds in root-derived medicines is unclear. Deoxymiroestrol can be oxidized *in vitro* to give miroestrol, a process which might occur naturally in *P. mirifica* as part of a biosynthetic sequence, or take place when the sliced tubers are being air-dried, powdered and processed into traditional Thai medicines. Thus, whilst deoxymiroestrol appears from *in vitro* tests to be more oestrogenic than miroestrol (Chansakaow *et al.*, 2000b), the activity of medicinal preparations *in vivo* may either reflect a predominance of miroestrol derived from deoxymiroestrol or, more probably, the combined effects of both compounds.

In its pure form, miroestrol does not appear to have any clear clinical advantages over steroid oestrogens such as 17$\beta$-oestradiol. Indeed, the side effects associated with taking even small quantities of miroestrol (Schoeller *et al.*, 1940; Cain, 1960) are a positive deterrent to its use, although these might be overcome, and activity enhanced, by chemical manipulation of the molecule. The satisfactory standardization of miroestrol and other compounds in traditional Thai medicine is clearly difficult to achieve but, assuming that the various constituents are present at reasonably constant levels, and bearing in mind the nausea and other symptoms caused by miroestrol in female patients, it may be the case that kwao keur preparations containing a complex mixture of many different interacting chemicals will prove generally to be of greater medicinal value than pure compounds individually administered.

## ACKNOWLEDGMENTS

We thank Mr R. Scarlett for his expertise in growing *P. mirifica* from seed, and Prof. T. Ishikawa (Chiba University) for supplying data on root constituents prior to publication. The help of Prof. E. Dagne (Addis Ababa University), Dr P. Friar (University of Reading), Dr G.P. Lewis (Kew Herbarium), Prof. W.M. Keung (Harvard Medical School), Prof. Y. Smitasiri (Mai Fah Luang University) and Prof. L.J.G. van der Maesen (Wageningen Agricultural University) is also gratefully acknowledged.

## REFERENCES

Benson, G.K., Cowie, A.T. and Hosking, Z.D. (1961) Mammogenic activity of miroestrol. *J. Endocrinol.*, **21**, 401–409.

Bickoff, E.M., Spencer, R.R., Witt, S.C. and Knuckles, B.E. (1969) Studies on the chemical and biological properties of coumestrol and related compounds. *U.S. Dep. of Agric. Tech. Bull.*, No. 1408.

Bounds, D.G. and Pope, G.S. (1960) Light-absorption and chemical properties of miroestrol, the oestrogenic substance of *Pueraria mirifica. J. Chem. Soc.*, 3696–3705.

Bunnag, K. (1938) A brief account of hua kwao. *J. Pharm. Assoc. Siam (Series II)*, **2**, 9–14.

Butenandt, A. (1940) Zur charakterisierung des oestrogen wirksamen tokokinins aus *Butea superba. Naturwissenschaften*, **28**, 533.

Cain, J.C. (1960) Miroestrol: an oestrogen from the plant *Pueraria mirifica. Nature*, **188**, 774–777.

Chansakaow, S., Ishikawa, T., Sekine, K., Okada, M., Higuchi, Y., Kudo, M. and Chaichantipyuth, C. (2000a) Isoflavonoids from *Pueraria mirifica* and their estrogenic activity: isolation of a new pterocarpene, puemiricarpene, and identification of kwakhurin as a new phytoestrogen. *Planta Med.*, **66**(6), 572–575.

Chansakaow, S., Ishikawa, T., Seki, H., Sekine, K., Okada, M. and Chaichantipyuth, C. (2000b) Identification of deoxymiroestrol as the actual rejuvenating principle of "kwao keur" *Pueraria mirifica*: a known miroestrol may be an artefact. *J. Nat. Prod.*, **63**, 173–175.

Corey, E.J. and Wu, L.I. (1993) Enantioselective total synthesis of miroestrol. *J. Am. Chem. Soc.*, **115**, 9327–9328.

Findlay, J.A., Mebe, P., Stern, M.D. and Givner, M.L. (1980) Total synthesis of 3,17$\beta$-dihydroxy-6-oxaestra-1,3 ,5,7-tetraen and related miroestrol analogues. *Can. J. Chem.*, **58**, 1427–1434.

Homans, A.L. and Fuchs, A. (1970) Direct bioautography on thin-layer chromatograms as a method for detecting fungitoxic substances. *J. Chromatogr.*, **51**, 325–327.

Ingham, J.L. (1983) Naturally occurring isoflavonoids (1855–1981). *Fortschr. Chem. Org. Naturst.*, **43**, 1–266.

Ingham, J.L., Tahara, S. and Dziedzic, S.Z. (1986a) A chemical investigation of *Pueraria mirifica* roots. *Z. Naturforsch.*, **41c**, 403–408.

Ingham, J.L., Markham, K.R., Dziedzic, S.Z. and Pope, G.S. (1986b) Puerarin-6″-O-β-apiofuranoside, a *C*-glycosylisoflavone *O*-glycoside from *Pueraria mirifica*. *Phytochemistry*, **25**, 1772–1775.

Ingham, J.L., Tahara, S. and Dziedzic, S.Z. (1988) Coumestans from the roots of *Pueraria mirifica*. *Z. Naturforsch.*, **43c**, 5–10.

Ingham, J.L., Tahara, S. and Dziedzic, S.Z. (1989) Minor isoflavones from the roots of *Pueraria mirifica*. *Z. Naturforsch.*, **44c**, 724–726.

Islam, Md. T. and Tahara, S. (2001) Repellent activity of estrogenic compounds toward zoospores of the phytopathogenic fungus *Aphanomyces cochlioides*. *Z. Naturforsch.*, **56c**, 253–261.

Jones, H.E.H. and Pope, G.S. (1960) A study of the action of miroestrol and other oestrogens on the reproductive tract of the immature female mouse. *J. Endocrinol.*, **20**, 229–235.

Jones, H.E.H. and Pope, G.S. (1961) A method for the isolation of miroestrol from *Pueraria mirifica*. *J. Endocrinol.*, **22**, 303–312.

Jones, H.E.H., Waynforth, H.B. and Pope, G.S. (1961) The effect of miroestrol on vaginal cornification, pituitary function and pregnancy in the rat. *J. Endocrinol.*, **22**, 293–302.

Kashemsanta, L., Suvatabandhu, K. and Airy Shaw, H.K. (1952) A new species of *Pueraria* (Leguminosae) from Thailand yielding an oestrogenic principle. *Kew Bull.*, 549–552.

Kashemsanta, L., Suvatabandhu, K., Bartlett, S. and Pope, G.S. (1957) The oestrogenic substance (miroestrol) from the tuberous roots of *Pueraria mirifica*. *Proc. Pacific Sci. Cong., Pac. Sci. Assoc. 9th Meeting, Bangkok, Thailand*, **5**, 37–40 (*Chem. Abstr.* **60**, 12368).

Kerr, A. (1932) A reputed rejuvenator. *J. Siam Soc. (Natural History Suppl.)*, **8**, 336–338.

Ketu-Sinh, O. (1941) Preliminary report on a pharmacologically active substance in *Butea superba*. *J. Med. Assoc. Thailand*, **24**, 81–82.

Kinjo, J., Furusawa, J., Baba, J., Takeshita, T., Yamasaki, M. and Nohara, T. (1987) Studies on the constituents of *Pueraria lobata*. III. Isoflavonoids and related compounds in the roots and the voluble stems. *Chem. Pharm. Bull. (Tokyo)*, **35**, 4846–4850.

Mebe, P.P., Findlay, J.A., Stern, M.D. and Givner, M.L. (1992) Uterotrophic activities of 6-oxa-steroids, analogues of miroestrol. *Bull. Chem. Soc. Ethiopia*, **6**, 105–108.

Miyano, M. and Dorn, C.R. (1972a) Miroestrol. I. Preparation of the tricyclic intermediate. *J. Org. Chem.*, **37**, 259–268.

Miyano, M. and Dorn, C.R. (1972b) Miroestrol. II. Synthesis of a new tricyclic system. *J. Org. Chem.*, **37**, 268–274.

Nilanidhi, T., Kamthong, B., Isarasena, K. and Shiengthong, D. (1957) Constituents of the tuberous roots of *Pueraria mirifica*. *Proc. Pac. Sci. Congr., Pacific Sci. Assoc. 9th Meeting, Bangkok, Thailand*, **5**, 41–47. (*Chem. Abstr.* **60**, 11041).

Niyomdham, C. (1992) Notes on Thai and Indo-Chinese Phaseoleae (Leguminosae-Papilionoideae). *Nordic J. Botany*, **12**, 339–346.

Pangsriwongse, K. (1938) A note on *Butea superba* Roxb. *J. Pharm. Assoc. Siam (Series II)*, **2**, 17–20.

Pendleton, R.F. and Suvatabandhu, K. (1952) (Kwao keur). Note on geology, soil and climate. *Kew Bull.*, 552.

Pope, G.S. (1959) A partition chromatographic column for fractionating 50 g batches of plant extract. *Lab. Pract.*, **8**, 416–417.

Pope, G.S., Grundy, H.M., Jones, H.E.H. and Tait, S.A.S. (1958) The oesterogenic substance (miroestrol) from the tuberous roots of *Pueraria mirifica*. *Proc. 69th Meeting Soc. Endocrinology, J. Endocrinol.*, **17**, XV.

Schoeller, W., Dohrn, M. and Hohlweg, W. (1940) Ubereine oestrogene substanz aus der knolle der siamesischen schlingpflanze *Butea superba*. *Naturwissenschaften*, **28**, 532–533.

Shutt, D.A. and Cox, R.I. (1972) Steroid and phyto-oestrogen binding to sheep uterine receptors *in vitro*. *J. Endocrinol.*, **52**, 299–310.

Smitasiri, Y. (1995) A new method in the controlling of pigeon fertility by using an antifertility plant (*Pueraria mirifica*). *Proc. 3rd Int. Conf. on Biodeterioration of Cultural Property, Bangkok*, pp. 135–142.

Smitasiri, Y., Fongkaew, B., Mon-Ing, A. and Supasai, S. (1990) *Pueraria mirifica*: the potent antifertility plant for dogs. In P. Poomvises and Ingkaninun (eds), *Proc. 7th Congr. Fed. Asian Veterinary Associations, Pattaya*, pp. 649–656.

Smitasiri, Y., Junyatum, U., Songjitsawad, A., Sripromma, P., Trisrisilp, S. and Anuntalab-hochai, S. (1986) Postcoital antifertility effects of *Pueraria mirifica* in rats. *J. Sci. Faculty Chiengmai Univ.*, 13, 19–28.

Sukhavachana, D. (1941) Oestrogenic principle of *Butea superba*. *J. Med. Assoc. Thailand*, 24, 92–94.

Tahara, S., Ingham, J.L. and Dziedzic, S.Z. (1987) Structure elucidation of kwakhurin, a new prenylated isoflavone from *Pueraria mirifica* roots. *Z. Naturforsch.*, 42c, 510–518.

Taylor, N.E., Hodgkin, D.C. and Rollett, J.S. (1960) The X-ray crystallographic determination of the structure of miroestrol. *J. Chem. Soc.*, 3685–3695.

Terenius, L. (1968) Structural characteristics of oestrogen binding in the mouse uterus: inhibition of $17\beta$-oestradiol binding *in vitro* by a plant oestrogen, miroestrol. *Acta Pharmacol. Toxicol.*, 26, 15–21.

Tirawat, P. and Smitasuwana, U. (1949) Preliminary study of the means for the identification and preservation of the tuberous roots, previously reported as the roots of *Butea superba*. *J. Pharm. Assoc. Siam (Series III)*, 3, 21–30.

van der Maesen, L.J.G. (1985) Revision of the genus *Pueraria* DC. with some notes on *Teyleria* Backer (Leguminosae). *Agric. Univ. Wageningen Papers (The Netherlands)*, 85–1, 1–132.

van der Maesen, L.J.G. (1994) *Pueraria*, the kudzu and its relatives. An update of the taxonomy. In M. Sorensen (ed.), *Proc. 1st. Internl. Symp. on Tuberous Legumes, Guadeloupe*, Jordbrugsfor-laget, Copenhagen, Denmark, pp. 55–86.

Vatna, S. (1939) A preliminary report on the presence of an oestrogenic substance and a poisonous substance in the strorage root of *Butea superba*. *Thai. Sci. Bull.*, 4, 3–9.

Wanandorn, P.W. (1933) A reputed rejuvenator. *J. Siam Soc. (Natural History Suppl.)*, 9, 145–147.

# 7 Biosynthesis and natural functions of *Pueraria* isoflavonoids

*Takashi Hakamatsuka and Yutaka Ebizuka*

## INTRODUCTION

*Pueraria lobata* Ohwi accumulates isoflavonoids as main secondary metabolites in various organs, e.g. roots, stems, leaves and flowers (Hakamatsuka *et al.*, 1994). Isoflavonoid is a subgroup of flavonoids whose basic carbon skeleton is constructed by a combination of shikimate pathway and acetate–malonate pathway (Figure 7.1). Biosynthetic pathway leading to isoflavonoid frameworks shares the early part of the pathway (often referred to general phenyl propanoid pathway) from phenylalanine to *p*-coumaroyl CoA with several other plant phenolics such as simple phenylpropanoids, lignins, coumarins and quinones. The construction of flavonoid skeleton from *p*-coumaroyl CoA and three molecules of malonyl CoA to form chalcone, and the isomerization of chalcone into flavanone are common biosynthetic reactions among all flavonoid subgroups. The first committed step of isoflavonoid-specific pathway is a migration of the side phenyl group (B-ring) of the flavanone molecule from C-2 to the adjacent C-3 position. One structural feature of *Pueraria* isoflavonoids is the lack of oxygen function at C-5 and this reduction process has been proven to occur in the course of chalcone formation (Stoessl and Stothers, 1979).

Biosynthesis of *Pueraria* isoflavonoids will be described in this chapter focusing on two key reactions, namely aryl migration and reduction at C-5, mainly based on our biosynthetic studies using *P. lobata* cell suspension cultures.

## TISSUE CULTURES

### Development of cell culture system

Plant tissue culture technology has contributed enormously toward the progress of biosynthetic studies in plant secondary metabolism. In cell culture systems, plant cells are maintained under controlled conditions independent of climatic or other environmental changes, and microbe-free plant materials are available whenever needed. The productivity of secondary metabolites in tissue cultures is not always equivalent to those in their mother plants. However, the establishment of cell cultures, capable of producing certain amount of metabolites of interest, is essential to an extensive investigation of the biosynthetic pathways of plant secondary metabolites at enzymatic and genetic levels.

Takeya and Itokawa (1982) induced callus from the stem of *P. lobata* on Murashige-Skoog's basal medium containing 0.1 mg/l of kinetin, 1.0 mg/l of 2,4-dichlorophenoxyacetic

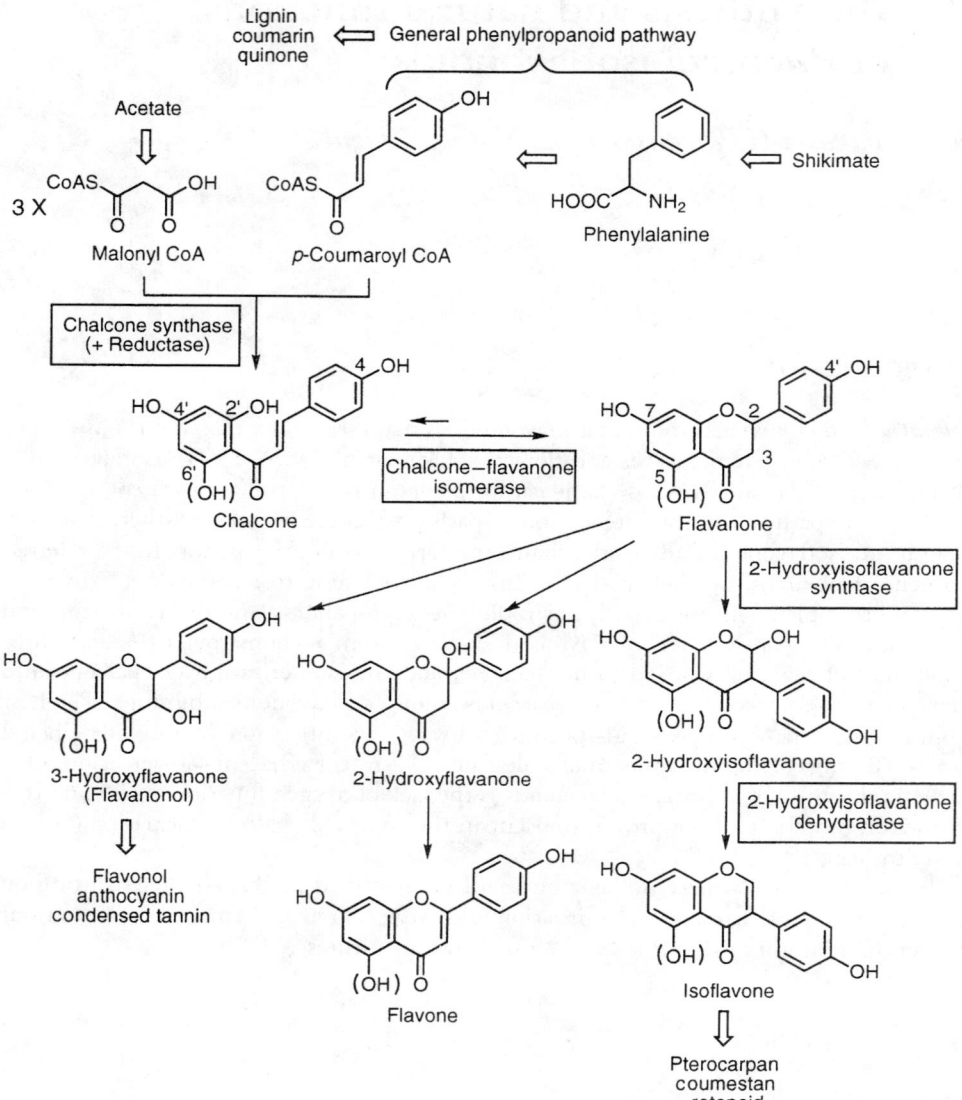

*Figure 7.1* General scheme of flavonoid biosynthesis.

acid (2,4-D), 3 per cent sucrose and 0.8 per cent Bacto agar. The callus tissues produced puerarin (1), daidzin (2), daidzein (3), genistein (4) and coumestrol (5), all of which are found in the mother plant (Figure 7.2). The callus cultures originally induced by Takeya and Itokawa (1982) were subsequently employed in our biosynthetic studies on *Pueraria* isoflavonoids. First, the callus from agar plate was transferred into liquid medium to initiate cell suspension cultures and they were maintained in Murashige-Skoog's liquid medium supplemented with kinetin (0.1 mg/l), 2,4-D (2 mg/l) and 3 per cent sucrose on a rotary shaker at 28 °C in the dark. They were rotated at a high speed

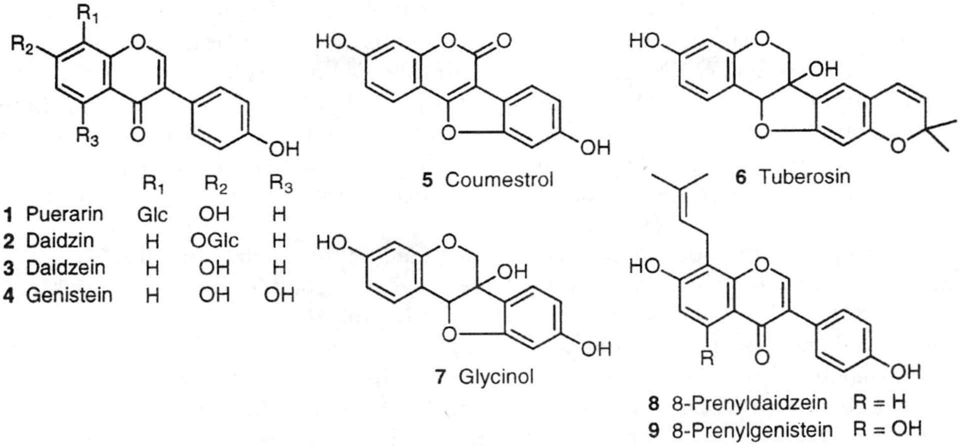

| | R$_1$ | R$_2$ | R$_3$ |
|---|---|---|---|
| **1** Puerarin | Glc | OH | H |
| **2** Daidzin | H | OGlc | H |
| **3** Daidzein | H | OH | H |
| **4** Genistein | H | OH | OH |

**5** Coumestrol

**6** Tuberosin

**7** Glycinol

**8** 8-Prenyldaidzein   R = H
**9** 8-Prenylgenistein   R = OH

*Figure 7.2* Isoflavonoids in *P. lobata* Ohwi.

(200 rpm), rather exceptional for most of plant cell cultures, since the cells tended to aggregate at lower speed. The main products identified in the later stage of suspension cultures were isoflavone glycosides, puerarin (1) and daidzin (2), most of them being as 6″-O-malonyl esters (Park *et al.*, 1992).

## Activation of isoflavonoid metabolism

Generally, enzymes involved in secondary metabolism are poorly expressed than those of primary metabolism even in the mother plants. Being able to activate the metabolic pathway of interest employing various experimental condition is important or even essential to the study of the enzymes that catalyze the pathway. Leguminous plants have been known to activate their isoflavonoid pathway and produce pterocarpan phytoalexins against microbial attacks. In *P. lobata*, tuberosin (6) and glycinol (7) have been identified as pterocarpan phytoalexins in infected leaves (Ingham, 1982). To study the biosynthesis of isoflavonoids at enzyme level in *P. lobata* cell cultures, we introduced new activation strategy taking advantage of the defense responses found in leguminous plants; despite the fact that cell suspension cultures of *P. lobata* produced comparable level of isoflavonoids with the mother plant (data not shown).

In plant–microbe interactions, cell wall components of pathogenic or non-pathogenic microorganisms, such as polysaccharides, proteins, glycoproteins and lipids, are recognized as the elicitors that can trigger phytoalexin accumulation. Besides these biotic elicitors, abiotic ones such as heavy metal ions, often cause phytoalexin induction. Glycoprotein elicitor from *Phytophthora megasperma* f. sp. *glycinea* (Keen and Legrand, 1980), which is specific for induction of glyceollin in soybean, was tested for its abilitiy of phytoalexin induction in cell suspension cultures of *P. lobata*, as none of specific pathogens for *P. lobata* had been identified. When the glycoprotein elicitor was added to the cell suspension cultures at a final concentration of 0.01 mg/ml, otherwise white cells turned into dark brown within 10 h with concomitant stop of cell growth. HPLC analysis indicated that pre-existed isoflavone glucosides rapidly disappeared from methanol extracts of the cells (Park *et al.*, 1995a). Instead, several less polar compounds appeared including tuberosin (6), 8-prenyldaidzein (8), and 8-prenylgenistein (lupiwighteone) (9) that had been identified as phytoalexins in the stems of *P. lobata* (Hakamatsuka *et al.*, 1991a). Furthermore, CuCl$_2$, at a final concentration of 1 mM induced almost the same response in Pueraria cell suspension cultures as that with fungal elicitor. Although both glycoprotein elicitor and CuCl$_2$ induced rapid accumulation of isoflavonoids, the response was so strong that the cells died soon after the treatment due to hypersensitive reaction(s). Therefore, these elicitors were not used in the activation of the metabolic pathways of interests in our studies.

In some plants, their own cell wall components are known to play parts in phytoalexin induction, and these elicitors are termed as endogenous elicitors (Bailey, 1980; Hahn *et al.*, 1981). Cell wall digests were prepared by hydrolyzing *P. lobata* cell walls with commercial endopolygalacturonidase from *Aspergillus niger* and tested for their capacity as endogenous elicitor. When the cell suspension cultures of *P. lobata* were treated with its own cell wall digests at a final concentration of 0.05 mg/ml, the cells turned into light brown but kept growing as well as untreated cells. This response was quite a contrast to the one with fungal glycoprotein elicitor or CuCl$_2$. HPLC analysis of methanol extracts of the treated cells indicated that the major constitutive glycosides such as daidzin (2) showed a rapid decline within 4 h after the treatment, as had been found with fungal elicitor. Feeding experiments with [$^{14}$C]-labelled isoflavone showed that the isoflavone glycosides, disappeared after elicitation, were incorporated into insoluble lignocellulose fractions of cell walls (Park *et al.*, 1995a). After 8 h of treatment, isoflavone glycosides began to accumulate and by 48 h, the amount of total isoflavone glycosides reached to more than five folds than that of untreated cells. The pterocarpan phytoalexins induced by the fungal glycoprotein elicitor or CuCl$_2$ were not induced under this condition. Subsequently, commercially available yeast extract (Bacto Yeast Extract from Difco) was found to have the same effect as the endogenous elicitor when it was added at a final concentration of 1 mg/ml. Furthermore, fungal glycoprotein elicitor, at concentrations <0.001 mg/ml, did not induce pterocarpan phytoalexins and exhibited the same effect as the endogenous elicitor (Park *et al.*, 1995a). Since the preparation of endogenous elicitors and fungal glycoprotein elicitor with reproducible activity was rather laborious, the easily available yeast extract has been routinely used to activate isoflavonoid biosynthesis in *P. lobata* cell cultures.


## BIOSYNTHESIS OF ISOFLAVONOIDS

### Formation of deoxy-type chalcone

As described above, both *P. lobata* plant and its cell suspension cultures produce mainly 5-deoxy-type isoflavonoids that lack an oxygen atom at C-5 (C-6' in chalcone numbering). The oxygen atom at this position corresponds to one of carbonyl oxygens in a triketide chain derived from three acetate units. Incorporation experiments with doubly labelled [1,2-$^{13}$C] sodium acetate were performed in *Pisum sativum* and $^{13}$C-NMR analysis of isoflavonoid end products indicated that the reduction occurred in the course of chalcone formation (Stoessl and Stothers, 1979). Enzymatic synthesis of 4,2',4'-trihydroxychalcone (6'-deoxychalcone) was first demonstrated in *Glycyrrhiza echinata* cell suspension cultures (Ayabe *et al.*, 1988). In the presence of NADPH, enzyme preparation from *G. echinata*

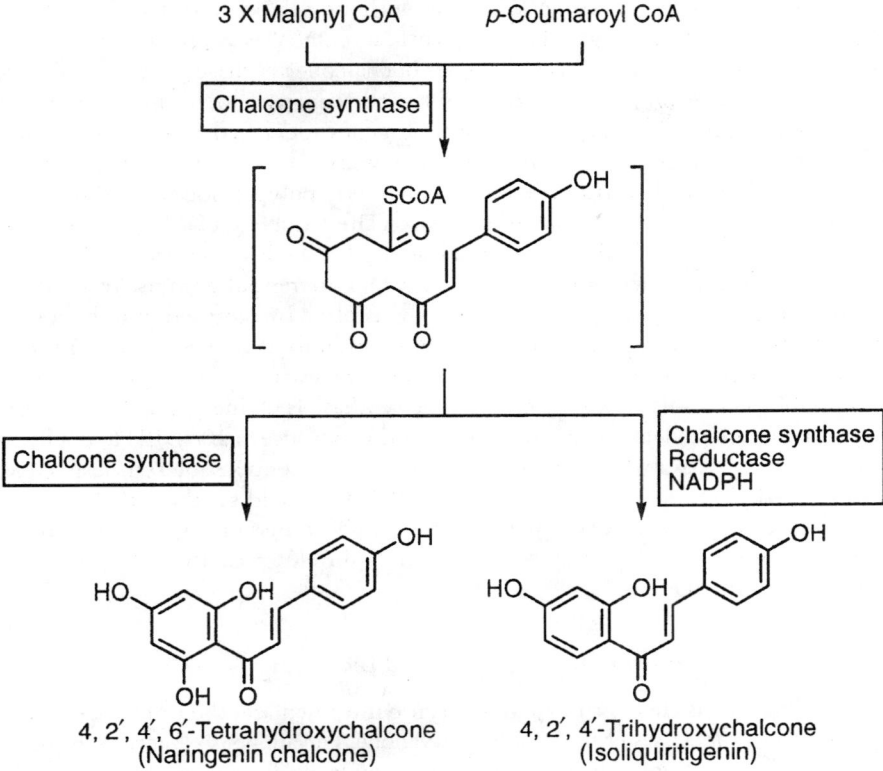

*Figure 7.3* Postulated reaction scheme for synthesis of 6-hydroxychalcone and 6-deoxychalcone.

produced 4,2′,4′-trihydroxychalcone from one molecule of *p*-coumaroyl CoA and three molecules of malonyl CoA. Characterization of deoxychalcone synthase of *Glycine max* (Welle and Grisebach, 1988a) indicated that this enzyme consists of two separable proteins, a chalcone synthase and a co-acting reductase. Using *p*-comaroyl CoA and malonyl CoA as substrates, the chalcone synthase alone catalyzes the formation of 4,2′,4′,6′-tetrahydroxy-chalcone (6′-hydroxychalcone) but the same enzyme, when accompanied with the reductase, produces 4,2′,4′-trihydroxychalcone (6′-deoxychalcone) in the presence of NADPH (Figure 7.3).

This enzyme system was also studied in cell suspension cultures of *P. lobata* (Hakamatsuka *et al.*, 1988). The subunit of the chalcone synthase of *P. lobata* had a molecular weight of 43 000 and its native form is dimeric. Two dimensional electro-phoresis of purified chalcone synthase indicated that it was composed of more than five isozymes with equal molecular weight. cDNA and genomic DNA of the chalcone synthase gene were cloned from *P. lobata* cell cultures using a chalcone synthase cDNA of *Phaseolus vulgaris* (Ryder *et al.*, 1984) as a probe, and the longest cDNA clone contained a complete open reading frame of 1170 bp encoding 389 amino acids that would predict a protein of almost the same size as the purified enzyme (Nakajima *et al.*, 1991, 1996). Within the translated regions, the cDNA sequences of chalcone synthase from

*P. lobata* and *Phaseolus vulgaris* shared 90.6 per cent homology at nucleotide level and 93.4 per cent homology at amino acid level without deletions or insertions. Southern blot analysis using full-length chalcone synthase cDNA as a probe showed that the genome of *P. lobata* cells contains 6–7 copies of chalcone synthase gene. The 5′ untranslated region of one genomic clone contains several sequences found in the promoters of stress-inducible phenylpropanoid biosynthetic genes identified in other plant species. The reductase co-acting with chalcone synthase was also purified to apparent homogeniety from *P. lobata* cell cultures. It was a monomeric polypeptide with MW of 35 000 and had at least two isozymes of the same size. The reductase cDNA was cloned from *P. lobata* cell cultures elicited with yeast extract by RT-PCR using primers based on the sequence of soybean reductase (Welle *et al.*, 1991). Differential response to yeast extract and $CuCl_2$ at transcriptional level has been demonstrated by Northern blot hybridization using *P. lobata* cDNA as a probe. Southern blot hybridization of genomic DNA fragments revealed the presence of four copies of reductase genes in *P. lobata* genome (Ebizuka *et al.*, unpublished results). The deoxychalcone synthase is a unique polyketide synthetic enzyme that affords a deoxy-type compound in the presence of NADPH and a hydroxy-type compound in its absence. In the reactions of other polyketide synthetic enzymes, such as 6-methylsalicyclic acid synthase as well as fatty acid synthase that seems to be the prototype of polyketide synthetic enzymes, the omission of NADPH from the reaction mixture results in the termination of chain elongation and the release of an incomplete intermediate.

### Flavanone: central intermediate of flavonoid biosynthesis

Although chalcone is the first flavonoid-specific intermediate, the immediate precursor toward different subclasses of flavonoids is flavanone. Flavanone is subjected to enzymatic modifications including reduction, oxidation, hydroxylation, and rearrangement to form various types of flavonoids such as flavones, flavonols, anthocyanidins, tannins and isoflavones (Figure 7.1). The stereospecific conversion of chalcone to (2*S*)-flavanone is catalyzed by chalcone–flavanone isomerase that is well studied in many plant systems at enzyme and gene levels (Heller and Forkmann, 1994). In *P. lobata* cell cultures, the chalcone–flavanone isomerase activity is constitutively detected, although it is enhanced several folds by the yeast-extract elicitor. It is quite a contrast to chalcone synthase and isoflavone synthase whose activities are hardly detectable without elicitation. The chalcone–flavanone isomerase in *P. lobata* cells was a monomer polypeptide with MW of 28 000 and was composed of two isozymes with equal molecular weight. cDNA cloning of chalcone–flavanone isomerase gene was also achieved in *P. lobata* and high homology with the one from *Phaseolus vulgaris* (Mehdy and Lamb, 1987) was observed at amino acid (88.9 per cent) and gene (89.9 per cent) levels (Terai *et al.*, 1996).

### Oxidative aryl migration in isoflavone formation

Flavonoids and isoflavonoids share the biosynthetic pathway until flavanone stage, i.e. general phenylpropanoid pathway, construction of chalcone and its isomerization to flavanone. The first reaction specific to isoflavonoid biosynthesis is oxidative aryl migration. The enzymatic activity was first detected in a microsomal preparation from cell suspension cultures of *Glycine max* (Hagmann and Grisebach, 1984). (2*S*)-Naringenin (5,7,4′-trihydroxyflavanone: 5-hydroxy type) was converted into genistein (5,7,4′-trihydroxyisoflavone) by the microsomal fraction in the presence of NADPH and molecular oxygen. (2S)-Liquiritigenin (7,4′-dihydroxyflavanone: 5-deoxy type) also served as a substrate for the enzyme, being converted to daidzein (7,4′-dihydroxyisoflavone). From its cofactor requirements and inhibition by low concentrations of cytochrome C, it appeared that the microsomal isoflavone synthase might be a cytochrome P-450-dependent monooxygenase. Furthermore, an intermediate in the transformation of (2S)-naringenin (flavanone) to genistein (isoflavone) was isolated and identified as 2,5,7,4′-tetrahydroxy-isoflavanone (2-hydroxyisoflavanone) by a careful analysis of mass and UV spectra (Kochs and Grisebach, 1986). The intermediate was easily converted into genistein by treatment with a soluble protein fraction from the same source. The conversion of (2S)-naringenin into 2,5,7,4′-tetrahydroxyisoflavanone required NADPH and molecular oxygen, but the formation of genistein from this intermediate required none of these cofactors. The former reaction was inhibited by typical inhibitors of cytochrome P-450, while the latter was not affected by these inhibitors. It is evident, therefore, that the reaction catalyzed by isoflavone synthase consists of two steps. The first step is an oxidative aryl migration of flavanone to yield 2-hydroxyisoflavanone, which is catalyzed by a cytochrome P-450 dependent-monooxygenase. The second step is catalyzed by a dehydratase in a soluble protein fraction to introduce a double bond between C-2 and C-3.

The enzymatic synthesis of isoflavonoid skeleton was also studied in a microsomal fraction of *P. lobata* (Hakamatsuka *et al.*, 1989, 1990). Because of the contamination of highly active chalcone–flavanone isomerase activity in the microsomal preparation, the transformation of both isoliquiritigenin (4,2′,4′-trihydroxychalcone: 6′-deoxychalcone) and liquiritigenin (7,4′-dihydroxyflavanone: 5-deoxyflavanone) into daidzein (7,4′-dihydroxyisoflavone: 5-deoxyisoflavone) was observed in the presence of NADPH and molecular oxygen. The competition experiments using [$^3$H]flavanone and [$^{14}$C]chalcone demonstrated that flavanone, and not chalcone, is the immediate substrate for the enzymatic synthesis of isoflavone. An intermediate of the reaction, which corresponds to 2,5,7,4′-tetrahydroxyisoflavanone described above, was also detected when the microsomal preparation was carefully washed by repeated ultracentrifugation to remove the contaminant soluble enzymes (Hashim *et al.*, 1990). Analysis of mass, UV and $^1$H-NMR spectra of the isolated intermediate rigorously established its structure as 2,7,4′-tri-hydroxyisoflavanone (2-hydroxyisoflavanone). Identification of the reaction intermediate as 2-hydroxyisoflavanone led us to look into the origin of the oxygen atom of the 2-hydroxy group. To clarify biosynthetic origin of the 2-hydroxy group in 2,7,4′-trihy-droxyisoflavanone, incorporation experiment using a washed microsomal fraction under an atmosphere of $^{18}$O gas was performed (Hashim *et al.*, 1990). Analysis of mass spectra of 2,7,4′-trihydroxyisoflavanone enzymatically synthesized in the presence of NADPH and $^{18}$O$_2$ indicated that $^{18}$O was incorporated solely into the 2-hydroxyl group of 2,7,4′-trihydroxyisoflavanone. From these results, we proposed a new mechanism for the cytochrome P-450 mediated hydroxylation associated with rearrangement in the conversion of flavanone to 2-hydroxyisoflavanone (Hakamatsuka *et al.*, 1991b) (Figure 7.4). Hydroxylation by cytochrome P-450 on non-activated carbon is normally initiated by abstraction of a hydrogen atom, which is followed by recombination with a hydroxy radical at the same carbon (Mansuy *et al.*, 1989). In the case of P-450 hydroxylation of flavanone to yield 2-hydroxyisoflavanone, the abstraction of a hydrogen atom at C-3 of flavanone is followed by 1, 2-shift of an aryl group leaving a carbon radical at C-2. Recombination of hydroxyl radical takes place at C-2 to give 2-hydroxyisoflavanone.

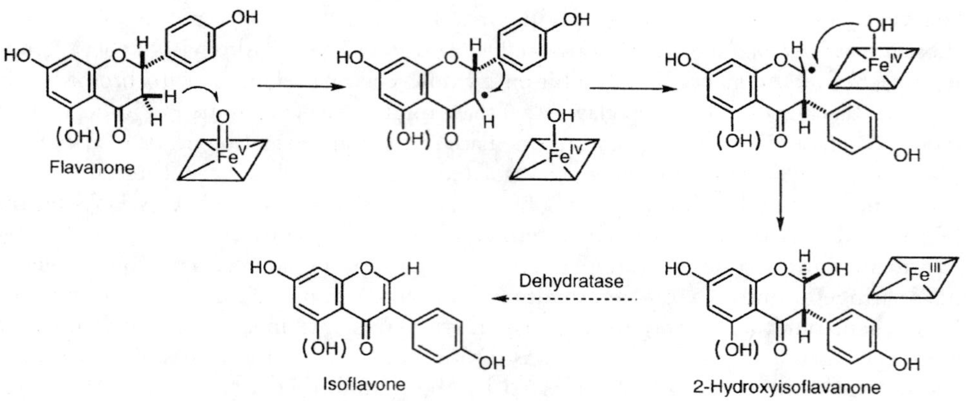

*Figure* 7.4 Proposed reaction mechanism for the oxidative aryl migration in isoflavone biosynthesis catalyzed by cytochrome P-450 enzyme system.

There are precedents in P-450 catalyzed reactions where bond formation between hydroxy radical and migrated radical center is involved, as exemplified in oxidation of cyclohexene, methylenecyclohexane, and β-pinene (Groves and Subramanian, 1984).

Purification of the P-450 protein involved in this reaction was attempted in order to obtain its amino acid sequence, expecting its potential use in cDNA cloning by reverse genetic approach (Hakamatsuka *et al.*, 1991b). The membrane-bound enzyme was first solubilized by treatment with non-ionic detergent Triton X-100 at a concentration of 2 per cent (w/v). Column chromatography on DEAE-Sepharose of the solubilized fraction gave a good separation of cytochrome P-450 from NADPH:cytochrome P-450 reductase. Further purification of cytochrome P-450 was carried out on aminooctyl-Sepharose and hydroxyapatite columns. Analysis of the final preparation from the hydroxyapatite column by SDS-PAGE showed the presence of three prominent bands with molecular weights range from 50 000 to 55 000. Because the cytochrome P-450 protein was highly unstable, further purification had to be abandoned. Recently, cDNA cloning of a P-450 encoding 2-hydroxyisoflavanone synthase was achieved by alternative approaches. In *Glycine max*, P-450 sequences were searched from an expression sequence tag database and the enzyme function of candidates were identified by heterologous expression in insect cells (Steele *et al.*, 1999). In *Glycyrrhiza echinata*, P-450 fragments were isolated by a PCR-based method and the expressed proteins in yeast cells were tested for their enzyme activities (Akashi *et al.*, 1999). The cloning of 2-hydroxyisoflavanone synthase gene will facilitate detailed studies on molecular mechanism of aryl migration and furthermore enable future metabolic engineering of isoflavonoid producing and/or non-producing plants.

The dehydratase catalyzing the conversion of 2-hydroxyisoflavanone to isoflavone was purified to apparent homogeneity from yeast extract-treated cell suspension cultures of *P. lobata* (Hakamatsuka *et al.*, 1998). The Km value of the purified enzyme for 2,7,4'-trihydroxyisoflavanone was appreciably low (7 μM) and treatment of yeast extract elicitor induced its activity. Therefore, the purified enzyme appears to possess the physiological function in isoflavonoid biosynthesis and is not a simple non-specific dehydratase, although non-enzymatic dehydration of 2-hydroxyisoflavanone gradually progresses even at physiological temperature. This enzyme was a monomeric polypeptide with MW 38 000. The stereochemistry of the dehydratase reaction is not clear, because the configuration of the 2,7,4-trihydroxyisoflavanone has not been established.

## Pterocarpans as antimicrobial phytoalexins

Biosynthesis of pterocarpans has been studied mostly in leguminous plants producing antimicrobial isoflavonoid phytoalexins in the context of plant–microbe interactions. In uninfected tissues of *P. lobata*, basic isoflavones, such as daidzein (3) and genistein (4), are subjected to simple modifications like hydroxylation, methylation and glycosylation (Hakamatsuka *et al.*, 1994). Coumestrol (5) is the only constituent of healthy tissues that elaborates at later stage of isoflavonoid biosynthesis. Although *P. lobata* produces pterocarpan phytoalexins, namely tuberosin (6) and glycinol (7), against microbial attacks, biosynthetic informations on the later stages after formation of isoflavone skeleton in this plant are rather limited. From the studies in other leguminous plants, a series of biosynthetic enzymes that are involved in the formation of pterocarpan from isoflavone have been detected and characterized (Figure 7.5). The first enzyme of this pathway, isoflavone 2′-hydroxylase, which is a P-450 monooxygenase, introduces a hydroxyl group at C-2′ on the side phenyl group (Hinderer *et al.*, 1987). cDNA of this P-450 gene was cloned from *Glycyrrhiza echinata* (Akashi *et al.*, 1998). The next step is catalyzed by isoflavone reductase, which stereospecifically reduces 2′-hydroxyisoflavone to 2′-hydroxyisoflavanone (Tiemann *et al.*, 1987). cDNA cloning of the isoflavone reductase has been achieved in chickpea (Tiemann *et al.*, 1991) and alfalfa (Paiva *et al.*, 1991). Then, a carbonyl group at C-4 of the 2′-hydroxyisoflavanone is reduced to form 2′-hydroxyisoflavanol (Guo *et al.*, 1994). The reductase utilizes NADPH as a cosubstrate and cDNA encoding this enzyme was cloned from *Medicago sativa* (Guo and Paiva, 1995). Finally, the ring closure associated with dehydration generates a fused furan ring of pterocarpan (Guo *et al.*, 1994). The last two steps were formerly proposed to be catalyzed by a single enzyme, pterocarpan synthase (Bleb and Barz, 1988; Fischer *et al.*, 1990). Enzymes catalyzing further modifications, such as 6a-hydroxylation of pterocarpan skeleton (Hagmann *et al.*, 1984), subsequent transfer of prenyl groups (Zahringer *et al.*, 1979) and cyclization of prenyl groups (Welle and Grisebach, 1988b) have also been reported.

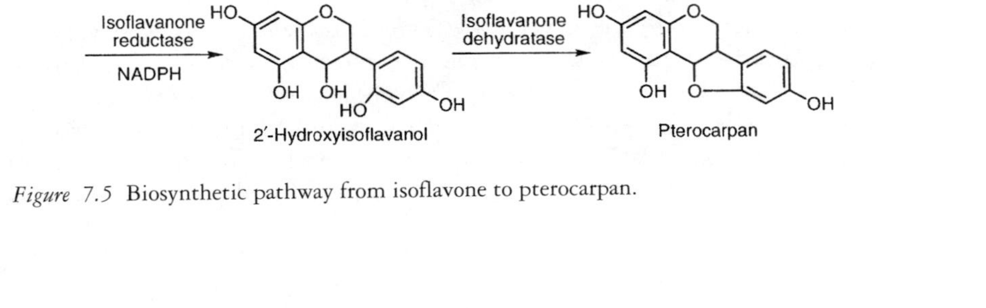

*Figure* 7.5 Biosynthetic pathway from isoflavone to pterocarpan.

## NATURAL FUNCTIONS OF *PUERARIA* ISOFLAVONOIDS

Like other isoflavonoids in leguminous plants, *Pueraria* isoflavonoids also play a major role in protection against microorganism. Isoflavonoids that accumulate in healthy tissues must act as constitutive defense substances. Upon treatment with elicitor to cell suspension cultures, the constitutive isoflavonoids were rapidly (within 4–6 h) incorporated into lignocellulose fractions of cell walls. They seem to contribute to physical and/or chemical defense by fortifying cell walls without gene activations. The experiments with exogenously applied isoflavones and flavones showed that the rapid response upon elicitation was selective for isoflavonoids and not for flavonoids (Park *et al.*, 1995a). Furthermore, exogenously added hydrogenperoxide stimulated the same response and rapid generation of $H_2O_2$ was detected within 10 min after elicitation at the cell surface. Pretreatment of some known inhibitors of NAD(P)H-oxidase, a $H_2O_2$ generating enzyme, resulted in almost complete inhibition of the early response upon elicitation. These results suggested that rapidly generated $H_2O_2$ plays a crucial role in the early defense response in *P. lobata* (Park *et al.*, 1995b). Soon after this early response involving pre-existing isoflavonoids and constitutive enzymes, *de novo* synthesis of biosynthetic enzymes starts and significant amount of isoflavonoids reaccumulate in the cells treated with elicitors if they are mild ones such as yeast extract and endogenous elicitor. On the other hand, as a response to violent elicitors such as fungal glycoprotein elicitor or $CuCl_2$, pterocarpan phytoalexins with strong antimicrobial activity accumulate in the treated cells. These inducible pterocarpans must be the major defense chemicals of *P. lobata* against microbial infections. However, like other plant metabolites, the true functions of constitutive and inducible isoflavonoids of *P. lobata* in natural habitat remain to be investigated or confirmed.

 *P. lobata* is important as a source of medicinals for human being. The dried roots of *P. lobata* are used, under the name of Ge Gen (Kakkon in Japanese), as an important crude drug in traditional Chinese medicine, and Ge Gen Tang (Kakkon-to in Japanese), a prescription containing Ge Gen, is one of the most commonly used medicines for the treatment of early symptoms of common cold. The main constituents of *P. lobata*, daidzein (3), daidzin (2) and puerarin (1) have been reported to stimulate cerebral and coronary blood circulation (Kinoshita, 1982). In addition, isoflavonoids have been shown to have various biological activities in animal cells. Genistein (4), another constituent of *P. lobata*, has been widely used by researchers in biochemical and pharmacological fields as a specific inhibitor of tyrosine kinase. Furthermore, many isoflavonoids containing plants exhibit estrogenic activities which may be the cause of infertility in livestocks, changing of bone density and sometimes disruption of the endocrine system. It is important to clarify the action mechanisms of these biological activities.

 Recent progress in biosynthetic studies on isoflavonoids will facilitate effective use of isoflavonoids for agricultural, medicinal and ecological purposes by producing transgenic plants with regulated isoflavonoid contents. Jung *et al.* (2000) expressed soybean 2-hydroxyisoflavanone synthase gene in *Arabidopsis thaliana*, which does not normally accumulate isoflavones, and the transgenic plant successfully produced isoflavone genistein (4). Because the biosynthetic pathway of flavonoids exist in most plant, introduction of single gene of 2-hydroxyisoflavanone synthase would engineer new transgenic plants producing isoflavonoids. Although development of novel isoflavonoid-producing transgenic crop species would become feasible from the agricultural point of view, it is necessary to observe and evaluate various biological activities of isoflavonoids produced carefully and continuously.

## REFERENCES

Akashi, T., Aoki, T. and Ayabe, S. (1998) CYP81E1, a cytochrome P450 cDNA of licorice (*Glycyrrhiza echinata* L.), encodes isoflavone 2 hydroxylase. *Biochem. Biophys. Res. Commun.*, 251, 67–70.

Akashi, T., Aoki, T. and Ayabe, S. (1999) Cloning and functional expression of a cytochrome P-450 cDNA encoding 2-hydroxyisoflavanone synthase involved in biosynthesis of the isoflavonoid skeleton in Licorice. *Plant Physiol.*, 121, 821–828.

Ayabe, S., Udagawa, A. and Furuya, T. (1988) NAD(P)H-dependent 6-deoxychalcone synthase activity in *Glycyrrhiza echinata* cells induced by yeast extract. *Arch. Biochem. Biophys.*, 261, 458–462.

Bailey, J.A. (1980) Constitutive elicitors from *Phaseolus vulgaris*: a possible cause of phytoalexin accumulation. *Ann. Phytopathol.*, 12, 395–402.

Bleb, W. and Barz, W. (1988) Isolation of pterocarpan synthase, the terminal enzyme of pterocarpan phytoalexin biosynthesis in cell suspension cultures of *Cicer arietinum*. *FEBS Lett.*, 235, 47–50.

Fischer, D., Ebenau-Jehle, C. and Griseback, H. (1990) Purification and characterization of pterocarpan synthase from elicitor challenged soybean cell cultures. *Phytochemistry*, 29, 2879–2882.

Groves, J.T. and Subramanian, D.V. (1984) Hydroxylation by cytochrome P-450 and metalloporphyrin models. Evidence for allylic rearrangement. *J. Am. Chem. Soc.*, 106, 2177–2181.

Guo, L., Dixon, R.A. and Paiva, L. (1994) Conversion of vestitone to medicarpin in alfalfa (*Medicago sativa* L.) is catalyzed by two independent enzymes. Identification, purification, and characterization of vestitone reductase and 7,2 dihydroxy-4 methoxyisofalvanol dehydratase. *J. Biol. Chem.*, 269, 22372–22378.

Guo, L. and Paiva, L. (1995) Molecular cloning and expression of alfalfa (*Medicago sativa* L.) vestitone reductase, the penultimate enzyme in medicarpin biosynthesis. *Arch. Biochem. Biophys.*, 320, 353–360.

Hagmann, M. and Grisebach, H. (1984) Enzymatic rearrangement of flavanone to isoflavone. *FEBS Lett.*, 175, 199–202.

Hagmann, M., Heller, W. and Grisebach, H. (1984) Induction of phytoalexin synthesis in soybean. Stereospecific 3,9-dihydroxypterocarpan 6a-hydroxylase from elicitor-induced soybean cell cultures. *Eur. J. Biochem.*, 142, 127–131.

Hahn, M.G., Darvill, A.G. and Albersheim, P. (1981) Host-pathogen interactions. XIX. The endogenous elicitor, a fragment of a plant cell wall polysaccharide that elicits phytoalexin accumulation in soybeans., 68, 1161–1169.

Hakamatsuka, T., Noguchi, H., Ebizuka, Y. and Sankawa, U. (1988) Deoxychalcone synthase from cell suspension cultures of *Pueraria lobata*. *Chem. Pharm. Bull.*, 36, 4225–4228.

Hakamatsuka, T., Noguchi, H., Ebizuka, Y. and Sankawa., U. (1989) Isoflavone synthase from cell suspension cultures of *Pueraria lobata*. *Chem. Pharm. Bull.*, 37, 249–252.

Hakamatsuka, T., Noguchi, H., Ebizuka, Y. and Sankawa., U. (1990) Isoflavone synthase from cell suspension cultures of *Pueraria lobata*. *Chem. Pharm. Bull.*, 38, 1942–1945.

Hakamatsuka, T., Ebizuka, Y. and Sankawa, U. (1991a) Induced isoflavonoids from copper chloride-treated stems of *Pueraria lobata*. *Phytochemistry*, 30, 1481–1482.

Hakamatsuka, T., Hashim, M.F., Ebizuka, Y. and Sankawa, U. (1991b) P-450-dependent oxidative rearrangement in isoflavone biosynthesis: reconstitution of P-450 and NADPH: P-450 reductase. *Tetrahedron*, 47, 5969–5978.

Hakamatsuka, T., Ebizuka, Y. and Sankawa, U. (1994) XXIII *Pueraria lobata* (Kudzu vine): *in vitro* culture and the production of isoflavonoids. In Y.P.S. Bajaj (ed.), *Biotechnology in Agriculture and Forestry*, Vol. 28, Medicinal and Aromatic Plants VII, Springer-Verlag, Berlin Heidelberg, pp. 386–400.

Hakamatsuka, T., Mori, K., Ishida, S., Ebizuka, Y. and Sankawa, U. (1998) Purification of 2-hydroxyisoflavanone dehydratase from the cell cultures of *Pueraria lobata. Phytochemistry*, 49, 497–505.

Hashim, M.F., Hakamatsuka, T., Ebizuka, Y. and Sankawa, U. (1990) Reaction mechanism of oxidative rearrangement of flavanone in isoflavone biosynthesis. *FEBS Lett.*, 271, 219–222.

Heller, W. and Forkmann, G. (1994) Biosynthesis. In J.B. Harborne (ed.), *The Flavonoids, Advances in Research since 1980*, Chapman and Hall, London and New York, pp. 399–425.

Hinderer, W., Flentje, U. and Barz, W. (1987) Microsomal isoflavone 2'- and 3'-hydroxylases from chickpea (*Cicer arietinum* L.) cell suspensions induced for pterocarpan phytoalexin formation. *FEBS Lett.*, 214, 101–106.

Ingham, J.L. (1982) Phytoalexins from the Leguminosae. In J.A. Bailey and J.W. Mansfield, (eds), *Phytoalexins*, Blackie, Glasgow and London, pp. 21–80.

Jung, W., Yu, O., Leu, S.-M.C., O eefe, D.P., Odell, J., Feder, G. and McGonigle, B. (2000) Identification and expression of isoflavone synthase, the key enzyme for biosynthesis of isoflavones in legumes. *Nature Biotech.*, 18, 208–212.

Keen, N.T. and Legrand, M. (1980) Surface glycoproteins: evidence that they may function as the race specific phytoalexin elicitors of *Phytophthora megasperma f.sp. glycinea. Physiol. Plant Pathol.*, 17, 175–192.

Kinoshita, T. (1982) Chemistry of *Puerariae Radix. Gendai Toyo Igaku*, 3, 58–62.

Kochs, G. and Grisebach, H. (1986) Enzymic synthesis of isoflavones. *Eur. J. Biochem.*, 155, 311–318.

Mansuy, D., Battioni, P. and Battioni, J.-P. (1989) Chemical model systems for drug-metabolizing cytochrome-P-450-dependent monooxygenases. *Eur. J. Biochem.*, 184, 267–285.

Mehdy, M.C. and Lamb, C.J. (1987) Chalcone isomerase cDNA cloning and mRNA induction by fungal elicitor, wounding and infection. *EMBO J.*, 6, 1527–1533.

Nakajima, O., Akiyama, T., Hakamatsuka, T., Shibuya, M., Noguchi, H., Ebizuka, Y. and Sankawa, U. (1991) Isolation, sequence and bacterial expression of a cDNA for chalcone synthase from the cultured cells of *Pueraria lobata. Chem. Pharm. Bull.*, 39, 1911–1913.

Nakajima, O., Shibuya, M., Hakamatsuka, T., Noguchi, H., Ebizuka, Y. and Sankawa, U. (1996) cDNA and genomic DNA clonings of chalcone synthase from *Pueraria lobata. Biol. Pharm. Bull.*, 19, 71–76.

Paiva, N.L., Edwards, R., Sun, Y., Hradina, G. and Dixon, R.A. (1991) Stress responses in alfalfa (*Medicago sativa* L.) 11. Molecular cloning and expression of alfalfa isoflavone reductase, a key enzyme of isoflavonoid phytoalexin biosynthesis. *Plant Mol. Biol.*, 17, 653–667.

Park, H.-H., Hakamatsuka, T., Noguchi, H., Sankawa, U. and Ebizuka, Y. (1992) Isoflavone glucosides exist as their 6″-O-malonyl esters in *Pueraria lobata* and its cell suspension cultures. *Chem. Pharm. Bull.*, 40, 1978–1980.

Park, H.-H., Hakamatsuka, T., Sankawa, U. and Ebizuka, Y. (1995a) Rapid metabolism of isoflavonoids in elicitor-treated cell suspension cultures of *Pueraria lobata. Phytochemistry*, 38, 373–380.

Park, H.-H., Hakamatsuka, T., Sankawa, U. and Ebizuka, Y. (1995b) Involvement of oxidative burst in isoflavonoid metabolism in elicited cell suspension cultures of *Pueraria lobata. Z. Naturforsch.*, 50c, 824–832.

Ryder, T.B., Cramer, C.L., Bell, J.N., Robbins, M.P., Dixon, R.A. and Lamb, C.L. (1984) Elicitor rapidly induces chalcone synthase mRNA in *Phaseolus vulgaris* cells at the onset of the phytoalexin defense response. *Proc. Natl. Acad. Sci. USA*, 81, 5724–5728.

Steele, C.L., Gijzen, M., Qutob, D. and Dixon, R.A. (1999) Molecular characterization of the enzyme catalyzing the aryl migration reaction of isoflavonoid biosynthesis in Soybean. *Arch. Biochem. Biophys.*, 367, 146–150.

Stoessl, A. and Stothers, J.B. (1979) The incorporation of [1,2-$^{13}$C$_2$]acetate into pisatin to establish the biosynthesis of its polyketide moiety. *Z. naturforsch.*, **34c**, 87–89.

Takeya, K. and Itokawa, H. (1982) Isoflavonoids and the other constituents in callus tissues of *Pueraria lobata. Chem. Pharm. Bull.*, **30**, 1496–1499.

Terai, Y., Fujii, I., Byun, S.-H., Nakajima, O., Hakamatsuka, T., Ebizuka, Y. and Sankawa, U. (1996) Cloning of chalcone-flavanone isomerase cDNA from *Pueraria lobata* and its overexpression in *Escherichia coli. Protein Exp. Purif.*, **8**, 183–190.

Tiemann, K., Hinderer, W. and Barz, W. (1987) Isolation of NADPH:isoflavone oxidoreductase, a new enzyme of pterocarpan phytoalexin biosynthesis in cell suspension cultures of *Cicer arietinum. FEBS Lett.*, **213**, 324–328.

Tiemann, K., Inze, D., Montagu, M. and van Barz, W. (1991) Pterocarpan phytoalexin biosynthesis in elicitor-challenged chickpea (*Cicer arietinum* L.) cell cultures. Purification, characterization and cDNA cloning of NADPH:isoflavone oxidoreductase. *Eur. J. Biochem.*, **200**, 751–757.

Welle, R. and Grisebach, H. (1988a) Isolation of a novel NADPH-dependent reductase which coacts with chalcone synthase in the biosynthesis of 6 deoxychalcone. *FBES Lett.*, **236**, 221–225.

Welle, R. and Grisebach, H. (1988b) Induction of phytoalexin synthesis in soybean: Enzymatic cyclization of prenylated pterocarpans to glyceollin isomers. *Arch. Biochem. Biophys.*, **263**, 191–198.

Welle, R., Schroeder, G., Schiltz, E., Grisebach, H. and Schroeder, J. (1991) Induced plant responses to pathogen attack. Analysis and heterologous expression of the key enzyme in the biosynthesis of phytoalexins in soybean (*Glycine max* L. Merr. cv. Harosoy 63). *Eur. J. Biochem.*, **196**, 423–430.

Zahringer, U., Ebel, J., Mulheirn, L.J., Lyne, R.L. and Grisebach, H. (1979) Induction of phytoalexin synthesis in soybean. Dimethylallylpyrophosphate:trihydroxypterocarpan dimethylallyl transferase from elicitor-induced cotyledons. *FEBS Lett.*, **101**, 90–92.

# 8 Pharmacological effects of *Pueraria* isoflavones on cardiovascular system

*Siang-shu Chai, Ai-ping Zhao, and Guang-Yao Gao*

## INTRODUCTION

In ancient China, *Radix puerariae* (RP, root of kudzu) was mostly used for the treatment of early symptoms of acute febrile diseases such as dizziness and stiffness in neck and back, lack of perspiration, and aversion to air draft (*Shang Han Lun, c.* AD 200). The study of the pharmacological action of RP on the cardiovascular system is a more recent adventure, dating back only to the 1970s when a medical group from the Chinese Academy of Medical Sciences went to Capital Iron and Steel Plant to conduct a research project on the prevention and treatment of hypertension among the factory workers. The major task was to devise a treatment method that would effectively relieve the symptoms that accompanied hypertension. The most prominent symptoms observed among these patients included headache, dizziness, stiffness and soreness in the neck, which persisted even after blood pressure was brought under control through conventional medications. This reminded the medical team of the indications of RP and led to a clinical trial testing the efficacy of RP in the management of the symptoms accompanied hypertension. The result was so impressive that it triggered a systematic and comprehensive research program that spanned across multiple disciplines from phytochemistry to pharmacology. Preliminary studies led to the belief that the active ingredients of RP were isoflavonoids. Later, some of these isoflavones were isolated and identified and their pharmacological activities on cardiovascular system were studied (Fang, 1980).

Early pharmacological studies of RP and its active principles concentrated mainly on their antihypertensive property. Fan and his colleagues (Fong *et al.*, 1974; Zeng *et al.*, 1974) confirmed its antihypertensive effect in dogs. Later, Lu and his colleagues demonstrated that RP is an antagonist of $\beta$-adrenergic receptor and attributed RP's antihypertensive effect to its $\beta$-blocking activity (Lu *et al.*, 1980). In the meantime, other beneficial effects of RP on cardiovascular diseases such as myocardiac infarction and angina pectoris were also discovered (Zeng *et al.*, 1974; Zhou *et al.*, 1977). In 1984, a systematic study of puerarin on ischemic myocardium and hemodynamics was carried out by Fan *et al.* (1984). Further, Lu and Chai using classical screening test (Turner, 1971) confirmed that puerarin, the major active constituent of RP, exhibits $\beta$-adrenergic blocking activity (Lu and Chai, 1987).

In recent years, emphasis on the study of the pharmacological actions of RP has shifted from cardio- to cerebral vascular system. Chai and his colleagues discovered that puerarin could promote cerebral blood circulation (Chen *et al.*, 1993) and blood metabolism (Zhao and Chai, 1995). Meanwhile, Zhao and Shen also showed that it improves cerebral microcirculation (Zhao *et al.*, 1999) and hemorrheology (Shen *et al.*, 1996). Such results have been used to explain, at least in part, its efficacy in the treatment of headache and acute deafness, etc. On the basis of these results, Wang and Zhao further inferred that RP in some cases could prevent cerebral ischemia and stroke (Wang *et al.*, 1997; Zhao *et al.*, 1999). To date, RP and RP-based medications have been used not only for the treatment of cardiovascular diseases but also for the improvement of cerebral vascular function in China.

In this chapter, we intent to provide a brief overview on the pharmacological effects of RP, its total isoflavonoid fraction and puerarin on cardiovascular system and their therapeutic potential in the management of cardiovascular diseases.

## EFFECTS OF CRUDE RP EXTRACT AND ITS ISOFLAVONES ON HYPERTENSION

*Radix pueraria* was shown to lower blood pressure in both normal and hypertensive laboratory animals. In normal anesthetized dog, intravenous administration of crude extract of RP, its total isoflavones or a pure preparation of puerarin elicited a short and quick reduction in blood pressure. In normal unanesthetized dogs, orally administered a crude water extract (2 g/kg), an alcohol extract (2 g/kg), a total isoflavone fraction or a puerarin preparation produced similar effect on blood pressure (Zeng *et al.*, 1974; Fan, 1977).

When puerarin was administered intraperitoneally (100 mg/kg) to spontaneous hypertension rat (SHR), both blood pressure and heart rate (Figure 8.1) were lowered as compared to the controls. Such effect was more potent in SHR than in normal Wistar rat (Song *et al.*, 1988). It has also been shown that puerarin significantly reduced plasma renin activity (PRA) in SHR by radioimmunoassay, reduced it by 67 per cent from $2.1 \pm 0.45$ ng/ml·h to $0.73 \pm 0.20$ ng/ml·h. However, the effect of puerarin on PRA was not observed in normal rats.

Other investigators showed that a *pueraria* isoflavone daidzein, when administered orally as a PVP (Polyvinylpyrrolidone)-dispersed product to SHR (300 mg/kg), reduced blood pressure of the test animals. Maximal reduction was observed 2 h after administration. The systolic pressure was reduced by 12.8 per cent, whereas no obvious change was found in control animals receiving a simple mixture of daidzein and PVP in amounts equivalent to those in daidzein–PVP dispersion preparation. The daidzein–PVP dispersion reduced PRA by 52.4 per cent and 24.7 per cent in SHR and Wistar rat, respectively. It has also been shown to lower systolic and diastolic pressure by $18.6 \pm 10.6$ per cent and $43.0 \pm 19.2$ per cent, respectively, when administered intravenously (25 mg/kg) in anesthetized cat (Guo *et al.*, 1995). In anesthetized rabbits, daidzein-PVP preparation elicited an early but weak antihypertensive effect (Guo *et al.*, 1995).

The mechanism by which RP elicits its antihypertensive effect is still largely obscure. *Radix pueraria* might antagonize blood pressure elevation caused by isoproterenol, and decrease or eliminate blood pressure elevation effect of epinephrine. It has been shown that RP extract antagonized elevated heart rates induced by isoproterenol in various animal models. On the basis of these results, it was suggested that RP may act as a $\beta$-adrenergic blocker (Lu *et al.*, 1980).

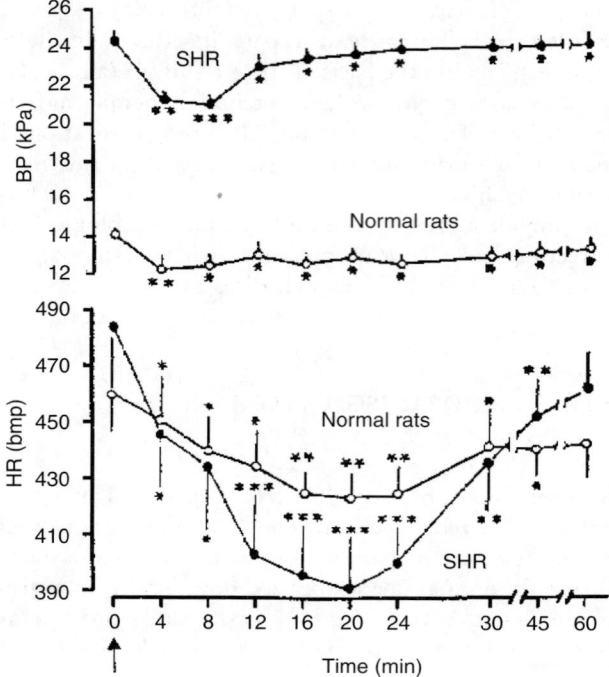

*Figure 8.1* Effect of puerarin (↑) on blood pressure (BP) and heart rate (HR) of spontaneous hypertensive rat. $N = 8$, * $p > .05$, ** $p < .05$, *** $p < .01$.

## EFFECTS OF RP ISOFLAVONES ON BLOOD VESSELS AND CORONARY CIRCULATION

The total isoflavonoid fraction of RP and a pure puerarin preparation were shown to cause significant relaxation in normal and experimentally induced constricted coronary blood vessels. Intravenous injection of the total isoflavonoid fraction (30 mg/kg) to normal dogs increased coronary blood flow (CBF) by 40 per cent and decrease vessel resistance (VR) by 29 per cent. When injected directly into the coronary artery (1 mg/kg), isoflavonoids increased CBF by 92 per cent and decreased VR by 42 per cent (Fan, 1977). Both effects were dose-dependent. Similar results on CBF reduction and VR increase have also been shown by other investigators using pharmacological models simulated with vasopressin (Yue and Hu, 1996).

Effects of puerarin on cardiac hemodynamics and coronary circulation have been studied using polygraph system and electromagnetic flowmeter, etc. (Fan *et al.*, 1982; Wang *et al.*, 1984) in open chest anesthetized dogs at two different doses, 25 mg/kg and 50 mg/kg. Intravenous injection of puerarin decreased blood pressure (BP), heart rate (HR), cardiac output (CO), systolic index (SI), left ventricular pressure (LVP), velocity of maximal change ($dp/dt_{max}$), total peripheral resistance (TPR) and left

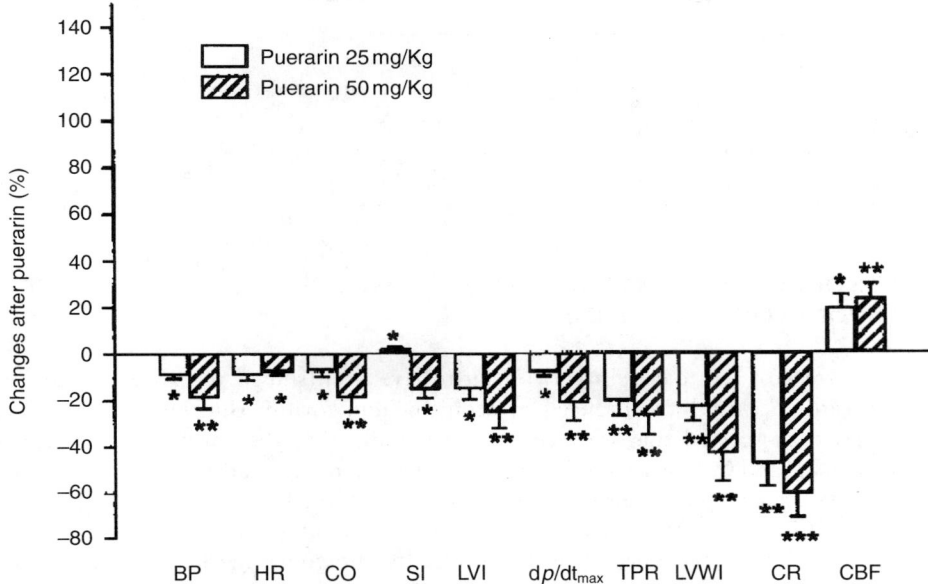

*Figure 8.2* Effect of puerarin on cardiac hemodynamics and coronary circulation in anesthetized dogs. $N = 6$, * $p > .05$, ** $p < .05$, *** $p < .01$.

ventricular work index (LVWI). All of these effects appeared to be dose-dependent (Figure 8.2).

It appeared that both the total isoflavonoid fraction of RP and the pure puerarin preparation exhibit beneficial effects on coronary vessels and circulation. The mechanism by which RP or puerarin increase coronary circulation may be related to responsive heterogeneity of vessels other than $\beta$-adrenergic receptor blocking effect.

## EFFECTS OF RP ISOFLAVONES ON ISCHEMIC MYOCARDIUM

Intravenous injection of the total isoflavonoid fraction of RP to anesthetized dog significantly increased content of blood oxygen in coronary node. Content of lactate and pyruvate also increased but the lactate–pyruvate ratio remained unchanged (Fan, 1977). This suggested that the metabolism of glucose was not affected. Another study also confirmed that RP isoflavones improved metabolism in both normal and ischemic myocardium (Lou and Qin, 1995).

The protective effects of puerarin on ischemic myocardium in dog with acute myocardial infarction (AMI) were demonstrated by pericardial ECG, plasma activity of creatine kinase (CK), radiocardiogram and myocardial staining (Li *et al.*, 1984; Liu *et al.*, 1998). Following intravenous injection of puerarin (20 mg/kg) at 5 min after acute coronary ligation there were a marked decrease in AMI size in the puerarin group (Table 8.1) and significant improvement on all the above indication of AMI.

*Table 8.1* Effect of intravenous puerarin on acute myocardial
infarction (AMI) in dogs

| Treatment | Number of animals | MI size (in %) | p value |
|---|---|---|---|
| Control | 9 | 14.4 | |
| Puerarin (20 mg/kg) | 9 | 5.8 | <.01 |
| Propranolol (0.5 mg/kg) | 6 | 55 | <.01 |

## EFFECTS OF CRUDE RP EXTRACTS AND PURIFIED RP ISOFLAVONES ON ARRHYTHMIA

Alcohol extract of RP, daidzein and puerarin have been shown to exhibit protective effect on cardiac arrhythmia induced by aconitine and barium chloride (Yue and Hu, 1996). It was suggested that these agents act by affecting the permeation of cations such as $K^+$, $Na^+$ and $Ca^{2+}$, across cell membrane, decreasing myocardium excitability and as a consequence preventing cardiac arrhythmia (Guo *et al.*, 1995).

Puerarin (Chai *et al.*, 1985) has also been shown to antagonize cardiac arrhythmia induced by chloroform–epinephrine in rabbits. Its antagonizing effect on ventricular extrasystole and ventricular tachycardia induced by ouabain in guinea pigs were also reported (Table 8.2, Table 8.3). Puerarin was less effective in alleviating arrhythmia induced by aconitine (Table 8.4).

The effects of puerarin on cellular electrophysiology of cardiac muscle in guinea pig were studied using intracellular microelectrode (Tian *et al.*, 1986). Puerarin decreased

*Table 8.2* Effect of puerarin on cardiac arrhythmia induced
by chloroform–epinephrine in rabbits

| Treatment | Dose (mg/kg) | Number of animals | Duration of arrhythmia |
|---|---|---|---|
| Control | – | 5 | 237±36 |
| Puerarin | 100 | 10 | 106±50* |
| Propranolol | 1 | 5 | 44±42* |
| Practolol | 3 | 5 | 44±57* |

Note
* $p < 0.01$.

*Table 8.3* Effect of puerarin on arrhythmia induced by ouabin in Guinea pigs

| Treatment | Dose (mg/kg) | Number of animals | Dose (μg/kg) of ouabin | | |
|---|---|---|---|---|---|
| | | | VE | VT | VF |
| Control | – | 5 | 173±15 | 164±32 | 258±45 |
| Puerarin | 100 | 10 | 178±29*** | 208±40** | 280±45* |
| Propranolol | 1 | 5 | 233±19*** | 281±48*** | 379±25*** |
| Practolol | 3 | 5 | 224±31*** | 313±14*** | 413±17*** |

Notes
* $p > .05$, ** $p < .05$, *** $p < .01$.

*Table 8.4* Effect of puerarin on arrhythmia induced by aconitine in rabbits

| Treatment | Dose (mg/kg) | Number of animals | Latency before arrhythmia |
|---|---|---|---|
| Control | – | 5 | $89 \pm 23$ |
| Puerarin | 100 | 10 | $99 \pm 20^*$ |
| Propranolol | 1 | 5 | $326 \pm 29^*$ |
| Practolol | 3 | 5 | $186 \pm 50^*$ |

Note
* $p < .01$.

both the amplitude and duration of action potential and reduced effective refractory period of cardiac muscle. While resting membrane potential was not changed, the spontaneous discharge of cardiac muscle attenuated.

Chloroform–epinephrine induces cardiac arrhythmia by stimulating $\beta$-adrenergic receptor. In the process of arrhythmia induced by ouabain the sympathetic nerve tonus played an important role. The aconitine-induced arrhythmia may be caused by direct effect on heart muscle. Therefore, purerain was able to antagonize only the chloroform–epinephrine and ouabain induced arrhythmia.

## $\beta$-ADRENERGIC RECEPTOR BLOCKING EFFECT OF PUERARIN

Since many pharmacological activities of RP and RP-based medications on the cardiovascular system were, in one way or another, related to their $\beta$-blocker activities, significant amounts of efforts have been devoted to the elucidation of the $\beta$-blocker effect of puerarin, the major active principle of RP.

Antagonizing effect of puerarin on $\beta$-adrenergic receptor was first discovered during experiments using isolated organs containing $\beta$-adrenergic receptors such as the atria muscle of rabbit and trachea of guinea pig. Lu and Chai (1986) showed that puerarin produced a parallel rightward shift of the dose–response curves for its positive chronotropic effect on isoproterenol-induced spontaneous contraction of atria muscle and negative inotropic effect on trachea strips (Figure 8.3).

The $\beta$-adrenergic receptor blocking effect has also been confirmed in vascular smooth muscle (Wang *et al.*, 1994). In the cases of renal arteries and femoral veins of cat, puerarin inhibited the relaxation response of the strips to isoproterenol in a dose dependent manner. This response resembled that of the propranolol. Puerarin did not alter the relaxation response to nitroglycerine (Figure 8.4). It was demonstrated that puerarin was able to antagonize the effects induced by isoproterenol and showed $\beta$-adrenergic receptor blocking effect.

The $\beta$-blocker effect was also confirmed *in vivo* (Lu and Chai, 1987) in anesthetized cats and monkeys. After intravenous injection of isoproterenol (10 mg/kg), heart rate was increased by $46 \pm 7$ per cent and blood pressure reduced by $26 \pm 4$ per cent. When puerarin (40 mg/kg) was given intravenously in advance, its maximal effect occurred in 10–15 min, then isoproterenol was administered at the same dose as before, the response of heart rate and blood pressure induced by isoproterenol was diminished (Figure 8.5).

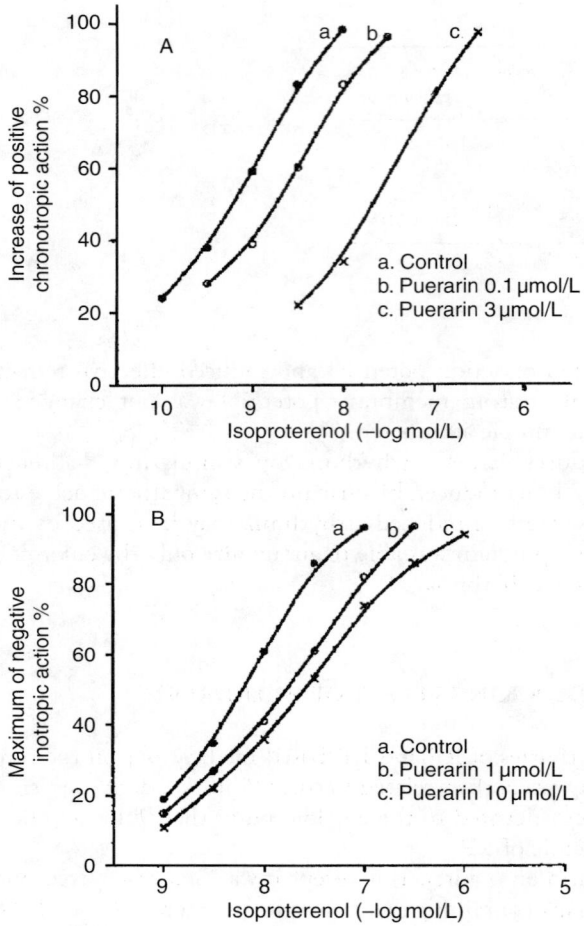

*Figure 8.3* Cumulative dose–response curve for isoproterenol in the presence of puerarin. (A) Heart atria of 6 rabbits, (B) Trachea of 6 guinea pigs.

The $\beta$-adrenergic receptor blocking effect of puerarin was also studied in detail in terms of its effect on radioligand binding capacity to membranes prepared from rat cardiac muscle and on the activity of adenylate cyclase of erythrocyte membranes of Beijing duck (Lou *et al.*, 1985). Results showed that intravenous injection of puerarin significantly decreased the maximal binding capacity of $\beta$-adrenergic receptors to [3]H-DHA and the binding capacity of those $\beta$-adrenergic receptor containing membranes which had been preincubated with increasing concentrations of puerarin to radioligand was markedly inhibited (Figure 8.6). Further, the increased activity of adenylate cyclase (AC) induced by epinephrine was completely abolished when the erythrocyte membranes containing AC were pre-incubated with puerarin (Figure 8.7). On the basis of these results, it was suggested that puerarin has $\beta$-adrenoceptor blocking effect.

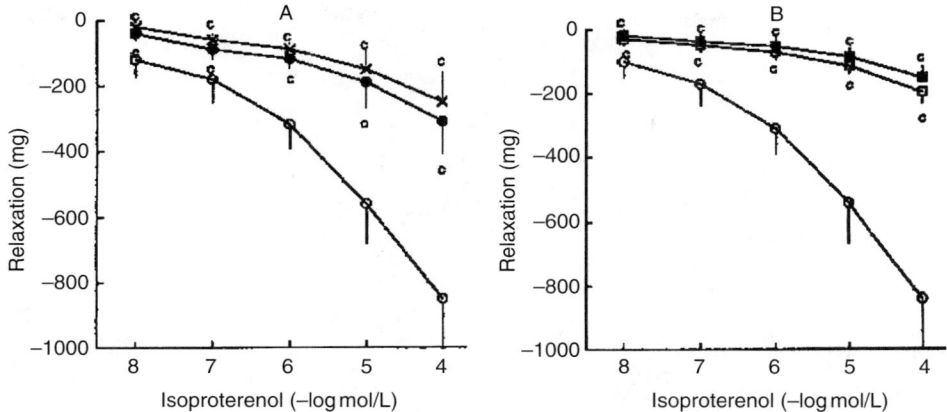

*Figure 8.4* Effects of puerarin 0.01(●), 0.1(×) mM; propranolol 0.1 (□), 1(■) mM and control (○) on isoproterenol-induced relaxation of methanoxamine contraction of renal artery in cats. N=6, "c"=$p < .05$ vs. control.

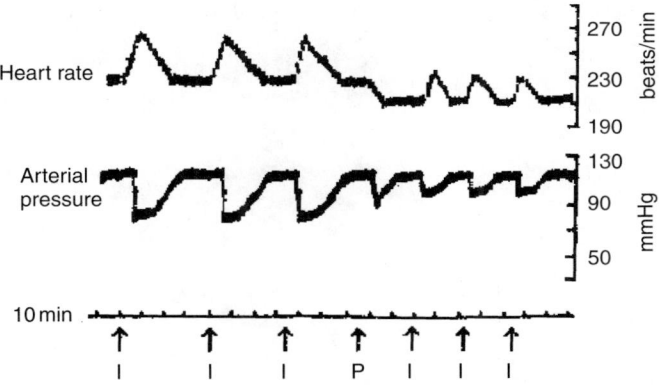

*Figure 8.5* Influence of iv puerarin (P) 40 mg/kg on the effect of isoproperenol (I) 10 µg/kg on heart rate and carotid arterial pressure of anesthetized cats.

Puerarin, as a β-adrenoceptor blocking agent, can be categorized by using conventional schemes as being cardioselective, having of certain intrinsic sympathemimetic activity and weak membrane stabilizing and local anesthetic properties.

The cardioselectivity of puerarin was conferred by the fact that its effect on atria muscle $\beta_1$ receptor was stronger than that on the trachea muscle $\beta_2$ receptor (Figure 8.3).

Puerarin could induce a decrease in normal heart rate and blood pressure in cats, but this effect was attenuated or reversed by the use of reserpine prior to the administration of puerarin. This fact showed that puerarin has certain intrinsic sympathemimetic action (Lu and Chai, 1986).

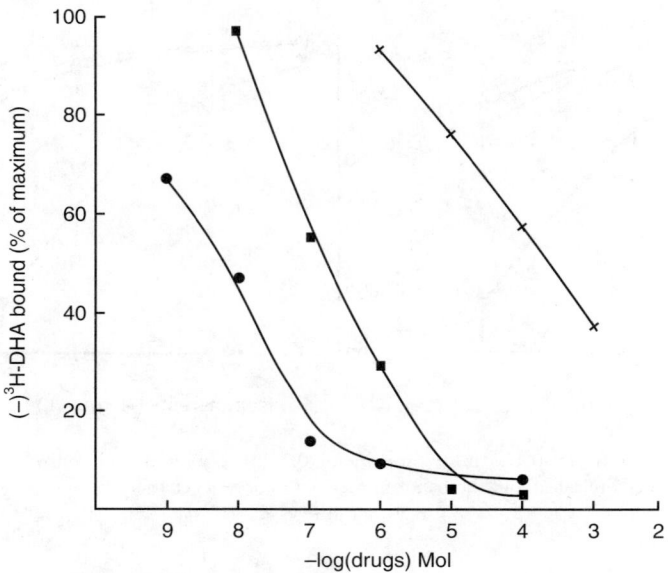

*Figure 8.6* Inhibition curve of the binding of $^3$H-DHA to rat cardiac muscle membrane for different β-adrenergic ligands, (●—●) propranolol (■—■) isoproterenol and (×—×) puerarin.

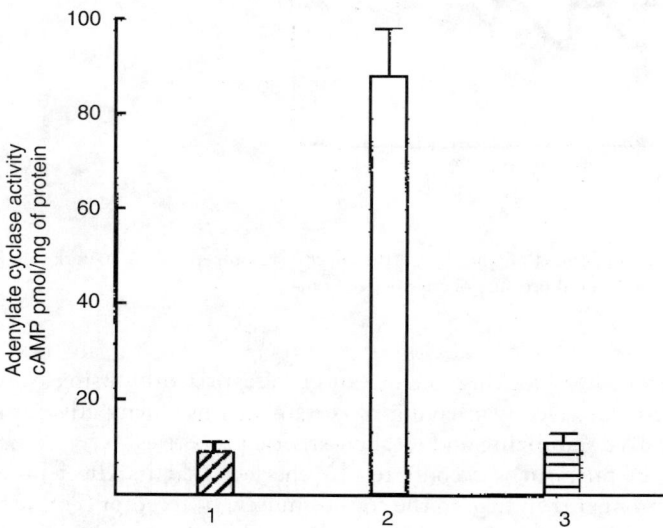

*Figure 8.7* Effect of puerarin on adenylate cyclase (AC) activity: (1) Basal AC activity of the membrane preparation, (2) AC activated by epinephrine, (3) AC activity of the membrane pre-incubated with puerarin.

The experiments on rabbit's corneal reflex and guinea pig's skin test showed that puerarin has weak local anesthetic properties (Chai *et al.*, 1985).

## CLINCAL USE OF KUDZU FOR CADIOVASCULAR DISEASES

The clinical application of RP and RP-based medications for cardiovascular disease was widespread in China. They are used mainly in the treatment of hypertension, angina pectoris and myocardial infarction.

### Treatment for hypertension

In human studies, conflicting results were obtained for the antihypertensive effect of RP. It was reported by some that water and alcoholic extract of RP, its total isoflavonoid fraction, daidzein or puerarin did not significantly reduce blood pressure in hypertensive patients (Lou and Qin, 1995). However, all of these preparations appeared to be effective in the relief of hypertension related symptoms such as headache, dizziness, stiffness in the neck and tinnitus. Such actions have been attributed to their abilities to improve cerebral blood circulation (Yue and Hu, 1996). However, another report showed that puerarin exhibited some antihypertensive effect. Puerarin was given (i.v.) to 50 cases of hypertension at a daily dose of 300 mg for 5 days, 76 per cent of them exhibited various degree of decrease in blood pressure and heart rates. Under this condition, no significant cardiac performance changes have been observed. It has been shown that rennin-angiotension system was suppressed by puerarin and this played an important role in its antihypertensive effect (Chen *et al.*, 1987).

### Treatment for angina pectoris

Since RP has been shown to improve heart function in multiple ways, RP and RP-based medications have also been tested in patients suffering from angina pectoris. A clinical study included 191 cases of angina pectoris and 177 cases of coronary heart diseases were treated with alcoholic extract of RP and daidzein independently, the results showed that the effective rates were 69–91 per cent and 79.1 per cent, respectively (Lou and Qin, 1995).

Zeng confirmed that symptoms of angina pectoris could be relieved by intravenous injection of puerarin. It was also shown that high level of plasma catecholamine found in these patients decreased simultaneously. It was suggested that reduction of catecholamine was a result of decreased lactic acid level in the myocardium (Zeng and Zhang, 1979). Another clinical study on 30 cases of angina pectoris patients reported that the total episode of chest pain experienced by these patients decreased from 325 per week among controls to 160 ($p < .01$) per week among puerarin (500 mg plus 5 per cent glucose 500 ml, i.v.) treated patients. In no cases were the conditions of these tested patients deteriorated during trial period (Yue and Hu, 1996).

### Treatment for myocardial infarction

Puerarin was tested in 67 cases of acute myocardial infarction (AMI) and shown to decrease the scope of AMI and improve several indications of the patients. The efficacy was similar to that observed in laboratory dogs (Li *et al.*, 1985). In addition, the modulation of puerarin on extrasystole was also studied in 12 patients. The effective rate was reported to be 53 per cent.

## REFERENCES

Chai, S.S., Wang, Z.X., Chen, P.P. and Wang, L.Y. (1985) Anti-arrhythmic action of puerarin. *Acta Pharmacol. Sin.*, 6, 166–168.

Chen, H.S., Wang, P.R. and Shao, J.F. (1987) The antihypertensive effect and mechanisms of puerarin. *Acta Acad. Med. Shandong*, 35, 28–33.

Chen, L.B., Chai, Q., Zhao, A.P. and Chai, S.S. (1993) Effects of β-adrenergic receptor blockers (Puerarin, propranolol) on cerebral blood flow in Dog. *J. Cereb. Blood Flow Metab.*, 13 (Suppl.1), S186.

Fan, L.L. (1977) Effect of kudzu (*Radix pueraria*) on coronary artery, dynamics of cardiac blood flow and myocardium metabolism in dog. *Chin. Med. J.*, 12, 724–725.

Fan, L.L., Zeng, G.Y., Zhou, Y.P., Zhang, L.Y. and Chen, Y.S. (1982) Pharmacologic studies on *Radix puerariae*. *Chin. Med. J.*, 95, 145–150.

Fan, L.L., Powel, W.W., O'Keefe, D.D. (1984) Effect of puerarin on regional myocardial blood flow and cardiac hemodynamics in dogs with acute myocardial ischemia. *Acta Pharmaceutica Sin.*, 19, 801–807.

Fang, Q.C. (1980) Some current study and research approaches relating to the use of plants in the traditional Chinese medicine. *J. Ethnopharmacol.*, 2, 57–63.

Fong, Q.Z., Lin, M., Sun, Q. M. *et al.* (1974) The studies on puerarin isoflavonoids. *Chin. Med. J.*, 54, 271.

Guo, J.P., Shun, Q.R. and Zhou, H. (1995) The pharmacological progress of *Radix puerariae*. *Chin. Tradit. Herbal Drugs*, 26, 163–165.

Li, X.I., Wang, P.R., Shao, J.F., Zue, X.R. and Zue, J.F. (1984) The effect of intravenous injection of puerarin on size of experimental AMI in dogs. *Acta Academiae Medicinae Shandong*, 22, 9–17.

Li, X.I., Wang, P.R. and Shao, J.F. (1985) The effect of puerarin on size of acute myocardial infarction patient. *Chin. J. Cardiovasc. Dis.*, 13, 175–178.

Liu, Q.G., Wang, L. and Lu, Z.Y. (1998) The protective effect of puerarin on ischemic myocardium in dogs and its possible mechanism. *J. Clin. Cardial. (China)*, 14, 292–295.

Lou, Z.Q. and Qin, B. (1995) Researches of Commonly Used Chinese Crude Herb medicines on Quality and Species (Northern part of China), Beijing Medical University and Peking Union Medical College Publishers, Beijing, pp. 380–420.

Lou, B.Z., Gao, E. and Shan, J.R. (1985) Effects of puerarin on the binding capacity of β-adrenergic receptor and the activity of adenylate cyclase. *Med. J. Chinese Peoples Liberation Army*, 10, 97–100.

Lu, X.R. and Chai, S.S. (1986) Blocking effect of puerarin on β-adrenoceptor of isolated organs and whole animal. *Acta Pharmacol. Sin.*, 7, 529–531.

Lu, X.R. and Chai, X.S. (1987) Puerarin β-adrenergic receptor blocking effect. *Chin. Med. J.*, 100, 25–28.

Lu, X.R., Chen, S.M. and Sun, T. (1980) Study of β-adrenergic receptor blocking effect of extracum of root *Pueraria hirsute*. *Acta Pharmacol. Sin.*, 15, 218–222.

Shen, X.L., Witt, M.R., Nielsen, M. and Sterner, O. (1996) Inhibition of $^3$H-flunitrazepam binding to rat brain membranes *in vitro* by puerarin and daidzein. *Acta Pharmaceutica Sin.*, 31, 59–62.

Song, X.P., Chen, P.P. and Chai, X.S. (1988) Effects of puerarin on blood pressure and plasma rennin activity in spontaneunsly hypertensive rats. *Acta Pharmacol. Sin.*, 9, 55–58.

Tian, J.Y., Wang, Z.X., Wang, L.Y. and Chai, S.S., (1986) The effect of puerarin on electrophysiology of cardiac muscle. *News Communication CPS*, 5, 46.

Turner, R.A. (1971) $\beta$-adrenergic blocking agents. In *Screening Methods in Pharmacology*, Vol. 2, Academic press, New York, p. 21.

Wang, L.Y., Zhao, A.P. and Chai, X.S. (1994) Effects of puerarin on cat vascular smooth muscle *in vitro*. *Acta Pharmacol. Sin.*, 15, 1–3.

Wang, L.Y., Zhao, A.P., Chai, Q. and Chai, X.S. (1997) Protective effect of puerarin on acute cerebral ischemia in rats. *China J. Chin. Mater. Med.*, 22, 752–754.

Wang, L.Y., and Chai, X.S. (1987) Effects of puerarin on cardiac hemodynamics. *News Communication CPS*, 3, 66.

Yue, H.W. and Hu, X.Q. (1996) Medicinal evaluation of kudzu and puerarin on cardiovascular system. *Chinese J. Integr. Tradit. West Med.*, 16, 382–384.

Zeng, G.Y., Zhou, Y.P., Zhang, L.Y. and Fan, L.L. (1974) Effect of kudzu (*Radix pueraria*) on blood pressure, reaction ability of blood vessel, cerebral and peripheral circulation in dog. *Chin. Med. J.*, 4, 265–267.

Zeng, G.Y. and Zhang, L.Y. (1979) Affection of *pueraria* flavonoids on the content of catecholamine in the patients of hypertension and coronary heart disease. *Chin. Med. J.*, 59, 479–480.

Zhao, A.P. and Chai, S.S. (1995) Cerebrovascular and cerebrometabolic effects of puerarin and propranolol in rabbits. *J. Cereb. Blood Flow Metab.*, 15 (Suppl.1), S528.

Zhao, A.P., Chai, Q. and Chai, X.S. (1999) Effect of topical application of puerarin on pial microcirculation in rabbits. *J. Cerebral Blood Flow Metab.* 19(Suppl.1), S731.

Zhao, A.P., Chai, Q., Liu, E.X., Wang, L.Y. and Chai, S.S (1999) Puerarin prolonged the onset time of brain ischemic seizure and death in SHRSP. *China J. Chin. Mater. Med.*, 23, 431–433.

# 9 Preclinical studies of kudzu (*Pueraria lobata*) as a treatment for alcohol abuse

*Wing Ming Keung*

## INTRODUCTION

Alcohol abuse is the third most common disease of man, following only heart disease and cancer. It is particularly prevalent in Western countries. In the United States recent surveys estimated the life time prevalence of alcohol abuse at about 20 per cent of the adult population, i.e. about 38 million, and a total economic cost to the nation at more than $100 billion per year (US Secretary of Health and Human Services, 1997). Therefore, safe and effective pharmaceutical agents that would selectively suppress the desire to drink alcohol (antidipsotropic agent) are urgently needed. To date, only two agents[1] – disulfiram (acts by aversion) and naltrexone (acts by blocking reward) – have been approved for the treatment of alcohol abuse/alcoholism in the United States. While these agents can be helpful as adjuncts to comprehensive therapy with problem drinkers who are responsible, socially stable and motivated to maintain complete abstinence (Ogborne, 2000), better agents and novel approaches remain to be found.

The earliest treatments for alcohol abuse were invariably based on psychological aversion: pairing alcohol consumption with unpleasant and/or horrifying experiences. For instance, Pliny the Elder, the Roman encyclopedist living in the first century AD, described in his *Historical Naturalis* (Rackham, 1938) that the Romans placed spiders

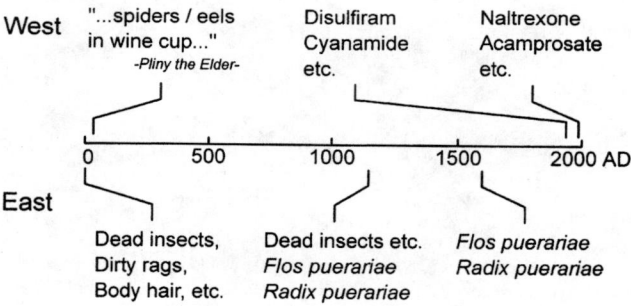

*Figure 9.1* Treatments for alcohol abuse/dependence – a historical perspective.

---

1 Two other agents, calcium cyanimide (Temposil™) and acamprosate (Aotol™), have been approved and used for the treatment of alcoholism in some European countries and Canada with limited success.

or eels in the bottom of wine cups to be found by patients while finishing their drinks (Figure 9.1). For more than two millennia, therapies based on this ancient concept have remained the most prescribed treatments for alcohol abuse in the West. However, both ancient and modern applications of this treatment method, including disulfiram (Antabuse™) and cyanimide (Temposil™), have remained controversial to date (American Medical Association, 1987; Howard *et al.*, 1991; Banys, 1998).

Like the Romans, the ancient Chinese had also discovered, presumably independently[2], psychological aversion as a way to deter alcohol consumption. Indeed, a thorough literature search of the vast collections of ancient Chinese pharmacopoeias has revealed that most remedies listed for the treatment of "alcohol addiction" were based on psychological aversion. For instance, Sun Simiao (*c.* AD 600) compiled 16 "cease drinking" treatments in his classic work *Beiji-Qianjin-Yaofeng*. In these treatments, herbalists placed repugnant substances, e.g. body hair, burned rags, dead insects, etc. in patients' alcoholic beverages (Figure 9.2). It was not until the thirteenth century when a kudzu (*Pueraria lobata*)-based

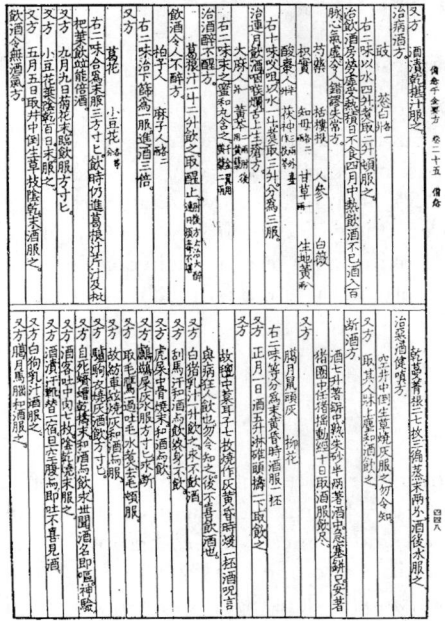

*Figure 9.2* Sun Simiao's "stop drinking" formula (highlighted) in *Beiji-Qianjin-Yaofeng*.

2  The first recorded diplomatic contact between the Roman and Chinese empires was in the year 166 when an envoy dispatched by Roman Emperor Marcus Aurelius arrived Luoyang, the imperial capital of China. However, the presence of Romans in ancient China could be long before Aurelius' Roman. As the legend goes, an ancient band of Roman legionnaires fought and worked their way to China more than 2000 years ago, settling in the village of Zhelaizhai on the edge of the Gobi Desert. According to the historian Pliny the Elder, the soldiers, under the command of Marcus Licinius Crassus to conquer Parthia, were defeated and taken to Central Asia as POW to help Parthia guard its eastern frontier. These soldiers were then disappeared from Western history. Some historians believe that some of them escaped and joined Jzh-Jzh, the mighty leader of Hun, who later invaded and was defeated by Chinese army. The Roman soldiers eventually ended up in China and built the so-called "The Lost City of Liqian" (Chu, H. (2000) Digging for Romans in China, *The Times of Central Asia*, 2(36)).

medication that apparently did not act by aversion was listed in the medical text *Pi Wei Lun* of Li Dongyuan (*c*. AD 1200) for the treatment of "alcohol addiction." Perhaps more intriguing is that virtually all remedies based on psychological aversion were slowly weeded out in ancient Chinese pharmacopoeias, presumably because of their ineffectiveness and/or undesirable side effects. By AD 1600, the only medications that have survived historical trial-and-error scrutiny are those based on the flower (*Flos puerariae*, FP)[3] and root (*Radix puerariae*, RP) of kudzu (Figure 9.1). On the basis of this historical backdrop, we started our search for safe and effective antidipsotropic (alcohol intake suppressive) agents with RP.

In this chapter, I intend to briefly review the conventional pharmaceuticals currently used for the treatment of alcohol abuse/alcoholism and provide a detail account on the discovery of and recent progress on the pharmacological and biochemical studies of the antidipsotropic action of RP and its active principles.

## CURRENT TREATMENTS FOR ALCOHOL ABUSE

### Disulfiram and calcium carbimide – by aversion

Disulfiram was first proposed as an aversive agent for the treatment of alcohol abuse based on the observation that workers in the rubber industry who had been exposed to thiuram compounds experienced a wide spectrum of unpleasant reactions after consuming alcohol (disulfiram–alcohol-reactions, DAR). Such reactions were later attributed to the fact that disulfiram irreversibly inactivated aldehyde dehydrogenases (ALDHs), the major family of isozymes involved in the detoxification of acetaldehyde, the primary and highly reactive metabolic intermediate of ethanol metabolism (Scheme 9.1).

$$\text{Ethanol} \xrightarrow{\text{ADH}} \text{Acetaldehyde} \xrightarrow{\text{ALDH}} \text{Acetate} \qquad \text{(Scheme 9.1)}$$

The rationale for the treatment with disulfiram is that fear of DAR will deter further drinking. Disulfiram had been approved by FDA for the treatment of alcoholism in 1948 before results of extensive experimental and clinical studies became available. Since then disulfiram has been shown to inhibit not only ALDHs but also a wide spectrum of enzymes critical in drug metabolism and detoxification, neurotransmitter metabolism, and multiple pathways of intermediary metabolism. DAR can be so severe that adverse conditions such as respiratory depression, cardiovascular collapse, cardiac arrhythmia, myocardial infarction, acute congestive heart failure, unconsciousness, convulsions and shock could be precipitated. Calcium carbimide, a similar aversive agent approved in Canada and some European countries, was introduced a year later as a less toxic version of disulfiram. The efficacies of both disulfiram and calcium carbimide

---

3 Abbreviations used are: FP, *Flos puerariae*; RP, *Radix puerariae*; ADH, alcohol dehydrogenase; ALDH, aldehyde dehydrogenase; AR, aldehyde reductase; DA, dopamine; DOPAC, 3,4-dihydroxyphenylacetic acid; DOPAL, 3,4-dihydroxyphenyl-acetaldehyde; 5-HIAL, 5-hydroxyindole-3-acetaldehyde; 5-HIAA, 5-hydroxyindole-3-acetic acid; 5-HTOL, 5-hydroxytryptophol; 5-HT, 5-hydroxytryptamine (serotonin); MAO, monoamine oxidase; i.p., intraperitoneally; P rat, alcohol preferring rat; AUC, area under concentration-time curve.

have since been re-evaluated in double-blind, placebo-controlled clinical trials. While, under strictly controlled clinical testings, these alcohol-sensitizing agents were shown to be no more effective in reducing alcohol intake than the placebo control, they can be helpful as adjuncts to comprehensive therapy with problem drinkers who are motivated to receive treatment (For further details, see Peachey, 1981; Banys, 1988; Ogborne, 2000).

### Acamprosate and naltrexone – via central reward system

Neuropharmacological and behavioral studies have led to the belief that the compulsion of taking an addictive drug is driven primarily by factors underlying its positive reinforcing properties such as various forms of euphoria and subjective sensations of stress reduction. Negative reinforcing actions are also believed to participate in drug seeking behavior. In this case, a drug is taken to avoid the unpleasant consequences of abstinence, the withdrawal syndromes. Recent studies suggested that the reinforcing actions of ethanol is mediated by multiple neural pathways, including dopamine (DA), serotonin (5-HT), γ-aminobutyric acid (GABA), glutamate and glycine (NMDA), and opioids, of the cerebral reward system (Koob, 1994). Therefore, it is believed that ethanol seeking behavior could be intervened by agents that modulate the activities of these neural pathways. The observation that opioid agonists increase whereas its antagonists decrease ethanol intake in laboratory animals (Hubbell *et al.*, 1986) led to the clinical evaluation of naltrexone (an opioid antagonist), and for similar reasons acamprosate (a GABAergic and/or glutamatergic agent), as anti-alcohol craving agents (Hubbell *et al.*, 1986; Littleton, 1995). Results indicate that these agents could be beneficial to some (about 15–30 per cent) when used in combination with psychotherapy (Lhuintre *et al.*, 1990; O'Malley *et al.*, 1992; Volpicelli *et al.*, 1992).

## ALCOHOL INTAKE SUPPRESSIVE EFFECT OF KUDZU ROOT

### Golden hamster – the alcohol craving rodent

The search for new antidipsotropic agents requires a laboratory animal that consumes alcohol voluntarily and, preferably, in measurable quantities. We found that Syrian golden hamsters meet this minimal requirement. Mature male hamsters (130 g body weight) consume very little water (about 4 ml per day) because they are desert adapted animals with both renal and respiratory mechanisms for water conservation (Schmidt-Nielsen and Schmidt-Nielsen, 1950). When provided with free choice between water and an ethanol solution (from 5 to 60 per cent), they acquire most, if not all, of their fluid intake from the ethanol solution (Arvola and Forsander, 1961; Keung and Vallee, 1993a). In about 7 to 10 days, hamster ethanol intake gradually increases and reaches a relatively constant level which is 5 to 10 times higher than that consumed by human heavy drinkers (Stibler, 1991). To consume such large amounts of ethanol, hamsters must increase their total fluid intake 2- to 3-fold and, as a consequence, their urine outputs also increase significantly from about <2 ml to 5–10 ml per day. Although, the amounts of ethanol consumed by different hamsters vary from 8 to 20 g/kg/day, the daily ethanol intake of each individual hamster is remarkably constant. Furthermore, the golden hamsters exhibit a predictive validity[4] just as good as those found in the most commonly used animal models of alcoholism (Keung and Vallee, 1994). Yet unlike other animal models, golden hamsters can be maintained easily and used without prior genetic manipulation or extensive conditioning or training. This feature, together with the high volumes of ethanol intake, has been crucial to monitoring the search for new antidipsotropic agents from RP (Keung and Vallee, 1993a).

### Kudzu root extract suppresses hamster alcohol intake

The putative antidipsotropic activity of RP was first verified using alcohol drinking golden hamsters under a continuous access, two-bottle free-choice (water vs. 15 per cent ethanol) condition. At 195 mg/hamster/day (i.p.), a crude alcoholic extract of RP suppressed hamster ethanol intake by more than 50 per cent (Keung and Vallee, 1993a). Ethanol intake remained suppressed during treatment but gradually returned to pre-treatment level after treatment was terminated. The effect of RP was dose-dependent with an $EC_{50}$ value of 110 mg per hamster per day (Keung *et al.*, 1996).

### Daidzin and daidzein – the active principles

The active principles of RP were isolated by chromatography of its crude extract on BioGel P4 and C18 HPLC columns, and identified by mass spectrometry (MS) and nuclear magnetic resonance (NMR) as the isoflavone daidzin and its aglycone daidzein (Figure 9.3). Daidzin, at 20 mg/hamster/day (i.p.), suppressed hamster ethanol intake by about 50 per cent (Keung and Vallee, 1993a). Repeated free-choice trial have confirmed that daidzin, both isolated from RP and synthesized in our laboratory, suppresses ethanol consumption in this animal species. The effect of daidzin was

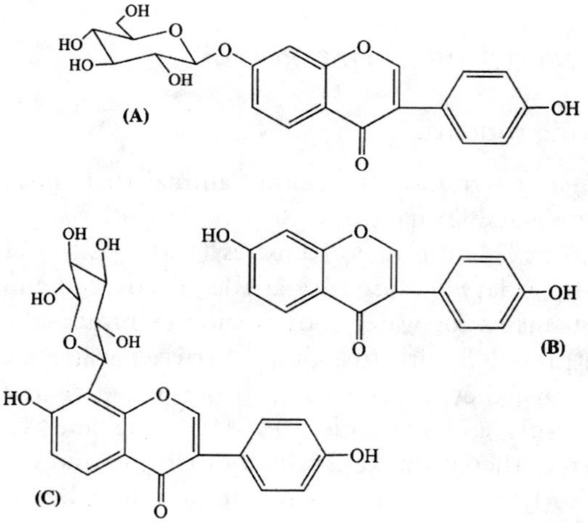

*Figure 9.3* Structure of (A) daidzin, (B) daidzein, and (C) puerarin.

4  The ability to predict the effectiveness of a drug in human based on animal data.

dose-dependent with an estimated $EC_{50}$ value of 23 mg/hamster/day. Over a period of more than 5 years, we have tested daidzin in 186 hamsters and in 174 of them ethanol intake was suppressed (29–83 per cent). Suppression data approximates a normal distribution with a mean±S.D. $= 52 \pm 11$ per cent.

Daidzein, the aglycone of daidzin, also suppressed ethanol intake but was less potent. Thus, a higher dose of daidzein (30 mg per hamster per day) was needed to produce ~50 per cent suppression. An equivalent dose of puerarin, another isoflavone found in RP (Figure 9.3), did not suppress hamster ethanol intake. However, in a later study, Overstreet *et al.* (1996) showed that puerarin, though less potent than daidzin, was antidipsotropic in the P and Fawn Hooded rats. The difference in ethanol drinking response to puerarin between rats and golden hamsters will be discussed later in light of the differences in sensitivity of their monoamine metabolizing enzymes to puerarin inhibition (Keung and Vallee, 1998). Hamsters received RP extract, daidzin, daidzein or puerarin remained healthy and did not exhibit any significant changes in water, total caloric intake or body weight throughout these experiments (Keung and Vallee, 1993a, 9994).

To test the generality of daidzin's antidipsotropic effect, we extended our study to the rat. In this study, a two-lever choice procedure developed by Heyman (1993) was adopted. Wistar rats used in these experiments were trained to consume large and relatively constant amounts of ethanol in a short period of time (2.5–3 g/kg in a 30-min session). Training and experimental sessions were carried out in an operant chamber equipped with two levers. At one lever, responses earned 10 per cent ethanol whereas at the other, responses earned an isocaloric polycose solution. In this experimental setting, daidzin suppressed both ethanol and polycose solution intake, but the suppression in ethanol intake was significantly larger (Heyman *et al.*, 1996). This result not only shows that the antidipsotropic effect of daidzin also pertain to the rat but also indicates that daidzin remains efficacious in an experimental setting in which ethanol is a potent reinforcer. Further, it demonstrates that reduction in ethanol intake caused by daidzin is not simply a reduction in feeding behavior. The latter point is of particular importance because calories probably play a very minor role in the regulation of human drinking. Puerarin, under this experimental paradigm, did not suppress rat ethanol intake (unpublished result).

## Other animal models of alcoholism

Since our discovery, the antidipsotropic activities of daidzin and various RP extracts have been confirmed independently by other investigators in all laboratory animal models tested including the Wistar rats, Fawn Hooded rats, the genetically bred alcohol-preferring P rats and African green monkeys under various experimental conditions including two-lever choice, two-bottle free-choice, limited access, and ethanol-deprived paradigms (Heyman *et al.*, 1996; Lin *et al.*, 1996; Overstreet *et al.*, 1996, 1998). Results of these studies are summarized in Table 9.1.

### POTENCY: PURE DAIDZIN VS. CRUDE KUDZU ROOT EXTRACT

HPLC analysis showed that each g of RP extract contained 22 and 2.6 mg of daidzin and daidzein, respectively. This, together with the fact that daidzein is at least 3 times less potent than daidzin, suggests that daidzein contributes little to the total antidipsotropic

*Table 9.1* Effects of extracts of RP, daidzin and puerarin on ethanol intake in various animal models

| Animal | RP extract | Daidzin | Puerarin |
| --- | --- | --- | --- |
| Golden hamster[a] | + | + | − |
| Rats | | | |
|   Wistar rats[b] | nd[e] | + | − |
|   Fawn hooded rats[c] | + | + | + |
|   P rats[c,d] | + | + | + |
| African green monkeys[c] | + | nd[e] | nd[e] |

Notes
a Keung and Vallee, 1993a.
b Heyman *et al.*, 1996.
c Overstreet *et al.*, 1996, 1998.
d Lin *et al.*, 1996.
e nd = not determined.

activity of the crude extract of RP. Comparison of the antidipsotropic activity of RP extract to that of pure daidzin, however, shows that daidzin alone could not account for the total activity of the crude extract. The $EC_{50}$ values estimated from the dose–response curves of pure daidzin and daidzin administered as a constituent of the crude extract are 23 and 2.3 mg per hamster per day, respectively. Thus, crude daidzin appears to be 10 times more potent than the pure compound. This discrepancy can be explained by one or both of the following possibilities: (1) in addition to daidzin, crude extract of RP contains other major active constituents that could either potentiate the effect of or act additively/synergistically with daidzin, and/or (2) the extract of RP contains one or more constituents that could increase the bioavailability of daidzin.

## Daidzin is the major active principle

Crude RP extract contains a number of isoflavones that are analogs of daidzin and hence could be antidipsotropic. HPLC analysis of this extract has identified 7 isoflavones with puerarin being most abundant (160 mg/g extract), followed by daidzin (22 mg/g), genistin (3.7 mg/g), daidzein (2.6 mg/g), daidzein-4′,7-diglucoside (1.2 mg/g), genistein (0.2 mg/g), and formononetin (0.16 mg/g). At suboptimal doses, daidzin and daidzein did not potentiate each other's activity, but rather they acted additively. Further, at a dose corresponding to the amounts present in an $EC_{50}$ dose of RP extract, neither purified nor synthetic forms of daidzin nor daidzein exhibited any measurable antidipsotropic effect. None of the other isoflavones identified in the crude extract of RP suppressed ethanol intake, either alone of in combination at the same amounts and proportions as found in an $EC_{50}$ dose of the extract. This lack of synergism among these compounds makes it unlikely that the small amounts of these isoflavones, including daidzein, could account for the enhanced antidipsotropic activity of the crude extract (Keung *et al.*, 1996).

## Daidzin in crude kudzu root extract is more potent

Physicians practicing traditional Chinese medicine generally believe that crude drugs are superior to pure ones for a number of reasons. Among them, crude drugs are thought to be less toxic and have better pharmacokinetic properties. To determine whether or not daidzin in the crude extract of RP constitutes a more bioavailable form than pure daidzin, the pharmacokinetic properties of the two dosage forms were evaluated in hamsters (Keung *et al.*, 1996). Results showed that rates of bioavailability ($t_{max}$) of daidzin given as a pure form and as a component of crude RP extract were 63 and 33 min, respectively. In both cases, $t_{max}$ values obtained were independent to the doses administered. The maximal blood daidzin concentration ($C_{max}$) in hamsters received the extract (containing 3.3 mg daidzin) is 10 times higher than that found in hamsters which received 6 mg of pure daidzin. The extents of bioavailability (area under plasma daidzin concentration–time curves AUC) estimated for both pure daidzin- and RP extract-treated hamsters was directly proportional to the amounts of daidzin administered. Within the dose range studied, daidzin administered as crude extract yields AUC values approximately ten times those obtained with the same dose of pure daidzin. Apparently, daidzin given as a component of the crude extract is more readily available to the hamsters than that given in pure form. Pure, synthetic daidzin added to a crude extract of RP acquired the bioavailability of the endogenous daidzin that exists naturally in the extract. These results strongly suggest that daidzin is the major, if not the only, antidipsotropic principle in the crude extract of RP and that other constituents in the crude extract promote the bioavailability of daidzin.

## MECHANISM OF ACTION OF DAIDZIN

### Daidzin is a potent, selective and reversible inhibitor of ALDH-2

The molecular mechanism by which daidzin suppresses ethanol intake in laboratory rodents is still unknown. In early studies, we have shown that daidzin potently, selectively and reversibly inhibits human, rat and hamster liver mitochondrial ALDH isozyme (ALDH-2) (Keung and Vallee, 1993b; Klyosov *et al.*, 1996), the major ALDH isozyme that catalyzes the oxidation of ethanol derived acetaldehyde. This, together with the fact that alcohol abuse is extremely rare among individuals who have inherited an inactive variant form of ALDH-2 (Harada *et al.*, 1982), would seem to suggest that daidzin might act by mimicking the consequence of this natural mutation of the ALDH-2 gene. As an ALDH-2 inhibitor daidzin, in principle, could affect ethanol intake by at least two routes. On the one hand, it might serve as an ethanol-sensitizing agent by inhibiting ALDH-2 subsequent to drinking and thereby allow acetaldehyde to reach toxic levels. On the other hand, it could disrupt an as-yet-undefined physiological pathway involving ALDH-2 and alter the concentration of one or more intrinsic metabolites that control ethanol drinking behavior.

### Daidzin does not act by aversion

To determine whether or not daidzin suppresses hamster ethanol intake by inhibiting acetaldehyde metabolism, we studied acetaldehyde clearance in daidzin-treated hamsters after ethanol administration. Acetaldehyde clearance in disulfiram (a known broad acting ALDH inactivator)-treated hamsters were also studied for comparison. Results showed that daidzin, given at a dose that would suppress hamster ethanol intake by >50 per cent, had no effect on either acetaldehyde or ethanol metabolism as compared

*Table* 9.2  Acetaldehyde oxidizing capacities of major human, hamster and rat liver ALDH isozymes

| Source | Isozyme | Oxidizing capacity nM/min/g *liver* | Inhibition *by daidzin* |
|---|---|---|---|
| Golden Hamster | | | |
|    Mitochondria | ALDH-2 | $1.9 \times 10^3$ | Yes |
|    Cytosol | ALDH-1 | $1.6 \times 10^3$ | No |
|    Cytosol | ALDH-3 | 0 | No |
| Human | | | |
|    Mitochondria | ALDH-2 | 430 | Yes |
|    Cytosol | ALDH-1 | 20 | No |
| Rat (Wistar) | | | |
|    Mitochondria | ALDH-2 | 590 | Yes |
|    Cytosol | ALDH-1 | 50 | No |

to that of the controls. On the other hand, disulfiram treatment greatly impaired hamsters' ability to metabolize acetaldehyde. In disulfiram-treated hamsters, plasma acetaldehyde concentration rose rapidly after ethanol administration, reached 0.6 mM within 50 min and 0.9 mM after 4 h. Ethanol metabolism was also inhibited in disulfiram-treated hamsters. On the basis of these results, we concluded that the mechanism by which daidzin suppresses hamster ethanol intake is different from that proposed for the classic ALDH inhibitors such as disulfiram and calcium carbimide (Hald and Jacobsen, 1948), and proposed that it may act by modulating the activity of an as-yet-undefined physiological pathway catalyzed by ALDH-2 (Keung *et al.*, 1995).

In the same study, we also showed that: (1) acetaldehyde metabolism catalyzed by isolated hamster liver mitochondria was very sensitive to daidzin inhibition ($IC_{50} = 0.4 \mu M$), and (2) liver mitochondria isolated from daidzin-treated golden hamsters contained high concentrations of daidzin (~40–70 μM). The apparently conflicting results – daidzin was delivered to the mitochondira at concentrations (~40–70 μM) high enough to inhibit acetaldehyde metabolism ($IC_{50} = 0.4 \mu M$) yet it did not affect the overall acetaldehyde metabolism *in vivo* – were latter attributed to the presence of a cytosolic ALDH isozyme in hamster liver which is not sensitive to daidzin inhibition but catalyzes acetaldehyde oxidation efficiently and with high capacity (Table 9.2). This unusual genetic endowment of the ALDH isozymes of Syrian golden hamsters has provided a unique system in which the physiological function of ALDH-2 and its role in the detoxification of acetaldehyde during ethanol metabolism can be differentiated and studied independently.

## Daidzin may act via the mitochondrial MAO/ALDH-2 pathway

It has long been postulated that ALDH-2 is involved in the oxidation of biogenic aldehydes that derive from biologically active monoamines such as serotonin (5-HT) and DA in mammalian tissues. Therefore, as a selective and potent inhibitor of ALDH-2, daidzin may act by affecting the metabolism of one or more of the monoamines (5-HT, DA) that mediate important neural pathways of the central reward system. Oxidative deamination of these monoamines, catalyzed by mitochondrial monoamine oxidase (MAO), generates reactive aldehyde intermediates. These aldehydes are either oxidized to their corresponding acid metabolites by ALDH-2 (Tank *et al.*, 1981) or are reduced to their corresponding alcohols by NADH-dependent alcohol dehydrogenase (ADH) and/or NADPH-dependent aldehyde reductase (AR) (Feldstein and Wong, 1961). In the liver and brain, 5-HT and DA are primarily converted to their respective acid products 5-hydroxyindole-3-acetic acid (5-HIAA) and 3,4-dihydroxyphenyl-acetic acid (DOPAC) indicating that the mitochondrion is probably an important subcellular compartment in which 5-HT and DA metabolism occur *in vivo*. Isolated liver mitochondria contain both MAO and ALDH-2 but not ADH and AR (Keung and Vallee, 1998). Therefore, they provide a simple yet physiologically relevant system in which the effects of daidzin on monoamine metabolism can be examined.

When supply with a monoamine such as 5-HT of DA, both hamster and rat liver mitochondria effectively metabolized these substrates to their respective acid metabolites, 5-HIAA and DOPAC. Under these conditions, daidzin potently inhibited the formation of 5-HIAA and DOPAC but had no effect on 5-HT and DA depletion. Inhibition was concentration-dependent ($IC_{50}$ ~2–3 µM) and was accompanied by a concomitant increase in the accumulation of their respective metabolic intermediates 5-hydroxyindole-3-acetaldehyde (5-HIAL) and 3,4-dihydroxyphenylacetaldehyde (DOPAL) (Keung and Vallee, 1998). On the basis of these results, we proposed that the anti-dipsotropic action of daidzin might not be mediated by the monoamines themselves but rather by their aldehyde intermediates, 5-HIAL or DOPAL, that accumulated in the presence of daidzin (Figure 9.4).

While the effects of daidzin on rat and hamster mitochondria-catalyzed monoamine metabolism were identical, those of puerarin were markedly different. Puerarin, at 30 µM, had no effect on 5-HT depletion, 5-HIAA formation and 5-HIAL accumulation in hamster liver mitochondria during 5-HT metabolism. However, at the same concentration, it significantly inhibited 5-HIAA formation and increased 5-HIAL accumulation in rat liver mitochondria (Keung and Vallee, 1998). Puerarin has been shown to suppress ethanol intake in P rats and Fawn Hooded rats but not in golden hamsters (Keung and Vallee, 1993a; Overstreet *et al.*, 1996). Hence, the difference in ethanol drinking response to puerarin between these two animal species could be attributed to the difference in

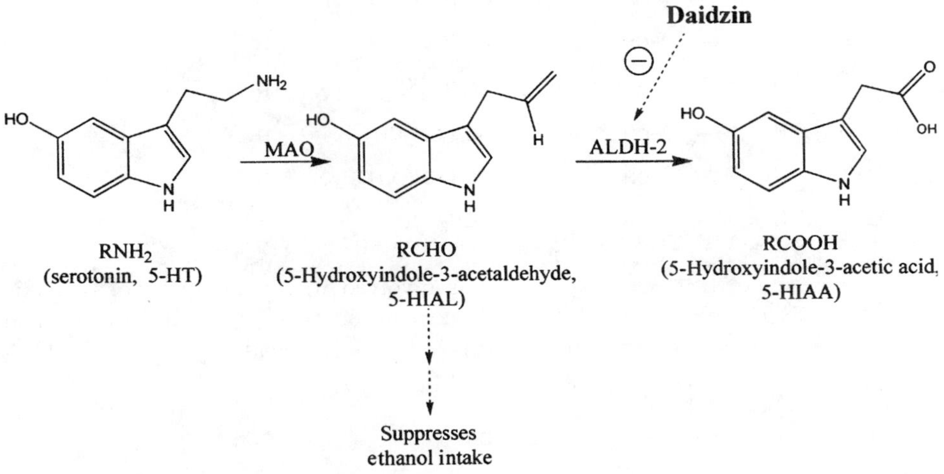

*Figure 9.4* The mitochondrial MAO/ALDH-2 pathway: a potential site of action for daidzin.

sensitivity of their mitochondrial MAO/ALDH-2 pathway to puerarin inhibition. While these results by no means prove a mechanism of action for daidzin but lend support to the proposition that it may suppress ethanol intake via the MAO/ALDH-2 pathway (Figure 9.4).

To further evaluate this hypothesis, we synthesized a series of structural analogs of daidzin (Figure 9.5), tested and compared their antidipsotropic activities with their abilities to increase 5-HIAL accumulation in isolated hamster liver mitochondria during 5-HT metabolism. Results from these studies revealed a positive correlation between the two (Figure 9.6). Concentrations of 5-HIAL attained during 5-HT metabolism in isolated mitochondria are determined by the relative catalytic efficiencies of MAO and ALDH-2. Thus, analogs of daidzin that increase the MAO/ALDH-2 activity ratio by inhibiting ALDH-2 activity increase 5-HIAL accumulation, whereas those that decrease

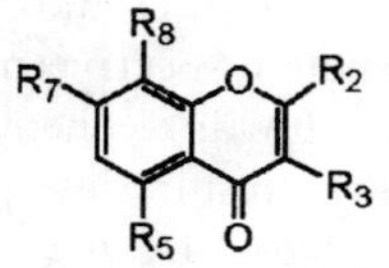


| Cpd | $R_2$ | $R_3$ | $R_5$ | $R_7$ | $R_8$ |
|---|---|---|---|---|---|
| **1** | H | $C_6H_4$-OH | H | OGlc | H |
| **2** | H | $C_6H_4$-OH | H | OH | H |
| **3** | H | $C_6H_4$-OH | H | $O(CH_2)_5CO_2H$ | H |
| **4** | H | $C_6H_4$-OH | H | $O(CH_2)_6CO_2H$ | H |
| **5** | H | $C_6H_4$-OH | H | $O(CH_2)_9CO_2H$ | H |
| **6** | H | $C_6H_4$-OH | H | $O(CH_2)_{10}CO_2H$ | H |
| **7** | H | $C_6H_4$-OH | H | $O(CH_2)_3Br$ | H |
| **8** | H | $C_6H_4$-OH | H | $O(CH_2)_4Br$ | H |
| **9** | H | $C_6H_4$-OH | H | $O(CH_2)_6Br$ | H |
| **10** | H | $C_6H_4$-OH | H | $O(CH_2)_2$-(1,3-dioxan-2-yl) | H |
| **11** | H | $C_6H_4$-OH | H | $OCH_2CH{=}CH_2$ | H |
| **12** | H | $C_6H_4$-OH | H | $OCH_2CH(OH)CH_2OH$ | H |
| **13** | H | $C_6H_4$-OH | H | $OCH_2CH_3$ | H |
| **14** | H | $C_6H_4$-OH | OH | OH | H |
| **15** | $C_6H_5$ | H | OH | OH | H |
| **16** | $C_6H_5$ | H | H | OH | OH |
| **17** | H | $C_6H_4$-$OCH_3$ | OH | OH | H |
| **18** | H | $C_6H_4$-OH | H | OH | CGlc |

*Figure 9.5* Structures of daidzin and its analogs.

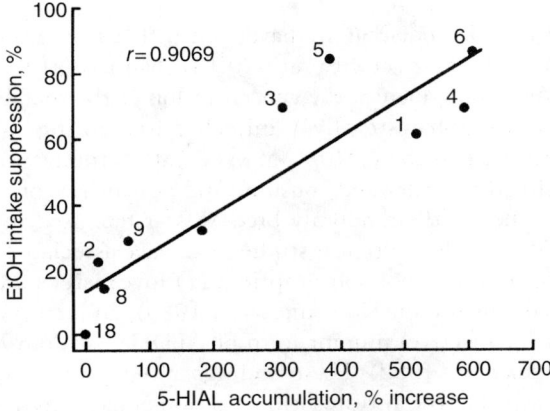

*Figure 9.6* Scatter of suppression of ethanol intake versus increase in 5-HIAL accumulation by daidzin and its structural analogs.

*Table 9.3* Inhibition of MAO and ALDH-2 activities and suppression of ethanol intake by daidzin and its structural analogs

| Compound[a] | IC$_{50}$, µM | | Ethanol intake suppression, % |
|---|---|---|---|
| | MAO | ALDH-2 | |
| 1 | n.i.[b] | 0.04 | 62±4 |
| 2 | 14 | 9 | 22±5 |
| 3 | 3 | 0.009 | 69±12 |
| 4 | 2.1 | 0.009 | 69±8 |
| 5 | 10 | 0.004 | 84±5 |
| 6 | 13 | 0.003 | 86±7 |
| 7 | 0.15 | 0.26 | 0 |
| 8 | 4 | 0.27 | 14±5 |
| 9 | 2 | 0.3 | 29±5 |
| 10 | 7 | 0.4 | 32±10 |
| 11 | 0.45 | 0.8 | 0 |
| 12 | 1.7 | 0.1 | 0 |
| 13 | 0.3 | 0.04 | 0 |
| 14 | 0.9 | n.i. | 0 |
| 15 | 1.6 | n.i. | 0 |
| 16 | 17 | n.i. | 0 |
| 17 | 0.4 | n.i. | 0 |
| 18 | n.i | n.i. | 0 |

Notes
a  See Figure 9.5 for structures of compounds.
b  n.i. = no inhibition up to 10 µM.

the MAO/ALDH-2 activity ratio by inhibiting MAO decrease 5-HIAL accumulation (Table 9.3). It appears that the antidipsotropic activity of daidzin, and its active analogs as well, stems from it's ability to increase the MAO/ALDH-2 activity ratio and that 5-HIAL, or an as-yet-unknown aldehyde intermediate derived from the action of MAO, is involved in the regulation of the ethanol drinking behavior of the golden hamster (Rooke *et al.*, 2000).

In this context, it is of interest to point out that we have shown that hamster liver mitochondria exhibit a lower MAO:ALDH-2 activity ratio (0.18) than that of the rat (1.6) (Keung and Vallee, 1998). As a consequence, the concentration of the metabolic intermediate 5-HIAL found in isolated hamster liver mitochondria during 5-HT metabolism is also much lower than that in the rat (0.2 μM vs. 2.3 μM). Interestingly, golden hamsters are by nature inclined to prefer and consume large quantities of ethanol (Arvola and Forsander, 1961) whereas the randomly bred Wistar rats used in our studies avoid ethanol (Li *et al.*, 1979). Epidemiological studies also have associated low MAO:ALDH-2 activity ratio with high ethanol consumption: (1) low platelet MAO activity correlates with type II alcoholism (von Knorring *et al.*, 1985), and (2) Asians who have inherited a low activity (or inactive) mutant form of ALDH-2 seldom have a problem with alcohol abuse (Harada *et al.*, 1982). These findings, although inconclusive, are consistent with the hypothesis that the antidipsotropic action of daidzin may be mediated by a biogenic aldehyde intermediate of the mitochondrial MAO/ALDH pathway such as 5-HIAL.

## Biogenic aldehyde(s) and alcohol consumption

Early interest in biogenic aldehydes in relation to alcohol research stems largely from the belief that acetaldehyde, the reactive intermediate of ethanol metabolism, interferes with their oxidative metabolism (Deitrich and Erwin, 1980). It was postulated that levels of biogenic aldehydes increase during ethanol metabolism because of competitive inhibition of ALDH-2 by acetaldehyde. Biogenic aldehydes, accumulated under such conditions, can be diverted to a reductive pathway leading to the formation of their alcohol metabolites (Feldstein and Wong, 1961; Feldstein, 1971), and/or undergo non-enzymatic condensation reactions forming adducts with biogenic amines, proteins and phospholipids (Deitrich and Erwin, 1980; Nilsson and Tottmar, 1987). While the physiological implication of shifting from an oxidative to a reductive metabolic pathway is completely unknown at this time, the condensation products have been shown to affect ethanol drinking behavior in laboratory animals (Myers, 1978). Moreover, biogenic aldehydes themselves could be physiologically active (Alivisatos and Tabakoff, 1973) and may play a more direct role in regulating ethanol intake. Our results suggest that daidzin, and its active antidipsotropic analogs, could act by mimicking the effect of acetaldehyde on biogenic aldehyde metabolism and provided a strong case for further pursuit of the role biogenic aldehydes in the regulation of alcohol use and abuse.

## PROSPECT

Unlike other agents described in the literature, daidzin was discovered based solely on empirical clinical information rather than any preconceived theories or hypotheses based on data derived from animal experiments. This unique property of daidzin not only has further strengthened our belief that it and/or more of its congeners could be developed into safe and efficacious agents for the clinical treatment of alcohol abuse/dependence but also provide a very powerful tool to identify and characterize the elusive biochemical pathways(s) that regulate the use and abuse of alcohol.

# REFERENCES

Alivisatos, S.G.A. and Tabakoff, B. (1973) Formation and metabolism of "biogenic" aldehydes. In H. Sabelli (ed.), *Chem. Modulation Brain Funct.*, Raven Press, New York, pp. 41–66.

American Medical Association, Council on Scientific Affairs (1987) Aversion therapy. *J. Am. Med. Assoc.*, **258**(18), 2562–2566.

Arvola, A. and Forsander, O. (1961) Hamsters in experiments of free choice between alcohol and water. *Nature*, 191, 819–820.

Banys, P. (1988) The clinical use of disulfiram (Antabuse®): A review. *J. Psychoactive Drugs*, 20(3), 243–260.

Deitrich, R.A. and Erwin, V.G. (1980) Biogenic amine-aldehyde condensation products: Tetra-hydroisoquinolines and tryptolines (β-carbolines). *Annu. Rev. Pharmacol.*, 20, 55–80.

Feldstein, A. (1971) Effect of ethanol on neurohumoral amine metabolism. In B. Kissin and H. Beleiter (eds), *The Biology of Alcoholism*, Plenum, New York, pp 127–159.

Feldstein, A. and Wong, K.-K. (1961) Enzymatic conversion of serotonin to 5-hydroxytryptophol. *Life Sci.*, 4, 183–191.

Hald, J. and Jacobsen, E. (1948) A drug sensitizing the organism to ethyl alcohol. *Lancet*, 255, 1001–1004.

Harada, S., Agarwal, D.P., Goedde, H.W., Tagaki, S. and Ishikawa, B. (1982) Possible protective role against alcoholism for aldehyde dehydrogenase isozyme deficiency in Japan. *Lancet*, 2, 827.

Heyman, G.M. (1993) Ethanol regulated preference in rats. *Psychopharmacology*, 112, 259–269.

Heyman, G.M., Keung, W.M. and Vallee, B.L. (1996) Daidzin decreases ethanol consumption in rats. *Alcohol. Clin. Exp. Res.*, 20, 1083–1087.

Howard, M.O., Elkins, R.L., Rimmele, C. and Smith, J.W. (1991) Chemical aversion treatment of alcohol dependence. *Drug Alcohol Depend.*, 29, 107–143.

Hubbell, C.L., Czirr, S.A., Hunter, G.A., Beaman, C.M., LeCann, N.C. and Reid, L.D. (1986) Consumption of ethanol solution is potentiated by morphine and attenuated by naloxone persistently across repeated daily administrations. *Alcohol.*, 3(1), 39–54.

Keung, W.M., Lazo, O., Kunze, L. and Vallee, B.L. (1995) Daidzin suppresses ethanol consumption by Syrian golden hamsters without blocking acetaldehyde metabolism. *Proc. Natl. Acad. Sci. USA*, **92**, 8990–8993.

Keung, W.M., Lazo, O., Kunze, L. and Vallee, B.L. (1996) Potentiation of the bioavailability of daidzin by an extract of *Radix puerariae*. *Proc. Natl. Acad. Sci. USA*, 95, 4284–4288.

Keung, W.M. and Vallee, B.L. (1993a) Daidzin and daidzein suppress free-choice ethanol intake by Syrian golden hamsters. *Proc. Natl. Acad. Sci. USA*, 90, 10008–10012.

Keung, W.M. and Vallee, B.L. (1993b) Daidzin: A potent, selective inhibitor of human mito-chondrial aldehyde dehydrogenase. *Proc. Natl. Acad. Sci. USA*, 90, 1247–1251.

Keung, W.M. and Vallee, B.L. (1994) Therapeutic lessons from traditional Oriental medicine to contemporary Occidental pharmacology. *EXS*, 71, 371–381.

Keung, W.M. and Vallee, B.L. (1998) Daidzin and its antidipsotropic analogs inhibit serotonin and dopamine metabolism in isolated mitochondria. *Proc. Natl. Acad. Sci. USA*, 95, 2198–2203.

Klyosov, A.A., Rashkovetsky, L.G., Tahir, M.K. and Keung, W.M. (1996) Possible role of liver cytosolic and mitochondrial aldehyde dehydrogenases in acetaldehyde metabolism. *Biochemistry*, 35, 4445–4456.

Koob, G.F., Rassnick, S., Heinrichs, S. and Weiss, F. (1994) Alcohol, the reward system and dependence. *EXS*, 71, 103–114.

Lhuintre, J.P., Moore, N., Tran, G., Steru, L., Langrenon, S., Daoust, M., Parot, P., Ladure, P., Libert, C., Boismare, F. *et al.* (1990) Acamprosate appears to decrease alcohol intake in weaned alcoholics. *Alcohol Alcohol*, 25, 613–622.

Li, T.-K., Lumeng, L., McBride, W.J. and Waller, M.B. (1979) Quantitative correlation of ethanol elimination rates *in vivo* with liver alcohol dehydrogenase activities in fed, fasted and food-restricted rats. *Drug Alcohol Depend.*, 4, 45–60.

Lin, R.C., Guthrie, S., Xie, C.-I. *et al.* (1996) Isoflavonoid compounds extracted from *Pueraria lobata* suppress alcohol preference in a pharmacogenetic rat model for alcoholism. *Alcohol. Clin. Exp. Res.*, **20**, 659–663.

Littleton, J. (1995) Acamprosate in alcohol dependence: how does it work? *Addiction*, **90**, 1179–1188.

Myers, R.D. (1978) Tetrahydroisoquinolines in the brain: the basis of an animal model of alcoholism. *Alcoholism: Clin. Exp. Res.*, **2**, 145–154.

Nilsson, G.E. and Tottmar, O. (1987) Biogenic aldehydes in brain: On their preparation and reactions with rat brain tissue. *J. Neurochem.*, **48**, 1566–1572.

Ogborne, A.C. (2000) Identifying and treating patients with alcohol-related problems. *Can. Med. Assoc. J.*, **162**(12), 1705–1708.

O'Malley, S.S., Jaffe, A.J., Chang, G. and Schottenfeld, R.S. (1992) Naltrexone and coping skills therapy for alcohol dependence: a controlled study. *Arch. Gen. Psychiatry*, **49**, 881–887.

Overstreet, D.H., Lee, D.Y.-W., Chan, Y.T. and Rezvani, A.H. (1998) The Chinese herbal medicine NPI-028 suppresses alcohol intake in alcohol-preferring rats and monkeys without inducing taste aversion. *Perfusion*, 11, 381–390.

Overstreet, D.H., Lee, Y.-W., Rezvani, A.H., Pei, Y.-Hm., Criswell, H.E. and Janowsky, D.S. (1996) Suppression of alcohol intake after administration of the Chinese herbal medicine, NPI-028, and its derivatives. *Alcohol. Clin. Exp. Res.*, **20**, 221–227.

Peachey, J.E. (1981) A review of the clinical use of disulfiram and calcium carbimide in alcoholism treatment. *J. Clin. Psychopharmacol.*, 1, 368–375.

Rackham, H. (1938) Pliny's Natural History. Harvard University Press, Cambridge, MA, USA.

Rooke, N., Li, D.J., Li, J. and Keung, W.M. (2000) The mitochondrial monoamine oxidase-aldehyde dehydrogenase pathway: a potential site of action of daidzin. *J. Med. Chem.*, 43(22), 4169–4179.

Schmidt-Nielsen, B. and Schmidt-Nielsen, K. (1950) Pulmonary water loss in desert rodents. *J. Cell Comp. Physiol.*, **162**, 31–36.

Stibler, H. (1991) Carbohydrate-deficient transferrin in serum: a new marker of potentially harmful alcohol consumption reviewed. *Clin. Chem.*, **37**, 2029.

Tank, A.W., Weiner, H. and Thurman, J.A. (1981) Enzymology and subcellular localization of aldehyde oxidation in rat liver. *Biochem. Pharmacol.*, **30**, 3265–3275.

Volpicelli, J.R., Alterman, A.I., Hayasgida, M. and O'Brien, C.P. (1992) Naltrexone in the treatment of alcohol dependence. *Arch. Gen. Psychiatry*, **49**, 876–880.

von Knorring, A.-L., Bohman, M., von Knorring, L. and Oreland, L. (1985) Platelet MAO activity as a biological marker in subgroups of alcoholism. *Acta Psychiatr. Scand.*, **72**, 51–58.

# 10 Human studies of kudzu as a treatment for alcohol abuse

*Scott E. Lukas*

## INTRODUCTION

In 1994, 3.4 million Americans (about 1.6 per cent of the population ages 12 and older) received treatment for alcoholism and alcohol-related problems; 26–34 year olds were most frequently treated (SAMHSA, 1994). Worldwide, the numbers are equally staggering. Coupled with the finding that more costly treatments are not necessarily more effective (NIAAA, 1993b), the fact that providing heavy drinkers (who are not yet alcohol-dependent) with any type of intervention yields positive outcomes, the need for an inexpensive, widely available treatment for alcoholism is clearly evident. Alcohol-related treatments for women are especially needed because of their greater vulnerability and the need to reduce the incidence of fetal alcohol syndrome that is secondary to *in utero* exposure to ethanol.

To date, there are only a few medically accepted treatments for alcohol abuse and alcohol dependence. Disulfiram (Antabuse™) was the leading medication for many years until 1995 when the Food and Drug Administration (FDA) approved the use of the opiate receptor antagonist, naltrexone (ReVia™), as a treatment for alcohol abuse. While behavioral therapies and self-help groups such as alcoholics anonymous (AA) are successful for some patients, they are not universally accepted by all alcoholics and many relapse to using alcohol. Thus, there is a definite need to provide alternative treatments to those who may need a medication either alone or in combination with psychotherapy. Also, medications without side effects are desirable because many patients cannot tolerate the side effects of currently available medicines. In addition, there are no currently accepted and safe medications for treating adolescent alcoholics or pregnant women.

Pueraria-based medicines have been used for centuries in ancient Chinese prescriptions to treat a variety of alcohol-related problems. As a result of a greater acceptance and desire for "natural" or herbal treatments for a variety of ailments and diseases, pueraria-based medicines have recently gained more widespread acceptance in Western cultures. Recently, Huang (1999) reported that kudzu flower was useful in treating alcohol addiction, but Shebek and Rindone (2000) reported that kudzu root did not alter craving or sobriety levels in chronic alcoholics.

Pueraria-based preparations have been shown to reduce alcohol drinking in a variety of animal models (see Keung, this volume) using well-established laboratory-based models. The strengths of such studies are obvious: (1) controlled setting, (2) controlled dose of alcohol, (3) controlled access to alcohol, (4) accurate identity and dose of pueraria-based medication, and (5) verifiable compliance with treatment. The conduct of human studies of alcohol drinking and assessing the efficacy of a new medication to alter that behavior does not enjoy the same level of assurances. Human alcohol drinking especially in the natural environment is a complex behavior that is subject to the influence of many factors, both biological and environmental. Further, as with nearly all herbal or alternative medicines, the preparations are unregulated and claims of efficacy cannot be made, only inferred. Other issues such as medication compliance, validity of self-reports, verification of alcohol use are all subject to error and can influence the interpretation of the results. Finally, understanding the complete pharmacodynamics, interaction between a medication and alcohol is extremely important from a safety perspective.

This chapter offers a brief overview of the alcohol problem, the current medications used to treat alcoholism and early studies designed to assess the safety and efficacy of pueraria-based medications to treat alcohol abuse. The common problems and pitfalls of conducting such clinical studies will be discussed as well.

## Alcohol abuse

Ethyl alcohol is the most widely used psychoactive drug in the world. Because of its unique pharmacological profile of disrupting psychomotor performance while enhancing mood, it contributes to about 100 000 deaths annually in the United States alone (McGinnis and Foege, 1993). This sobering statistic means that alcohol-related deaths trail only cancer and heart disease. However, the victims of alcohol-related deaths tend to be much younger (NIAAA, 1993a), thus robbing them of their most productive years. According to a NIAAA news release (1998), from 1985 to 1992, the economic costs of alcoholism and alcohol-related problems rose 42 per cent to $148 billion. A full-two thirds of the costs were related to lost productivity, either due to alcohol-related illness (45.7 per cent) or premature death (21.2 per cent). Most of the remaining costs are related to health care expenditures to treat alcohol use disorders and the medical consequences of alcohol consumption (12.7 per cent), property and administrative costs of alcohol-related motor vehicle crashes (9.2 per cent), and various additional costs of alcohol-related criminal activity (8.6 per cent). The costs in 1995 were estimated to be $166.5 billion.

The above numbers suggest that seven percent of the United States population, ages 18 and older (or nearly 13.8 million Americans) had problems with drinking, including the 8.1 million alcoholics. The distribution of problem drinkers is nearly 3:1 for 18–29 year old males to females (NIAAA, 1994), but women appear to be at greater risk for alcohol-induced liver damage (Schuckit, 1985; Frezza *et al.*, 1990) and women are more at risk for a variety of other alcohol-related problems such as reproductive, sexual, dependence and victimization by others (NIAAA, 1993c). It also appears that people who begin drinking before age 15 are four times more likely to develop alcoholism than those who begin at age 21 (NIAAA, 1998). In fact, 64 per cent of high school seniors report that they have been drunk and more than 31 per cent say that they have had five or more drinks in a row during the last two weeks (Johnston *et al.*, 1997).

## Pharmacological manipulation of alcohol effects

There are no uniformly effective pharmacotherapies for treating alcohol abuse/dependence (Jaffe *et al.*, 1992), and because of the multifaceted nature of alcoholism, it is not surprising that drugs from a number of different pharmacological classes have been used to treat alcohol abuse and dependence (Liskow and Goodwin, 1987; Kranzler and Orrok, 1989; Litten and Allen, 1991). Liskow and Goodwin (1987) and more recently Swift (1997, 1999) reviewed identified categories of agents used treat alcoholism: (1) agents to treat withdrawal, (2) anticraving agents, (3) aversive agents, (4) agents to treat concomitant psychiatric problems, (5) agents to treat concomitant drug abuse, and (6) amethystic agents. Pharmacotherapies for alcoholism continue to evolve (Litten *et al.*, 1996) and the following drugs have been used to treat ethanol withdrawal: benzo-diazepines, $\beta$-adrenergic blockers, $\alpha_2$-adrenergic agonists, dopamine receptor blockers such as haloperidol, diphenylhydantoin, oxygen/nitrous oxide combinations, NMDA antagonists, calcium channel blockers/GABA glutamate interactive agent (acamprosate), and carbamazepine (Palestine and Alatorre, 1976; Sellers and Kalant, 1976; Lichtigfeld and Gillman, 1982; Kraus *et al.*, 1985; Simon, 1988; Malcolm *et al.*, 1989; Nutt *et al.*, 1989; Leslie *et al.*, 1990; Paille *et al.*, 1995; Sass *et al.*, 1996). Craving for ethanol is now thought to be related to low serotonin levels so fluoxetine (Gorelick, 1986), zimelidine (Naranjo *et al.*, 1984), citalopram (Naranjo *et al.*, 1987), ondansetron (Johnson *et al.*, 1993, 2000; Swift *et al.*, 1996) and fluvoxamine (Linnoila *et al.*, 1987) have been studied for their utility as anticraving agents. Opiate receptor antagonists such as naloxone and naltrexone have been used in both animal (Altshuler *et al.*, 1980; Davidson and Amit, 1996) and clinical studies (O'Malley *et al.*, 1992; Volpicelli *et al.*, 1992) to reduce drinking, and it appears that one effect of naltrexone may be to increase the latency to drink (Davidson *et al.*, 1996). However, none of these medications is consistently effect-ive in reducing drinking, are available only by prescription and most have adverse side effects that limit their usefulness and safety especially in pregnant women and adolescents. Thus, a completely safe and effective medication for reducing alcohol intake remains unavailable.

## Isoflavones in the plant kingdom

As a group, the isoflavones are benzo-$\gamma$-pyrone derivatives that are found in all leguminous plants. About 500 varieties are known, many of which were studied extensively in the 1950s and found to have weak estrogenic activity (Cheng *et al.*, 1955). These polyphenolic compounds also possess a number of other pharmacological effects such as inhibiting enzymes (Havsteen, 1983; Keung and Vallee, 1993a), scavenging for free-radicals (Bors *et al.*, 1990), reducing inflammation (Di Perri and Auteri, 1988) and activating polymorphonuclear leukocytes. In addition, isoflavones appear to have antifebrile, anti-hypertensive, antioxidant, and antidysrhythmic properties (Harada and Ueno, 1975; Nakamoto *et al.*, 1977).

In 1993, at the Second International Conference on Phytoestrogens (Little Rock, Arkansas) scientists presented results on the pharmacodynamics and pharmacokinetics of these phytoestrogenic flavonoids (Kelly *et al.*, 1995; Lundh, 1995; Miksicek, 1995; Wähälä *et al.*, 1995). It also was recently reported that daidzein is one of the more bio-available isoflavones in adult women (Xu *et al.*, 1994). One such plant, kudzu (*Puerariae lobata*) was introduced to the United States in 1876 as a method of controlling soil erosion. As most Southerners know, kudzu rapidly spread throughout most of the south eastern states and its thick long roots (up to 20″ in length) dig deep into the soil and its large leaves overshadow other crops such that it has been branded "the vine that ate the South."

## *Pueraria*-based treatments for alcohol abuse and dependence

The use of herbal plants to treat alcohol-related diseases dates back to AD 600. One such Chinese herbal medicine, XJL (NPI-028), has long been used to reduce the inebriation that results from alcohol consumption. NPI-028 contains the extracts of several plants including *Puerariae lobata* (kudzu) and *Citrus reticulata*, which were recorded in an ancient Chinese materia medica entitled Ben Cho Gang Mu (Li, AD 1590–1596) and have long been used to lessen alcohol intoxication (antidrunkedness) (Sun, AD 600). A total of seven isoflavonoids have been isolated from *Puerariae lobata* including NPI-031G (puerarin), NPI-031D (daidzin), NPI-031E (daidzein), NPI-031F (3′-methoxy-puerarin), and NPI-031L (genistein).

In one of the first empirical studies of these plants, Niiho *et al.* (1989) found that plasma ethanol and acetaldehyde levels were lower in mice that had received oral doses of an isoflavonoid fraction of the flower of the kudzu plant, *Puerariae flos*. Ethanol's effects on spontaneous locomotor activity also were attenuated in the treatment group. Subsequently, it was shown that NPI-028 could significantly reduce alcohol intake in two strains of alcohol-preferring rats under a range of conditions without the development of tolerance. Low doses of NPI-028 also were effective in alcohol-preferring vervet monkeys in a 24 h free-choice drinking paradigm (Overstreet *et al.*, 1996, 1998). The effects of certain components of kudzu on suppressing alcohol intake have been reported by several laboratories. Keung and Vallee at Harvard Medical School, demonstrated that daidzin and daidzein were the active components isolated from *Radix pueraria* that suppressed alcohol intake in Syrian Golden hamsters (Keung and Vallee, 1993a, 1993b). Drinking resumed rather quickly once the treatment stopped. Interestingly, daidzin also decreases blood alcohol levels and shortens sleep time induced by ethanol (Xie *et al.*, 1994). In another study, plasma ethanol levels in daidzin-treated rats peaked 1 h later and achieved a significantly lower peak level than those attained in placebo-treated animals (Xie *et al.*, 1994). Plasma ethanol levels also declined more quickly. These authors noted that daidzin was ineffective if given as a single pretreatment just before ethanol challenge. Further, daidzin treatment failed to alter liver ADH or mitochondrial ALDH in these rats, a finding that was confirmed in the hamster (Keung *et al.*, 1995). Daidzin shortened ethanol sleep time in rats, but only if ethanol was given orally, not i.p., suggesting that daidzin's effects may be partially due to a delay in gastric emptying (Xie *et al.*, 1994).

The degree of reduction was over 50 per cent and the effects appeared within 1 day of treatment. When the treatment was stopped, ethanol consumption resumed rather quickly. These i.p. injections of these agents also decreased ethanol intake (Keung and Vallee, 1994). The experimental design involved a choice procedure of ethanol vs. water and as water intake by the hamsters was unaffected by the isoflavone treatment, the authors concluded that the isoflavones were selectively reducing ethanol consumption. The Chinese herbal medicine (NPI-028) and two of its derivatives suppressed ethanol intake in alcohol-preferring Fawn-Hooded rats (Overstreet *et al.*, 1996). This preparation contains a number of different herbs including kudzu. Ethanol intake was reduced rather abruptly after introduction of the NPI-028 while water and food intake were essentially unaffected.

Lin *et al.* (1996) found that three isoflavonoids isolated from kudzu (daidzin, daidzein and puerarin) decreased alcohol consumption by female alcohol-preferring rats by 75 per cent, 50 per cent and 42 per cent, respectively. These data were observed in the absence of any effects on liver ADH and ALDH. Heyman *et al.* (1996) used a two lever choice procedure to demonstrate that daidzin decreased ethanol consumption in rats.

In 1997, we reported at the American College of Neuropsychopharmacology Meeting (Lukas *et al.*, 1997) that kudzu treatment attenuated ethanol's subjective effects in female social drinkers. This preliminary report was the first double-blind and placebo-controlled study of kudzu's effects on alcohol-induced intoxication. We recently reported that the efficacy of kudzu may differ in individuals with different drinking patterns at the recent meeting of the College on Problems of Drug Dependence (Lukas *et al.*, 1999).

## Isoflavone preparations

One of the unique features of the kudzu plant is its ubiquitousness in the environment and its widespread availability via health, herbal medicine and grocery stores. This factor alone might make it more desirable as an aid to control drinking. United States species of kudzu contain only about 0.7–1.1 per cent total isoflavones, and with the advent of electronic purchasing on the web, many suppliers of pueraria-based preparations have surfaced to sell these directly to the public. In spite of the claims of high purity, potency and quality, the actual content of pueraria and related isoflavones is very low, highly variable and inconsistent even from the same manufacturer. Further, even the claim that a preparation is an "extract" does not ensure greater potency as these preparations usually contain the same small amounts of puerarin as the raw powder. Given the extremely low potency of OTC available kudzu products, extrapolation from the animal data suggests that doses of 7–10 g of raw kudzu root powder needs to be taken at least two times daily, to match the doses that have been effective in animals.

## Summary

Alcohol abuse and alcoholism remain the largest drug abuse problem facing all countries. There are a few drugs that have a modest success rate in treating alcoholism, but the lack of a universally effective medication continues to limit successful treatment of this disease. Although, the mechanism by which pueraria-based preparations reduce alcohol intake in animals is unknown, these compelling preclinical studies suggest that isoflavones may alter ethanol-induced effects and reduce drinking behavior in humans. Pueraria-based preparations are unique in that they are widely available and have little or no toxicity in these animal studies. Because the content, potency and quality of puerarain-based products (as are most alternative medicines) remain unregulated, clinical studies must include detailed assessments of the ingredients of the preparation used. Well-controlled double-blind clinical trials are now needed to test whether these preparations will be useful in treating alcohol abuse and dependence in human subjects.

## CLINICAL SAFETY TRIAL OF KUDZU AND ALCOHOL INTERACTIONS

One of the fundamental steps required to bring any new medication to clinical practice is to subject it to the rigors of preclinical and clinical study, not only to determine the new drug's efficacy, but to identify any adverse reactions, side effects or toxicity.

Because kudzu is considered a food product by the FDA, it is unregulated and thus does not fall under the strict guidelines for drugs established by this agency. In our first experience with this preparation we sought to conduct a thorough evaluation of kudzu, including giving a challenge dose of alcohol to subjects to document the interactions and determine the safety of kudzu in the presence of alcohol.

## Subjects

None of the subjects met criteria for Alcohol Dependence using the Cahalan quantity-frequency variable drinking practices or CQFV (Cahalan *et al.*, 1969), but their drinking patterns and histories were collected for future analyses. Subjects ranged from light (CQFV scores of 13–16) to heavy (CQFV scores of 2–6) drinkers. Family history of alcoholism was calculated as described in the human subjects section using a revised family density method (Hill, 1984; McCaul *et al.*, 1991). As tobacco smoking was very rare in these subjects, the data were recorded as packs per week. Because of the difference in body weight between males and females, the number of drinks were also corrected for body weight.

## Kudzu and alcohol preparations

Crushed kudzu root (*Pueraria lobata*) was obtained from a United States supplier and formulated into gelatin capsules. Each capsule contained 500 mg of crushed kudzu root and the isoflavone content was assayed using HPLC and found to be 0.77 per cent (38.5 mg/capsule). Because of the differences in subject weight and the need to standardize the number of capsules, the subjects were instructed to take 8, 10 or 15 capsules, 3 times a day for 2-1/3 days. This dosing regimen resulted in daily doses of 15–20 g/day of raw kudzu root. Virtually identical gelatin capsules were used for placebo treatment.

Beverage grade ethyl alcohol (86 proof vodka) was mixed with ice-cold orange juice and presented in a series of 4 cups, each containing 120 ml. Subjects were instructed to drink each cup over a 5 min period in order to ensure similar dosing. This practice prevented some subjects from "chugging" the drinks or "sipping" them over a much longer period. This technique of controlled drinking yields rather similar blood alcohol levels over time (Lukas *et al.*, 1986a,b, 1989, 1992). The alcohol dose was kept standard at either 0.56 or 0.7 g/kg.

## General methods

The procedure for all studies was as follows: After passing clinical laboratory physical and psychiatric examinations, informed consent was obtained from male and female occasional drinkers. Subjects were given pre-packaged envelopes containing capsules of crushed kudzu root or placebo and were instructed to take one packet three times a day for two consecutive days. Subjects then called the laboratory and left a message stating that they had taken the medication; the date and time of the calls were stamped on the tape. On the third day they took their morning dose, returned to the laboratory and participated in an ethanol challenge experiment as follows: An indwelling i.v. catheter was inserted into an antecubital vein for blood sampling, a blood pressure cuff affixed to the appointed arm, standard EKG leads were attached to the chest and a thermistor probe was attached to a fingertip to measure heart rate and skin temperature, respectively (Figure 10.1). See Lukas *et al.* (1986a,b, 1991) for further details.

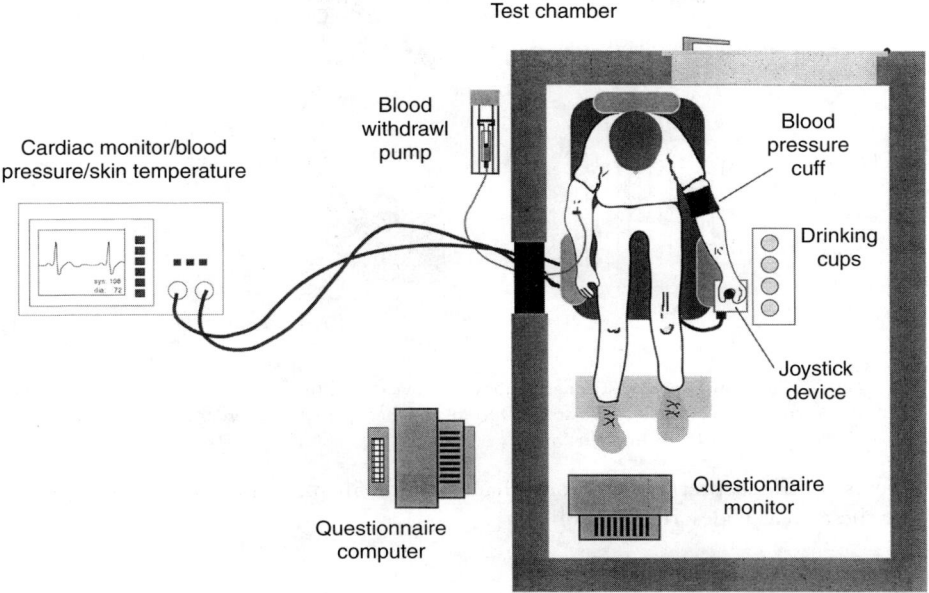

*Figure 10.1* Overview of experimental laboratory in which subjects are studied after active administration of alcohol. All data recording devices are located outside of the room that is devoid of other stimuli that could affect their subjective mood responses to the alcohol dose.

Subjects sat semi supine in a sound-attenuated chamber and were instructed to use a joystick device (Lukas *et al.*, 1986a) to answer computer-generated questionnaires including the Addiction Research Center Inventory (ARCI), Subjective High Assessment Scale (SHAS) and numerous Visual Analog Scales (VAS) that asked "How happy do you feel?", "How stimulated do you feel?", "How strong is your desire to use alcohol?", "How anxious do you feel?". In addition, subjects were queried regarding feelings of nausea, sweating, abdominal pain, headaches, etc. which might indicate a disulfiram-like reaction. The joystick device also served as a means of reporting "detection of alcohol effects" as well as episodes of intense good feelings ("euphoria") or bad feelings ("dysphoria") (Figure 10.2, close up of joystick). As this function of the joystick was continuously available, the subject could report rapid changes in mood state that occurred between the questionnaire sessions; this device has been used successfully in our laboratory for a number of years. After baseline measures were obtained, subjects were instructed to consume the beverage in the cup in a 20 min period. Measures of subjective reports of intoxication and blood pressure were sampled every 30 min for three hours after ethanol administration. Blood or breath samples were obtained every 30 min. Skin temperature and heart rate were measured every minute. Subjects returned a week later and repeated the experiment with the alternate pretreatment but received the same dose of ethanol.

The pilot studies had a number of specific aims: to assess the safety of kudzu itself, to ensure that there were no adverse events when subjects were challenged with ethanol, to explore the dose range of kudzu that might alter ethanol effects and to document medication compliance using added riboflavin. The main study was designed to compare

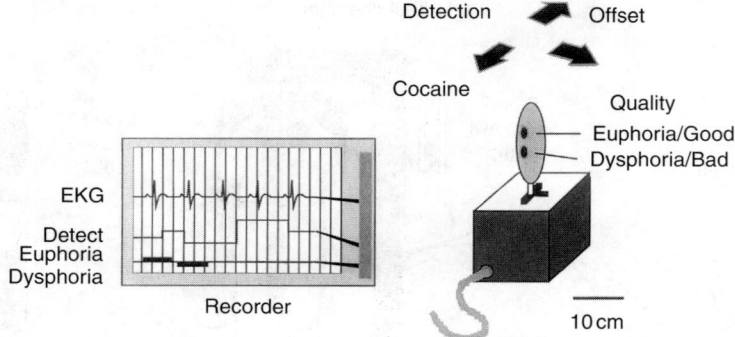

*Figure 10.2* Close up view of instrumental joystick device used by research subjects to continuously report changes in mood state. Output is either to a strip recorder (shown) or directly to computer for offline plotting.

the effects of kudzu pretreatment on ethanol effects in men and women. Progress in each of these areas is described as follows.

## Medication compliance

Medication compliance is always an issue in outpatient treatment studies. Generally, compliance is more problematic when the medication causes side effects or adverse

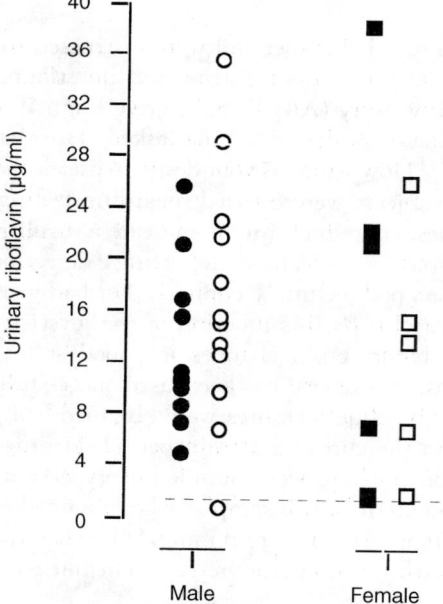

*Figure 10.3* Urinary riboflavin levels in subjects taking either placebo (open symbols) or kudzu (closed symbols) capsules in male and female research subjects. The dotted line represents normal riboflavin levels without supplement, suggesting that compliance for taking the medication was quite good.

events or can be distinguished from the placebo via taste, smell or texture. Fifty mg of riboflavin (vitamin B2) was added to each daily packet of kudzu and placebo capsules that the subjects were given to take over the 2 and 1/3 days. An analysis of the urinary riboflavin levels revealed equivalent and excellent compliance by the subjects (Figure 10.3). Urinary riboflavin levels ranged from 1–39 μg/ml regardless of whether subjects had taken kudzu or placebo. Normal dietary levels (dotted line) are in the <2 μg/ml range so it appears that this is an effective way of monitoring compliance. We are currently testing a urinary assay for puerarin to provide a more direct measure of compliance as well as the bioavailability of the puerarin-based preparation.

## Safety assessment of kudzu

A total of 49 male and female subjects have been treated on 1–6 occasions with the kudzu root preparation. Doses were taken as capsules, three times daily, for 2 and 1/3 days before they were challenged with either 0.7 g/kg ethanol or placebo. All subjects received both placebo and kudzu root preparation and no one was able to detect any effects of the kudzu preparation. There were no subjective reports of changes in mood state, sleep/wake profiles, appetite, diarrhea or nausea. Furthermore, a comprehensive battery of blood and urinalysis assessments was performed both before and after kudzu treatment (Table 10.1). Subjects' lab results were essentially unchanged from their baseline values. The only report that came back from the lab was the notation that the urine was bright yellow, which was due to the added riboflavin. The second safety issue related to the ethanol challenges. As we were initially concerned that, while a full

*Table 10.1* Blood chemistry, metabolic and urinalysis profiles after kudzu treatment

| Test | Pre kudzu | Post kudzu | Normal range | Units |
|---|---|---|---|---|
| *Metabolic* | | | | |
| GGT | 22.57 ± 10.91 | 20.29 ± 8.14 | 0–45 | μ/l |
| LDH | 146.43 ± 15.10 | 148.14 ± 19.28 | 0–250 | μ/l |
| SGPT | 21.71 ± 9.29 | 15.00 ± 4.73 | 0–48 | μ/l |
| Triglycerides | 124.14 ± 90.52 | 95.14 ± 39.28 | <200 | μ/l |
| Glucose | 87.83 ± 8.61 | 79.86 ± 14.87 | 70–115 | mg/dl |
| BUN | 13.57 ± 4.35 | 13.43 ± 2.7 | 7.0–25 | mg/dl |
| creatinine | 0.83 ± 0.10 | 0.76 ± 0.11 | 0.5–1.4 | mg/dl |
| BUN/creatinine ratio | 16.14 ± 4.18 | 17.86 ± 4.18 | 6.0–25 | (calc) |
| Sodium | 141.57 ± 2.23 | 140.00 ± 1.53 | 135–146 | meq/l |
| Potassium | 4.53 ± 0.35 | 4.06 ± 0.17 | 3.5–5.3 | meq/l |
| Chloride | 104.86 ± 2.48 | 105.00 ± 2.24 | 95–108 | meq/l |
| Calcium | 9.83 ± 0.61 | 8.61 ± 0.18 | 8.5–10.3 | meq/l |
| Protein, total | 7.64 ± 0.46 | 6.57 ± 0.24 | 6.0–8.5 | g/dl |
| Alkaline Phosphate | 71.00 ± 7.53 | 62.43 ± 6.43 | 20–125 | μ/l |
| SGOT | 23.43 ± 11.22 | 16.57 ± 3.95 | 0–42 | μ/l |
| *Hematology* | | | | |
| Hemoglobin | 15.87 ± 0.73 | 17.69 ± 9.94 | 13.8–15.6 | g/dl |
| Hematocrit | 46.65 ± 2.09 | 48.21 ± 20.79 | 41–46 | % |
| MCV | 92.58 ± 2.55 | 90.39 ± 3.23 | 80–100 | FL |
| Platelet count | 240.83 ± 48.75 | 211.00 ± 49.06 | 130–400 | thous/mcl |
| *Urinalysis* | | | | |
| Specific gravity | 1.02 ± 0.01 | 1.02 ± 0.01 | 1.001–1.035 | |
| pH | 5.64 ± 0.38 | 5.64 ± 0.69 | 4.6–8.0 | |

disulfiram-like reaction was unlikely, some of the isoflavones in kudzu would inhibit a variant of the ALDH enzyme and might precipitate nausea after drinking ethanol. This was an unwarranted concern as subjects did not report feeling nauseous and plasma acetaldehyde levels were not increased in kudzu-treated subjects (Figure 10.4).

Once the safety of the kudzu preparation, both alone and in combination with alcohol, had been established, we explored the efficacy of this preparation in altering acute alcohol effects. The following study was designed to study the effects of kudzu in a group of light and moderate drinking volunteers who did not meet criteria for alcohol dependence.

## Subjective mood effects

Using the joystick device, subjects report "feeling" the effects of ethyl alcohol rather quickly during the drinking phase (Figure 10.5, top). A lower dose of alcohol results in a delayed response and less intense magnitude of effects (Figure 10.5, middle) and placebo alcohol results in a highly variable response characterized by a delayed detection and short duration of effect (Figure 10.5, bottom). Kudzu pretreatment does not appreciably alter the behavioral profile of the higher dose of alcohol.

Figures 10.6–10.8 show the results of some of the subjective mood measures (e.g. SHAS, ARCI, and VAS) in female light drinkers. In this series of figures, both the area under the curve (AUC) of the subject's responses and the maximal response were analyzed. Kudzu pretreatment failed to change measures of how "drunk," "high," "confused," or whether subjects had "slurred speech" after consuming a 0.7 g/kg dose of ethanol (Figure 10.6). However, certain other, more negative, effects of alcohol appeared to be attenuated in this population; these included how "terrible," "anxious," "nausea," and "uncomfortable" (Figure 10.7). A similar attenuation in two of the scales in the ARCI were noted: the MBG or "morphine-benzedrine" scale and the LSD scale were attenuated in the kudzu-treated subjects (Figure 10.8). The MBG scale is a general indicator of euphoria or pleasant feelings and the LSD scale measures dysphoria. Figure 10.9 depicts the subjects' response to the question "What is your desire to NOT use alcohol?". Kudzu pretreatment markedly increased both the AUC and maximal response scores to this measure suggesting that it may have some effect on craving for alcohol. These data are preliminary and are in a limited number of female subjects who report drinking only 6–8 beers per week. Thus, the generalizability of these data to males as well as to individuals who have different drinking histories will remain to be seen.

The effects of kudzu pretreatment on blood alcohol levels is an area that we are currently pursuing. Although, the isoflavones in the kudzu root may inhibit certain enzymes responsible for alcohol's metabolism *in vitro*, it remains to be demonstrated that this translates to an altered blood alcohol level *in vivo*. However, some of our pilot data suggest that plasma alcohol levels are slightly higher in some subjects; the significance of this finding and the reason for this response in a selected group of subjects is currently under study.

## Summary and implications for future developments

One intention of these types of studies is to demonstrate that a new medication blocks or reverses the positive effects of alcohol and thus might be useful as a treatment medication much like the opiate receptor antagonists naloxone and naltrexone block heroin's

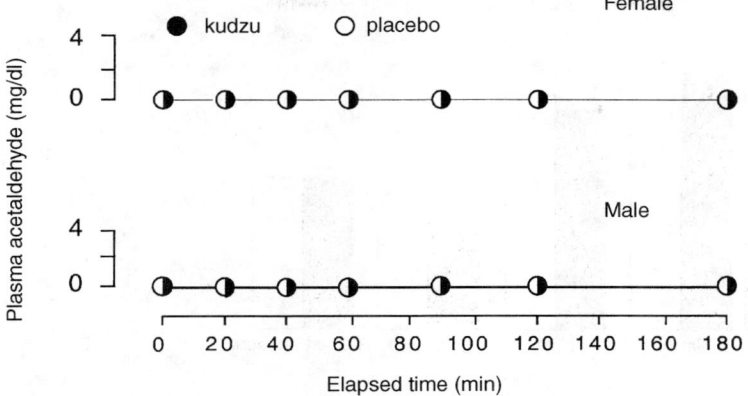

*Figure 10.4* Effects of kudzu pretreatment on ethanol-induced alterations in plasma acetaldehyde levels in a representative female (top) and male (bottom) social drinker. Ethanol dose was 0.7 g/kg and was consumed from 0–20 min.

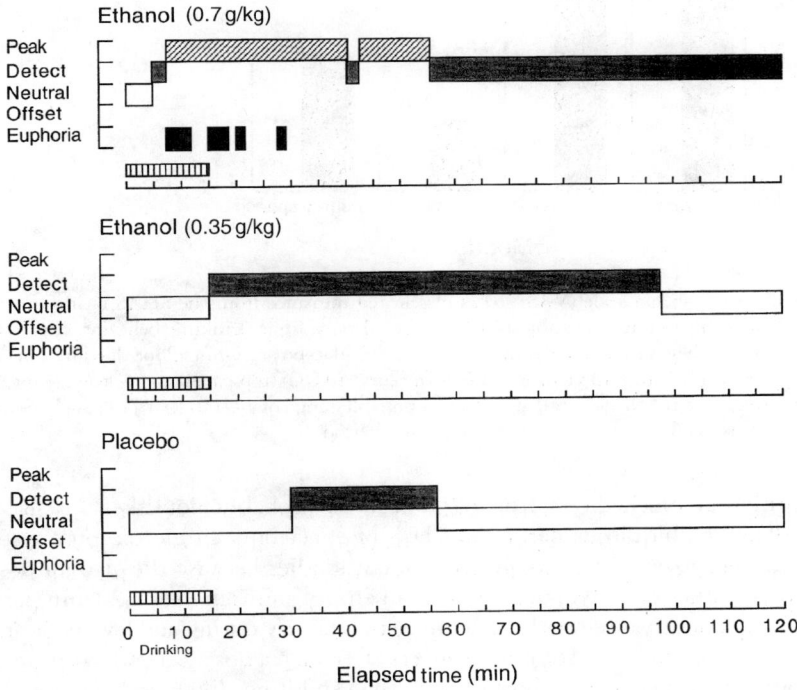

*Figure 10.5* Behavioral profile of the effects of acute alcohol administration using a continuously available joystick device. Data are from one representative male subject who consumed the beverages over a 15 min interval. Each visit was scheduled at least one week apart.

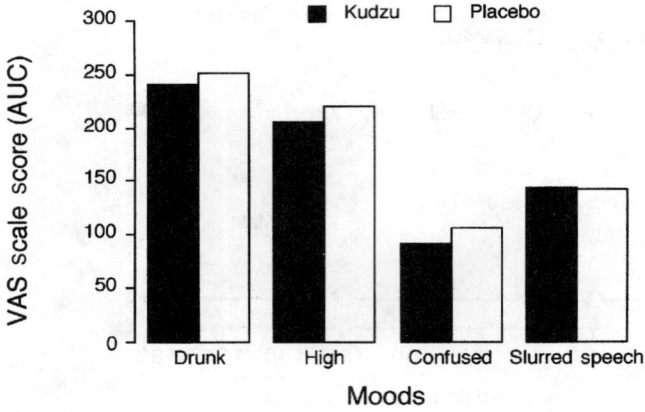

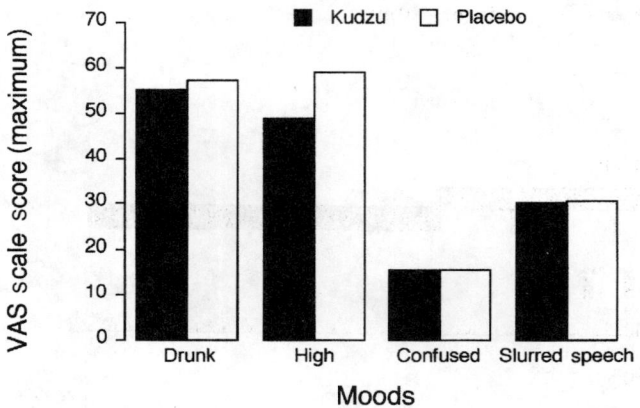

*Figure 10.6* Visual analog scale (VAS) scores of selected measures from the SHAS rating questionnaire in five female subjects who reported only light drinking behavior. Scores from
when they were treated with either kudzu or placebo are summed for the duration of the
study (180 min) to yield area under the curve (AUC) (top panel) or the average maximal
response to ethanol challenge (from a possible range of 1–100) (bottom panel). Subjects
consumed 0.7 g/kg ethanol over a 20 min period.

effects. Unequivocal blockade of ethanol's effects is probably not likely or feasible
because of ethanol's ubiquitous nature and lack of effect on a single receptor system.
Further, we did not expect blockade in the present studies because the alcohol dose is
equivalent to 6–7 shots of vodka, delivered on an empty stomach over a 20 min period.
This relatively high dose was selected to maximize the toxicity testing and few medications
would be expected to block the subjective effects after such a dose. Second, the duration
of kudzu treatment was relatively short (2–1/3 days) and is not likely to have been long
enough to cause significant changes that would alter ethanol's acute effects. Third, the
inability of a medication to block all of the acute effects of alcohol does not negatively
impact its potential use as a treatment for alcoholism. Blockade of positive effects is

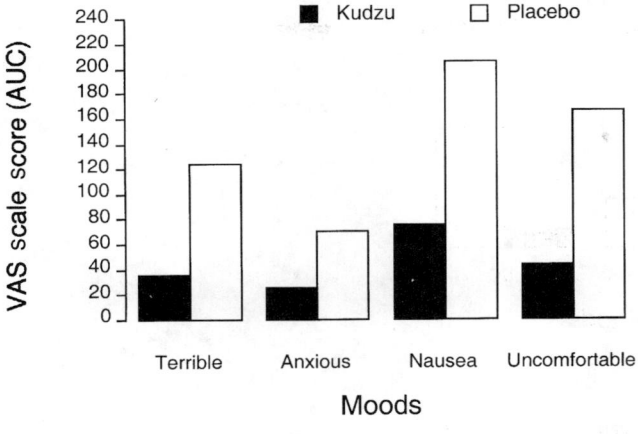

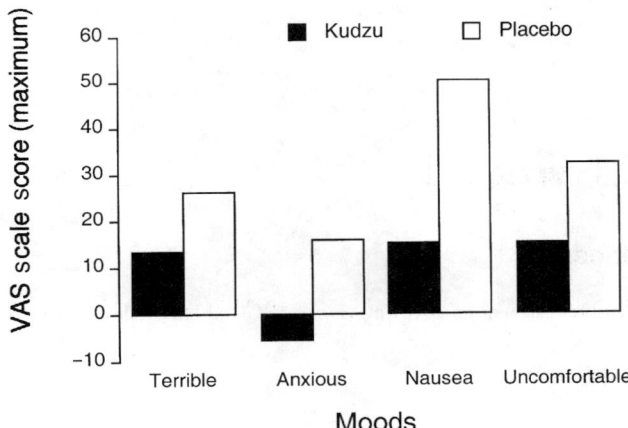

*Figure 10.7* Effects of kudzu pretreatment on another set of VAS and SHAS measures. Other details as in Figure 10.6.

only one of a number of strategies that is being explored for developing a medication for alcoholism. In contrast, we found preliminary evidence that kudzu pretreatment altered a number of ethanol's effects, both positive and negative. This complex profile of effects is intriguing and suggests that kudzu may have potential in managing some aspects of alcohol abuse and alcoholism. The final test of this hypothesis must be conducted in actual clinical trials of the medication in real world conditions.

Studies designed to measure drinking behavior in a patient's "home" environment have traditionally relied on subjective reports that are collected in daily diaries; such diaries are often completed at the end of the day. We are currently engaged in studies designed to measure such behavior using a real time data recorder in the form of a small wristwatch device called ActiWatch® Score (Mini-Mitter Co., OR). The device (see Figure 10.10) monitors sleep/wake activity and has a single button on its face that the subject uses to record when they have consumed an alcoholic beverage, tobacco cigarette, or other drugs as directed. In addition, the watch "beeps" about every 3 h to which the

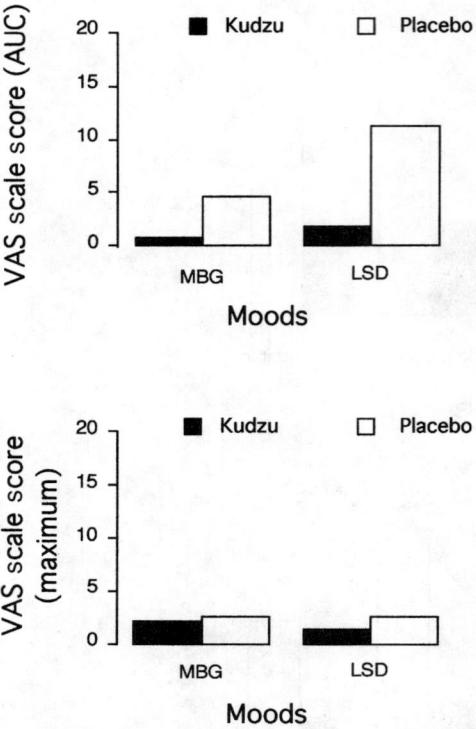

Figure 10.8 Effects of kudzu pretreatment on ARCI measures MBG and LSD scales. Other details as in Figure 10.6.

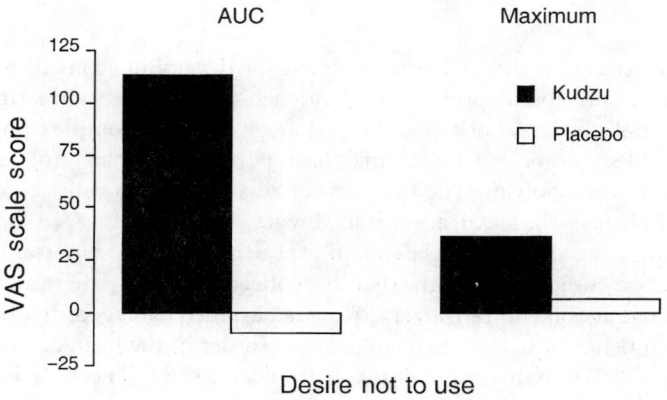

Figure 10.9 Effects of kudzu pretreatment on subjects' desire to NOT use alcohol. Other details as in Figure 10.6.

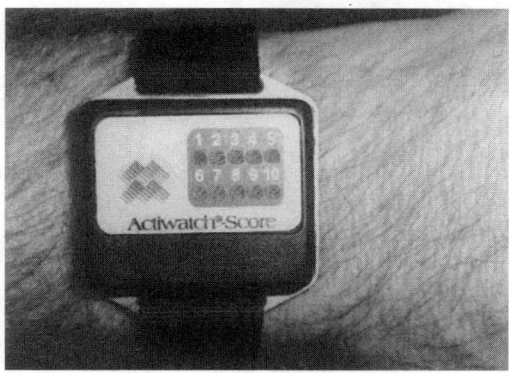

*Figure 10.10* Close up view of the ActiWatch-Score device that is used to monitor sleep/wake activity and provide a means for the subjects to enter data about their drinking behavior in real time.

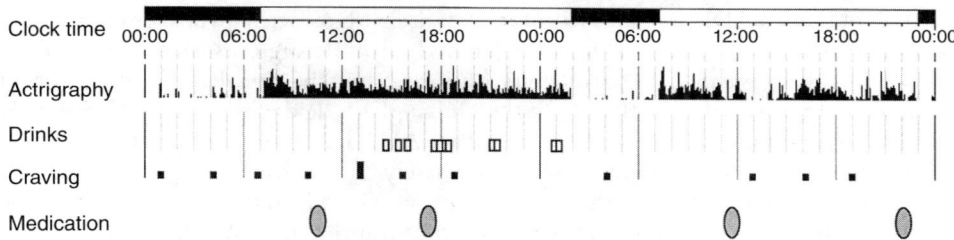

*Figure 10.11* Representative data from a single subject who wore the ActiWatch device over a 2-week period; only a 24 h period is shown. Clock time is depicted along the top panel and actigraphy is depicted in the next row as dense black vertical line. Each drink is recorded and is shown in the next row. Note the "binge" drinking pattern that occurred on a Friday night. Craving for alcohol is shown in the bottom panel; the magnitude of the craving is highest just prior to the binge drinking episode.

subject is required to enter a number from 1 to 10 indicating his/her desire to drink alcohol at that time. Typical data from this device is depicted in Figure 10.11. Sleep/wake activity is indicated by the density of vertical lines in the upper tracing. Note the cluster like "binge" drinking pattern in this female subject that began in late afternoon and continued into the evening. Her craving for alcohol peaked early in the evening before she began the binge. This technology will increase out ability to collect precise data on subjects' drinking patterns and so will serve as a solid foundation upon which to study new medications for treating alcoholism.

In summary, pueraria-based preparations have been used for centuries by Chinese herbalists to treat a variety of ailments, including alcohol-related problems in humans. With the recent demonstration that such preparations can reduce alcohol drinking in a number of animal species, further scrutiny of these preparations as possible medications for treating alcoholism are warranted. The results of pilot controlled laboratory studies reveals that kudzu/ethanol combinations do not result in increased acetaldehyde levels or in disulfiram-like reactions, attesting to the safety of this pueraria-based preparation in an alcohol-drinking population. As the highest ethanol dose used in these pilot studies is equivalent to about 5–6 shots of vodka, there is little chance that any adverse side effects will emerge from future studies. These results further show that kudzu treatment may attenuate low dose ethanol-induced intoxication, but has less of an effect on high dose ethanol challenges.

The mechanism by which these pueraria-based preparations reduce ethanol consumption in animals remains unknown. However, there are some interesting prospects that involve the dynamics of ethanol metabolism. Ethanol is first oxidized mainly by ADH to acetaldehyde. Further oxidation of acetaldehyde to acetate or its activated form, acetyl coenzyme A, is mainly catalyzed by the enzyme ALDH. As the *in vivo* oxidation of acetaldehyde is faster than ethanol, accumulation of the former does not usually occur under normal conditions and 85–90 per cent of this process occurs in the liver. As ALDH is present in nearly every tissue of the human body (von Wartburg, 1971) it is conceivable that it fulfills some, as yet unknown, metabolic function in other pathways. Biochemically, isoflavones also inhibit a number of enzymes including tyrosine-specific protein kinase (Akiyama *et al.*, 1987) and DNA topoisomerase II (Okura *et al.*, 1988), but the significance of this to alcohol abuse is unknown.

Although one isoflavone in the kudzu plant, daidzein, inhibits the class I (γ-type) isozymes of human ADH (Keung, 1993) and studies with 4-methylpyrazole, a compound that inhibits human class I ADH (Waller *et al.*, 1982), suggests that it suppresses ethanol intake in ethanol-preferring rats by inhibiting the metabolic elimination of ingested ethanol. Another isoflavone in the kudzu plant, daidzin, is a potent, reversible and selective inhibitor of human mitochondrial ALDH-2 (Keung and Vallee, 1993a). As alcohol abuse is rare among the approximately 50 per cent of Asians who have inherited an inactive mutant form of this enzyme (Goedde and Agarawal, 1990), we speculated that daidzin may suppress ethanol consumption by mimicking the effect of this apparently harmless natural mutation of the human mitochondrial ALDH gene. In marked contrast, disulfiram (Antabuse$^{TM}$) irreversibly inhibits the cytosolic isozyme ALDH-1 by covalently modifying sulfhydryl groups (Banys, 1988). It appears that daidzin and daidzein differ from disulfiram in their selective and reversible inhibition of ALDH-2 (Keung and Vallee, 1993a) and puerarin does not inhibit ALDH at all. However, they later reported that daidzin does not affect overall acetaldehyde metabolism in hamster (Keung *et al.*, 1995) and we found no increase in acetaldehyde levels after kudzu treatment so the suppression of alcohol intake is not likely an Antabuse-like effect.

There are two alternative hypotheses for the mechanism of action of these isoflavones. The first, proposed by Lin *et al.* (1996) suggests (although they did not have any data for this) that these isoflavones may act in the CNS to suppress or stimulate ethanol-responsive neurotransmitter or modulator systems in the reward pathway. Correlation studies using a series of daidzin analogs have led to the hypothesis that antidipsotropic isoflavones might act via the mitochondrial MAO/ALDH pathway and that a biogenic aldehyde such as 5-HIAL or DOPAL may be important in mediating their antidipsotropic action (Rooke *et al.*, 2000). There is a substantial rodent literature showing that there is a significant negative correlation between brain 5-HT activity and spontaneous ethanol intake. Lin and Li (1998) showed that puerarin, daidzin and daidzein suppressed voluntary alcohol consumption in alcohol-preferring rats. This effect was accompanied by an increase in water intake, but daily food consumption was unchanged. They postulated that these isoflavonoids are working via the brain reward pathways.

The second mechanism is that these isoflavones may alter cerebral blood flow that somehow affects the amount of ethanol that enters (or remains in) the brain. Evidence for this hypothesis is scarce although there is an approved intravenous preparation of one of the kudzu isoflavones, puerarin, that is used as a vasodilator in China to "expand coronary arteries and brain blood vessels, reduce myocardial oxygen consumption and improve micro-circulation."

The mechanism of action of kudzu in altering the effects of alcohol in animals can only be speculated at this time. Possible biochemical/metabolic effects are implied from animal studies and the human studies to date have not been designed to reveal a mechanism of action. Regardless of how it may work, the encouraging preclinical and clinical data, lack of toxicity and relatively high availability of this pueraria-based preparation suggests that further clinical studies of their utility in treating alcohol-related disorders is warranted.

## REFERENCES

Akiyama, T., Ishida, J., Nakagawa, S., Ogawara, H., Watanabe, B., Itoh, N. *et al.* (1987) Genistein, a specific inhibitor of tyrosine-specific protein kinases. *J. Biol. Chem.*, 262, 5592–5595.

Altshuler, H.L., Phillips, P.E. and Feinhandler, D.A. (1980) Alteration of ethanol self-administration by naltrexone. *Life Sci.*, 26, 679–688.

Banys, P. (1988) The clinical use of disulfiram (Antabuse): a review. *J. Psychoactive Drugs*, 20, 243–261.

Bors, W., Heller, W., Michel, C. and Saran, M. (1990) Flavonoids as antioxidants: determination of radical-scavenging efficiencies. *Methods Enzymol.*, 186, 343–355.

Cahalan, D., Cissin, I. and Crossley, H. (1969) *American Drinking Practices*. Rutgers Center for Alcohol Studies, New Brunswick, NJ.

Cheng, E.W., Yoder, L., Story, C.D. and Burroughs, C.D. (1955) *Ann. N.Y. Acad. Sci.*, 61, 637–736.

Davidson, D. and Amit, Z. (1996) Effects of naloxone on limited-access ethanol drinking in rats. *Alcohol. Clin. Exp. Res.*, 20, 664–669.

Davidson, D., Swift, R. and Fitz, E. (1996) Naltrexone increases the latency to drink alcohol in social drinkers. *Alcohol. Clin. Exp. Res.*, 20, 732–739.

Di Perri, T. and Auteri, A. (1988) Action of S-5682 on the complement system. *In vitro* and *in vivo* study. *Int. Angiol.*, 7(2 Suppl), 11–15.

Frezza, M., DiPadova, C., Pozzato, G., Terpin, M., Baraona, E. and Lieber, C.S. (1990) High blood alcohol levels in women: The role of decreased gastric alcohol dehydrogenase and first-pass metabolism. *N. Engl. J. Med.*, 322, 95–99.

Goedde, H.W. and Agarawal, D.P. (1990) Pharmacogenetics of aldehyde dehydrogenase (ALDH). *Pharmac. Ther.*, 45, 345–371.

Gorelick, D.A. (1986) Effect of fluoxetine on alcohol consumption. *Alcohol. Clin. Exp. Res.*, 10, 113.

Harada, M. and Ueno, K. (1975) Pharmacological studies on pueraria root. I. Fractional extraction of pueraria root and identification of its pharmacological effects. *Chem. Pharm. Bull.*, 23, 1798–1805.

Havsteen, B. (1983) Flavonoids, a class of natural products of high pharmacological potency. *Biochem. Pharmacol.*, 32, 1141–1148.

Heyman, G.M., Keung, W.M. and Vallee, B.L. (1996) Daidzin decreases ethanol consumption in rats. *Alcohol. Clin. Exp. Res.*, 20, 1083–1087.

Hill, S.Y. (1984) Vulnerability to the biomedical consequences of alcoholism and alcohol-related problems among women. In S.C. Wilsnack and L.J. Beckman (eds), *Alcohol Problems in Women. Antecedents, Consequences, and Interventions.* Guilford Press, New York, pp. 121–154.

Huang, K.C. (1999) *The Pharmacology of Chinese Herbs.* CRC Press, Boca Raton.

Jaffe, J.H., Kranzler, H.R. and Ciraulo, D.A. (1992) Drugs used in the treatment of alcoholism. In J.H. Mendelson and N.K. Mello (eds), *Medical Diagnosis and Treatment of Alcoholism*, McGraw Hill Book Co., New York.

Johnson, B.A., Campling, G.M., Griffiths, P. and Cowen, P.J. (1993) Attenuation of some alcohol-induced mood changes and the desire to drink by 5-HT3 receptor blockade: a preliminary study in healthy male volunteers. *Psychopharmacol.*, 112, 142–144.

Johnson, B.A. *et al.* (2000) Ondansetron for reduction of drinking among biologically pre-disposed alcoholic patients. *J. Am. Med. Assoc.*, 284, 963–971.

Johnston, L.D. *et al.* (1997) *Monitoring the Future Study.* Institute for Social Research, University of Michigan.

Kelly, G.E., Joannou, G.E., Reeder, A.Y., Nelson, C. and Waring, M.A. (1995) The variable metabolic response to dietary isoflavone in humans. *Proc. Soc. Exp. Biol. Med.*, 208, 40–43.

Keung, W.M. (1993) Biochemical studies of a new class of alcohol dehydrogenase inhibitors from *Radix puerariae. Alcohol. Clin. Exp. Res.*, 17, 1254–1260.

Keung, W.M. and Vallee, B.L. (1993a) Daidzin: a potent, selective inhibitor of human mito-chondrial aldehyde dehydrogenase. *Proc. Natl. Acad. Sci. USA.*, 90, 1247–1251.

Keung, W.M. and Vallee, B.L. (1993b) Daidzin and daidzen suppress free-choice ethanol intake by Syrian Golden hamsters. *Proc. Natl. Acad. Sci. USA.*, 90, 10008–10012.

Keung, W.M. and Vallee, B.L. (1994) Therapeutic lessons from traditional Oriental medicine to contemporary Occidental pharmacology. In B. Jansson, H. Jörnvall, U. Rydberg, L. Terenius and B.L. Vallee (eds), *Toward a Molecular Basis of Alcohol Use and Abuse*, Birkhäuser Verlag, Basel, pp. 371–381.

Keung, W.M., Lazo, O., Kunze, L. and Vallee, B.L. (1995) Daidzin suppresses ethanol con-sumption by Syria Golden hamsters without blocking acetaldehyde metabolism. *Proc. Natl. Acad. Sci. USA.*, 92, 8990–8993.

Keung, W.M. and Vallee, B.L. (1998) Daidzin and its antidipsotropic analogs inhibit seroto-nin and dopamine metabolism in isolated mitochondria. *Proc. Natl. Acad. Sci. USA*, 95, 2198–2203.

Kranzler, H.R. and Orrok, B. (1989) The pharmacotherapy of alcoholism. In A. Tasman, R.E. Hales and A.J. Frances (eds), *Review of Psychiatry*, American Psychiatric Press, Inc, Washington, D.C., pp. 359–379.

Kraus, M., Gottlieb, L.D., Horwitz, R.I. and Anscher, M. (1985) Randomized clinical trial of atenolol in patients with alcohol withdrawal. *N. Engl. J. Med.*, 313, 905–909.

Leslie, S.W., Brown, L.M., Dildy, J.E. and Sims, J.S. (1990) Ethanol and neuronal calcium channels. *Alcohol*, 7, 233–236.

Li, S.C. (AD 1590–1596) *Ben Cho Gang Mu.*

Lichtigfeld, F.J. and Gillman, M.A. (1982) The treatment of alcoholic withdrawal states with oxygen and nitrous oxide. *S. Afr. Med. J.*, 61, 349–351.

Lin, R.C., Guthrie, S., Xie, C.-Y., Lee, D.Y., Lumeng, L. and Li, T.-K. (1996) Isoflavonoid compounds extracted from *Pueraria lobata* suppress alcohol preference in a pharmacogenetic rat model of alcoholism. *Alcohol. Clin. Exp. Res.*, 20, 659–663.

Lin, R.C. and Li, T.K. (1998) Effects of isoflavones on alcohol pharmacokinetics and alcohol-drinking behavior in rats. *Am. J. Clin. Nutr.*, 68, 1512S-1515S.

Linnoila, M., Eckardt, M., Duncan, M., Lister, R. and Martin, P. (1987) Interactions of serotonin with ethanol: Clinical and animal studies. *Psychopharmacol. Bull.*, 23, 452–457.

Liskow, B.I. and Goodwin, D.W. (1987) Pharmacological treatment of alcohol intoxication, withdrawal and dependence: A critical review. *J. Stud. Alcohol*, 48, 356–370.

Litten, R.Z. and Allen, J.P. (1991) Pharmacotherapies for alcoholism: Promising agents and clinical issues. *Alcohol. Clin. Exp. Res.*, 15, 620–633.

Litten, R.Z., Allen, J. and Fertig, J. (1996) Pharmacotherapies for alcohol problems: A review of research with focus on developments since 1991. *Alcohol. Clin. Exp. Res.*, 20, 859–876.

Lukas, S.E., Benedikt, R., Mendelson, J.H., Kouri, E., Sholar, M. and Amass, L. (1992) Marihuana attenuates the rise in plasma ethanol levels in human subjects. *Neuropsychopharmacol.*, 7(1), 77–81.

Lukas, S.E., Daniels, S.L., Lundahl, L.H., Kern, B.J. and Wines, J. (1997) Kudzu, a Chinese herb, attenuates ethanol's subjective effects in women. In *American College of Neuropsychopharmacology (ACNP)*, Kamuela, Hawaii, December.

Lukas, S.E., Lex, B.W., Slater, J., Greenwald, N. and Mendelson, J.H. (1989) A microanalysis of ethanol-induced disruption of body sway and psychomotor performance in women. *Psychopharmacology*, 98, 169–175.

Lukas, S.E., Lundahl, L.H., Lachance, M., Duggan, S., Stull, M. and Lee, D.Y. (1999) The Chinese herb, kudzu, alters ethanol effects in male and female subjects. *The College on Problems of Drug Dependence*, Acapulco.

Lukas, S.E., Mendelson, J.H., Amass, L., Benedikt, R.A., Henry, J.H. and Kouri, E.M. (1991) Electrophysiologic correlates of ethanol reinforcement. In G.F. Koob, M.J. Lewis, R.E. Meyer and S. Paul (eds), *Neuropharmacology of Ethanol: New Approaches*, Boston: Birkhauser Boston Inc, pp. 201–231.

Lukas, S.E., Mendelson, J.H. and Benedikt, R.A. (1986a) Instrumental analysis of ethanol-induced intoxication in human males. *Psychopharmacology*, 89, 8–13.

Lukas, S.E., Mendelson, J.H., Benedikt, R.A. and Jones, B. (1986b) EEG alpha activity increases during transient episodes of ethanol-induced euphoria. *Pharmacol. Biochem. Behav.*, 25, 889–895.

Lundh, T. (1995) Metabolism of estrogenic isoflavones in domestic animals. *Proc. Soc. Exp. Biol. Med.*, 208, 33–39.

Malcolm, R., Ballenger, J.C., Sturgis, E.T. and Anton, R. (1989) Double-blind controlled trial comparing carbamazepine to oxazepam treatment of alcohol withdrawal. *Am. J. Psychiatry.*, 146, 617–621.

McCaul, M.E., Turkkan, J.S., Svikis, D.S. and Bigelow, G.E. (1991) Familial density of alcoholism: Effects on psychophysiological responses to ethanol. *Alcohol*, 8, 219–222.

McGinnis, J. and Foege, W. (1993) Actual causes of death in the United States. *J. Am. Med. Assoc.*, 270, 2208.

Miksicek, R.J. (1995) Estrogenic flavonoids: Structural requirements for biological activity. *Proc. Soc. Exp. Biol. Med.*, 208, 44–50.

Nakamoto, H., Iwasaki, Y. and Kizu, H. (1977) [The study of aqueous extract of *Puerariae radix*. IV. The isolation of daidzin from the active extract (MTF-101) and its antifebrile and spasmolytic effect (author's transl.)], *Yakugaku Zasshi*, 97, 103–105.

Naranjo, C.A., Sellers, E.M., Roach, C.A., Woodley, D.V., Sanchez-Craig, M. and Sykora, K. (1984) Zimelidine-induced variations in alcohol intake by nondepressed heavy drinkers. *Clin. Pharmacol. Ther.*, 35, 374–381.

Naranjo, C.A., Sellers, E.M., Sullivan, J.T., Woodley, D.V., Kadlec, K.E. and Sykora, K. (1987) The serotonin uptake inhibitor citalopram attenuates ethanol intake. *Clin. Pharmacol. Ther.*, 41, 266–274.

NIAAA (1993a) *Eighth Special Report to US Congress on Alcohol and Health*. US Department of Health and Human Services, Washington, D.C., p. 16.

NIAAA (1993b) *Eighth Special Report to US Congress on Alcohol and Health*. US Department of Health and Human Services, Washington, D.C., p. 261.

NIAAA (1993c) *Eighth Special Report to US Congress on Alcohol and Health*. US Department of Health and Human Services, Washington, D.C., p. 275.

NIAAA (1994) *Alcohol Health & Res. World*, 18, 243, 245.

NIAAA (1998) News Release.

Niiho, Y., Yamazaki, T., Nakajima, Y., Itoh, H., Takeshita, T., Kinjo, J. and Nohara, T. (1989) [Pharmacological studies on puerariae flos. I. The effects of *Puerariae flos* on alcoholic metabolism and spontaneous movement in mice.] *Yakugaku Zasshi*, 109, 424–431.

Nutt, D., Adinoff, B. and Linnoila, M. (1989) Benzodiazepines in the treatment of alcoholism. In M. Galanter (ed.), *Recent Developments in Alcoholism: Treatment Research*, vol. 7, Plenum Press, New York, pp. 283–313.

O'Malley, S.S., Jaffe, A.J., Chang, G., Schottenfeld, R.S., Meyer, R.E. and Rounsaville, B. (1992) Naltrexone and coping skills therapy for alcohol dependence. *Arch. Gen. Psychiatry*, 49, 881–887.

Okura, A., Arakawa, H., Oka, H., Yoshinari, T. and Monden, Y. (1988) Effect of genistein on topoisomerase activity and on the growth of [Val 12] ha-ras-transformed NIH 3T3 cells. *Biochem. Biophys. Res. Commun.*, 157, 183–189.

Overstreet, D.H., Lee, Y.W., Rezvani, A.H., Pei, Y.H., Criswell, H.E. and Janowsky, D.S. (1996) Suppression of alcohol intake after administration of the Chinese herbal medicine, NPI-028, and its derivatives. *Alcohol. Clin. Exp. Res.*, 20, 221–227.

Overstreet, D.H., Lee, Y.-W., Chen, Y.T. and Rezvani, A.H. (1998) The Chinese herbal medicine NPI-028 suppresses alcohol intake in alcohol-preferring rats and monkeys without inducing taste aversion. *J. Perfusion*, 11, 381–389.

Paille, F.M., Guelfi, J.D., Perkins, A.C., Royer, R.J., Steru, L. and Parot, P. (1995) Double-blind randomized multicentre trial of acamprosate in maintaining abstinence from alcohol. *Alcohol Alcohol.*, 30, 239–247.

Palestine, M.L. and Alatorre, E. (1976) Control of acute alcoholic withdrawal symptoms. A comparative study of haloperidol and chlordizepoxide. *Curr. Ther. Res.*, 20, 289–299.

Rooke, N., Li, D.-J., Li, J. and Keung, W.M. (2000) The mitochondrial monoamine oxidase-aldehyde dehydrogenase pathway-A potential site of action of daidzin. *J. Med. Chem.*, 43, 4169–4179.

SAMHSA (1994) *National Household Survey on Drug Abuse: Main Findings*, p. 138.

Sass, H., Soyka, M., Mann, K. and Zieglgansberger, W. (1996) Relapse prevention by acamprosate: Results from a placebo-controlled study on alcohol dependence. *Arch. Gen. Psychiatry.*, 53, 673–680.

Schuckit, M.A. (1985) Ethanol-induced changes in body sway in men at high alcoholism risk. *Arch. Gen. Psychiatry.*, 42, 375–379.

Sellers, E.M. and Kalant, H. (1976) Alcohol intoxication and withdrawal. *N. Engl. J. Med.*, 294, 757–762.

Shebek, J. and Rindone, J.P. (2000) A pilot study exploring the effect of kudzu root on the drinking habits of patients with chronic alcoholism. *J. Altern. Compliment. Med.*, 6, 45–48.

Simon, R.P. (1988) Alcohol and seizures. *N. Engl. J. Med.*, 319, 715–716.

Sun, S.-M. (c. AD 600) *Beiji-Quianjin-Yaofang*.

Swift, R.M. (1997) The pharmacological treatment of alcohol dependence. *Med. Health R.I.*, 80, 91–93.

Swift, R.M. (1999) Drug therapy for alcohol dependence. *N. Engl. J. Med.*, 340, 1482–1490.

Swift, R.M., Davidson, D., Wheliham, W. and Kuznetsov, O. (1996) Ondansetron alters human alcohol intoxication. *Biol. Psychiatry*, 40, 514–521.

Volpicelli, J., Alterman, A., Hayashida, M. and O'Brien, C. (1992) Naltrexone in the treatment of alcohol dependence. *Arch. Gen. Psychiatry.*, 49, 876–880.

von Wartburg, J.P. (1971) The metabolism of alcohol in normals and alcoholics: Enzymes. In B. Kissin and H. Begleiter (eds), *The Biology of Alcoholism*, Vol. 1, Plenum Press, New York, pp. 63–102.

Wähälä, K., Hase, T. and Adlercreutz, H. (1995) Synthesis and labeling of isoflavone phytoestrogens, including daidzein and genistein. *Proc. Soc. Exp. Biol. Med.*, **208**, 27–32.

Waller, M.B., McBride, W.J., Lumeng, L. and Li, T.-K. (1982) Effects of intravenous ethanol and of 4-methylpyrazole on alcohol drinking in alcohol-preferring rats. *Pharmacol. Biochem. Behav.*, **17**, 763–768.

Xie, C.I., Lin, R.C., Antony, V., Lumeng, L., Li, T.K., Mai, K., Liu, C., Wang, Q.D., Zhao, Z.H. and Wang, G.F. (1994) Daidzin, an antioxidant isoflavonoid, decreases blood alcohol levels and shortens sleep time induced by ethanol intoxication. *Alcohol. Clin. Exp. Res.*, **18**, 1443–1447.

Xu, X., Wang, H.J., Murphy, P.A., Cook, L. and Hendrich, S. (1994) Daidzein is a more bioavailable soymilk isoflavone than is genistein in adult women. *J. Nutr.*, **124**, 825–832.

# 11 Chemopreventive effects of isoflavones on estrogen-dependent diseases: osteoporosis and cancer of the breast, prostate and endometrium

*Chun-Kowk Wong*

## INTRODUCTION

Human vegetarian diets contain a complex array of naturally occurring bioactive non-nutrients called phytochemicals. The flavonoids exemplify one class of these naturally occurring phytochemicals in plants that cannot be synthesized by mammals. Dietary flavonoids are thought to have beneficial effects on human health. More than 4000 flavonoids have been identified. They can be classified as anthocyanidins, flavans, flavanones, flavones, flavonols, isoflavonols, isoflavanones and isoflavones (Das, 1994). The isoflavones also belong to a class of phytochemicals called "phytoestrogens" because they exhibit estrogenic and/or anti-estrogenic properties (Zava *et al.*, 1998).

Common North American and West European diets contain very little isoflavones. Hence, their daily isoflavone intake range from only 1 to 3 mg (Jones *et al.*,1989; Messina *et al.*, 1994). In contrast, people living in Asia (China, Indonesia, Korea, Japan etc.) consume relatively large amounts of isoflavones, an average of 25–100 mg each day, mostly from dietary soy products. Epidemiological studies have suggested that isoflavones may play a preventive role in many hormone-dependent diseases including cancer of the breast, prostate, and endometrium (Lamartiniere *et al.*, 1998; Griffiths *et al.*, 1999); and symptoms that usually associated with menopause such as hot flashes (Murkies *et al.*, 1995) and osteoporosis (Ishimi *et al.*, 1999). The discovery that dietary isoflavones such as daidzein, genistein, biochanin A and formononetin (Figure 11.1) can bind to estrogen receptors (ER) (Shutt and Cox,1972; Miksicek, 1995) raised the possibility that the phytoestrogens may exert their beneficial effect by modulating estrogenic activity *in vivo*.

While the putative beneficial effects of isoflavones remain to be verified, an enormous market for this class of chemicals has already been created for neutriceutical industries, particularly for those in soybean-producing countries such as United States and Australia. Many health foods, drinks and food supplements are now marketed based on the content and putative beneficial effects of isoflavones. Tablets, pills, and capsules containing extracts of isoflavones are being marketed as hormone-replacement therapies for post-menopausal women, available over the counter throughout the world. Such products have raised much public interests and debates, and initiated intensive researches on the true efficacy, indication, safety and mechanism of action of these commercial products.

Using dietary phytoestrogens has now become a popular practice in the prevention of cancer, particularly for people at high risk. Pueraria plant components are rich in phytoestrogens and may be developed into dietary supplements for human consumption.

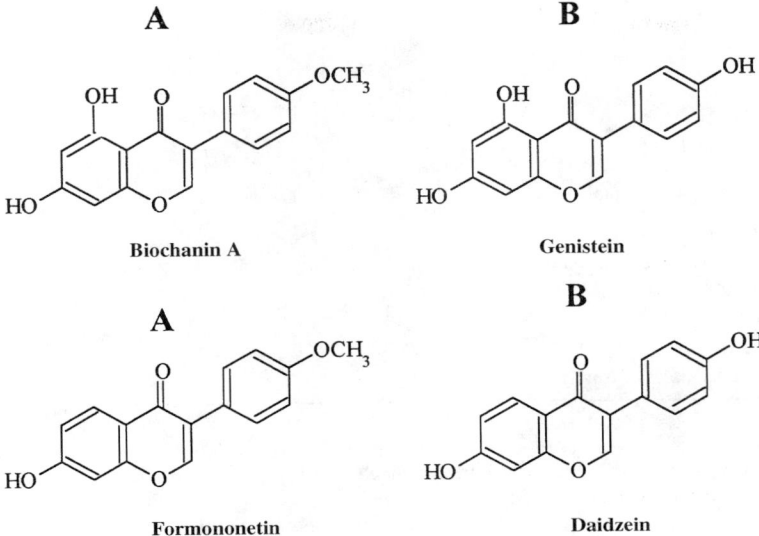

*Figure 11.1* Isoflavones in A, clover (*Trifolium* spp.) and B, soybean.

This article attempts to give a brief account on current researches in the chemopreventive effects of estrogenic isoflavones on estrogen-dependent disease ssuch as osteoporosis and cancer of the breast, prostate, and endometrium.

## ISOFLAVONES

### Chemical structure of isoflavones

The structural skeleton of an isoflavone (Figure 11.1) is strikingly similar to that of the mammalian estrogens (Setchell and Adlercreutz, 1988). A significant feature of the chemical structure of an isoflavone is the presence of a phenolic ring, a key structural element for ER binding (Leclerq and Heuson, 1979). In fact, the distance between the hydroxyl groups at the two ends of the estrogenic isoflavone metabolite equol is virtually identical to that between the 3- and 17$\beta$-hydroxylgroups of 17$\beta$-estradiol (Figure 11.2). For that matter, estrogenic isoflavones can compete with natural estrogens for binding to the ER and elicit weak estrogenic responses in estrogen-responsive tissues (Miksicek, 1994). In principle, they can act either as antagonists or partial agonists depending on whether or not there are natural estrogens around.

### Dietary sources of isoflavones

Isoflavones are mostly found in legumes, nuts and seeds (Table 11.1) (Franke *et al.*, 1994; Mazur *et al.*, 1996). Soybean, a major component of Asian diets, contains large amounts of the isoflavones daidzein and genistein. They are found either as the aglycone (unconjugated form) or as various types of sugar conjugates. The total amount of

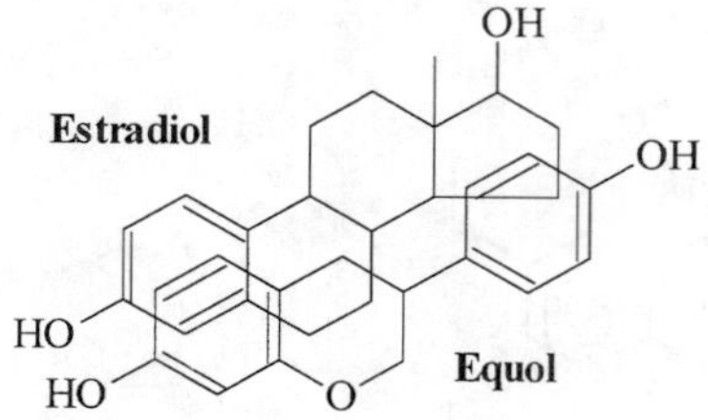

*Figure 11.2* Structures of equol and estradiol. Planar spatial arrangement shows striking similarity of the two molecules.

*Table 11.1* Food sources of isoflavones

| Isoflavones | Food |
| --- | --- |
| Daidzein and genistein | Black beans, Green split peas, Clover sprouts |
| Biochanin A and formononetin | Green beans, Chick peas, Lima beans, Split peas, Alfalfa sprouts, Clover sprouts, Sunflower seeds |

*Table 11.2* Total isoflavones in foods of Asian origin*

| Food | Isoflavones (mg/kg *fresh weight*) |
| --- | --- |
| Soy nuts | 2500 |
| Soybeans | 1800 |
| Soy powder | 1700 |
| Soybean flour | 1300–1650 |
| Soy milk | 800 |
| Miso | 800 |
| Akamiso soup | 800 |
| Shiromoiso soup | 700 |
| Soybean paste | 600 |
| Tofu | 500 |
| Soya milk | 450 |
| Tempeh | 400 |
| Soy cheese | 150 |
| Soya sauce | 150 |
| Dried beans | 10–70 |
| Fresh vegetables and fruit | 1 |

Note
* Data adopted from Mazur *et al.* (1996); Dwyer *et al.* (1994); Eldridge and Kwolek (1983); Seo and Morr (1984); Setchell *et al.* (1987); Barnes *et al.* (1994); Setchell *et al.* (1997).

isoflavones in soy proteins or soyfoods, and content of daidzein and genistein in soy food products are shown in Table 11.2 and Table 11.3, respectively. Besides dietary legumes, isoflavones are also found in medicinal plants. For example, *Radix puerariae* (root of *Pueraria*, a plant commonly used in traditional Chinese medicine) contains the isoflavones daidzin and daidzein which have been shown to exhibit antidipsotropic (alcohol intake suppressive) activity (Keung and Vallee, 1993).

*Table 11.3* Daidzein and genistein contents in soy food
products (µg/g)*

| Product | Daidzein | Genistein |
|---|---|---|
| Soy ingredient | | |
| Green soybean | 546 | 729 |
| Soy flour | 226 | 810 |
| Traditional soy food | | |
| Tofu | 146 | 162 |
| Bean paste | 272 | 245 |
| Fermented bean curd | 143 | 224 |
| Miso | 79 | 177 |

Note
* Data adopted from Wang and Murphy (1994).

## Metabolism of isoflavones after intake

Many naturally occurring isoflavones exist as sugar conjugates. The major glycosides found in soybean are daidzin, genistin and glycitin. After ingestion, the sugar moiety of these isoflavones are usually hydrolyzed first by intestinal glycosidases resulting in the release of the respective isoflavone aglycones daidzein, genistein and glycitein. The aglycones may be absorbed or further metabolized to many specific metabolites including equol and *p*-ethylphenol (Joannou *et al.*, 1995; Lundh, 1995). High carbohydrate can increase intestinal fermentation and encourage more extensive biotransformation of phytoestrogens and augment the formation of equol. After absorption in the small and probably large intestines, isoflavones undergo phase II metabolism forming sulfate and glucuronide conjugates in the liver, and then excreted through the kidney and bile. Although urinary and blood isoflavone levels increase rapidly after a soybean meal, only 7–30 per cent of ingested isoflavones are recovered in urine (Cassidy *et al.*, 1994; Xu *et al.*, 1995). Pharmacokinetic studies showed that blood isoflavones begin to increase within 30 min of consumption of a soybean supplemented meal, and begin to decline 5 h after ingestion. Blood isoflavone levels remain elevated 24 h after adminstration (Morton *et al.*,1997). The $t_{1/2}$ of plasma daidzein (and genistein) is 7.9 h in adults; $t_{max}$ ranges from 6 to 8 h after administration of pure compounds (King and Bursill, 1998; Setchell, 1998). In Japanese subjects who consumed about 50 mg of total isoflavones per day, plasma concentrations of daidzein, genistein and equol were found in the range between 50 and 800 ng/ml (Adlercreutz *et al.*, 1993) and genistein and daidzein were excreted in the urine at a rate of 12 µ mol per 24 h (Herman *et al.*, 1995). It appears that levels of plasma isoflavones observed in these individuals are much higher than that of normal plasma estradiol, which in men and women generally range from 40 to 80 pg/ml (Setchell *et al.*, 1984). In Western males, whose daily total isoflavones intake is less than 1 mg, plasma daidzein, genistein, equol and desmethylangolensin concentrations were generally less than 10 ng/ml (Adlercreutz *et al.*, 1993) and the amounts of genistein and daidzein excreted thourgh urine were about 0.22 µ mol per 24 h (Herman *et al.*, 1995).

## Biochemical activities of isoflavones

Isoflavones have been shown to have a range of beneficial effects seemingly related to their mild estrogenic and/or anti-estrogenic activity (Adlercreutz *et al.*, 1993). Isoflavones exhibit a wide spectrum of biochemical and pharmacological activities such as anti-cancer (Uckun *et al.*, 1995), anti-angiogenic (Fotsis *et al.*, 1995), estrogenic and anti-estrogenic (Pelissero *et al.*, 1991; Miksicek *et al.*, 1995), anti-oxidant (Wei *et al.*, 1993); anti-microbial (Elgammal and Mansour, 1986); anti-hypertensive (Fan *et al.*, 1985); anti-hyperthermic (Nakamoto *et al.*, 1977); amethystic (Xie *et al.*, 1994); and anti-dipsotropic activity (Keung and Vallee, 1993). They have also been shown to have anabolic effect on bone metabolism (Yamaguchi and Gao, 1998); induce leukemic cell differentiation (Finlay *et al.*, 1994; Constantinous and Huberman, 1995) and apoptosis (Spinozzi *et al.*, 1994); inhibit enzymes involved in the regulation of cell growth such as topoisomerase (Kondo *et al.*, 1991) and tyrosine protein kinase (Watanabe *et al.*,1991), and in steroid metabolism such as aromatase/estrogen synthetase (Adlercreutz *et al.*,1993), 17$\beta$-hydroxy-steroid oxidoreductase Type I (Mäkelä *et al.*,1995, 1998), sterol sulfatase (Wong and Keung, 1997), 3$\beta$-hydroxysteroid dehydrogenase/5-ene-4-ene-isomerase (3$\beta$-HSD/isomerase) (Wong and Keung,1999) and 5$\alpha$-reductase (Weber *et al.*, 1999).

## DIETARY ISOFLAVONES: PREVENTION OF OSTEOPOROSIS AND CANCER OF THE BREAST, PROSTATE AND ENDOMETRIUM

Osteoporosis and cancer of the breast, prostate and endometrium are common diseases of Western societies with serious clinical, social and economic repercussions. However, prevalence of such diseases is relatively low in Asia. Epidemiological studies have attributed the low incidence of these estrogen-dependent diseases among Asian popula-tions to their high dietary phytoestrogen intake, presumably from soy products. The fact that Westernization and abandoning traditional diet in some Asian countries is associated with an increase in prevalence of estrogen-dependent diseases lends further support to the hypothesis.

### Osteoporosis

Osteoporosis is defined as a condition in which the amount of bone per unit volume is decreased, but the composition remains unchanged, causing structural failure and pre-disposition to fracture. It is the most common metabolic bone disease in the United States. Recent studies estimated that about 23 per cent of American women over the age of 50 have osteoporosis. In fact, by the age of 50, between 40 and 56 per cent of them have osteopenia, typified by abnormally lowbone density, and may eventually deterio-rate into osteoporosis if left untreated. These rates are significantly higher than those found in Asia or Africa. It is estimated that 50 per cent of women over the age of 50 and living in the West will suffer a fracture of the hip, wrist or vertebra. On the contrary, hip fracture rates are significantly lower among Asian and African populations (World Health Organization, 1994).

The potential beneficial effects of phytoestrogens on the prevention of osteoporosis in populations consuming large amounts of soy products has been a subject of intensive investigation. Recent reports have shown that genistein, the most studied isoflavone, may prevent bone loss in mice (Ishimi *et al.*, 1999) and rats (Yamaguchi and Gao, 1998) by its estrogenic and anabolic actions. One clinical study indicated that high isoflavone-containing diets could protect post-menopausal women against spinal bone loss (Potter *et al.*, 1998*).* Further, ipriflavone, an isoflavone analog of daidzein holds great promise in the prevention and treatment of osteoporosis and other metabolic bone diseases in human (Head, 1999).

## Cancer of the breast

Recently, the WHO reported that breast cancer has become the most common cancer in women throughout the world, affecting one in every 15 between the age of 60 and 79 (WHO, 1997). The risk of breast cancer increases exponentially after age 30 with an average diagnosing age of 60. The incidence of breast cancer is generally four times lower in women living in Japan than those in North America. Epidemiological studies predicted that 12 per cent American women will develop breast cancer at some point in life comparing with only about 4 per cent women in Hong Kong (Parkin *et al.*, 1992). Established genetic factors BRCA1, BRCA2, p53 and ATM account for less than 10 per cent of all breast cancers diagnosed (McPherson *et al.*, 1994; Berrino *et al.*, 1996). In fact, less than half of all women with breast cancer have established risk factors including inherited genetic defects, menarche before 12, meno pause after age 55, having either no pregnancy or delayed age at first pregnancy, no lactation, early or repeated exposures to radiation, prolonged use of hormone replacement therapy, increased breast density, higher socio economic status and post-menopausal obesity. It appears that most of women with breast cancer can be related to increased lifetime exposure to estrogen and other hormones like progesterone and testosterone (Yu, 1999).

Five epidemiological studies were conducted to evaluate the potential link between risk of developing breast cancer and amounts of soy products consumed. Among them, three showed decreased risk with high soy product consumption but two showed no correlation between the two (Messina *et al.*, 1994). Another recent case-control study showed a decreased risk of breast cancer in women with high amounts of urinary phytoestrogens (Ingram *et al.*, 1997). In animal studies, genistein was shown to enhance mammary gland differentiation and as a consequence these animals have significantly less proliferating gland that is significantly less susceptible to mammary cancer (Lamartiniere *et al.*, 1998). It appears that breast cancer protection observed in Asian women consuming traditional soy-containing diets may stem from exposure to isoflavone-containing diets during their earlier years.

## Cancer of the prostate

Prostate cancer is the most prevalent hormone-related cancer in men and the most common cause of death from cancer in men over age 75. Prostate cancer is rarely found in men younger than 40 years old (Ho *et al.*, 1997). Highfat and meat diets are thought to increase the risk of developing prostate cancer. Like breast cancer, it is relatively rare in Asia where most men consuming phytoestrogen rich soy products. The lowest incidence is found among Japanese men and vegetarians. In the United Kingdom, the incidence of prostate cancer has increased rapidly in recent years, about 3–4 per cent per year (Coleman *et al.*, 1993).

Dietary isoflavones have been proposed to act as preventive agents for prostate cancer in Asian men (Griffiths *et al.*, 1999) and recurrence of prostate cancer after radical prostatectomy (Lee and Fair, 1999). The levels of daidzein and its metabolite equol in prostatic fluids are much higher in Hong Kong and Chinese men consuming soybean than in Portuguese and British men (Morton *et al.*, 1997). In four epidemiological studies conducted, three suggested that the protective effects of soy products on prostate cancer were not significant (Messina *et al.*, 1994) but one recent, case-control study indicated genistein had aslight protective effect (Strom *et al.*, 1999). Clinical trial of soybean in men with abnormally high prostate specific antigen is in progress (Barnes *et al.*, 1996).

## Cancer of the endometrium

Endometrial cancer is the most common type of uterine cancer. The incidence rate of endometrial cancer in women in the United States is 1–2 per cent. In North America, women at age of 60 is about five times more likely to develop endometrial cancer than in Japanese women (Parkin *et al.*, 1992). Increased risk of developing endometrial cancer has been noted in women with increased levels of natural estrogen. Early experimental studies showed a potential link between high dietary phytochemical intake and low incidence of endometrial cancer (Cline and Hughes, 1998).

## CHEMOPREVENTIVE EFFECTS OF ISOFLAVONES ON ESTROGEN-DEPENDENT DISEASES: EXPERIMENTAL STUDIES

### Modulating effects of isoflavones on estrogenic activities

#### *Dual effects of estrogen*

Estrogen and estrogenic steroids (estradiol, estrone, androstenediol) play dual roles in mammals. On the one hand, they play critically important physiological roles such as, programming the development of breast, uterus and brain, and in reproduction. On the other hand, they exhibit negative effects including the promotion of cancer of the breast, uterine lining and prostate (Armstrong and Doll, 1975; Winter *et al.*, 1995; Persson *et al.*, 1999). The effects of estrogen and estrogenic steroids on cell proliferation are mainly mediated by their interaction with ER.

While the precise role of these steroids in the development and growth of human breast, prostate and endometrial cancer remains to be detailed, three sites of action are thought to be important (Lippman, 1989). First, estrogens could serve as carcinogenic agents inducing mutations that may eventually lead to tumor formation. Second, estrogenic steroids may act as promoters of a pre-existing carcinogenic event. Third, they may play a permissive role in allowing a carcinogenic process to proceed by stimulating the uncontrolled proliferation of cells via the activation of *c*-fos and *c*-jun early genes or genes involved in cell cycle control and the stimulation of the src/p21ras/mitogen-activated protein kinase pathway (Tesarik *et al.*, 1999).

Selective ER modulators (SERMs) behave like estrogen in some tissues but block estrogen action in others. In a recent case-controlled study, the SERM Tamoxifen was shown to be beneficial in the treatment of breast cancer and reduction of incidence of breast cancer among women belong to the high-risk group. However, in the same study, the drug was also implicated in an increased incidence of cancer of the endometrium (Bernstein *et al.*, 1999). Only about a third of breast cancer patients respond to Tamoxifen. Apparently, besides hormonal factors, other factors like genetic, environmental, socioeconomic, obesity, exercise and alcohol intake also contribute to the risk of the genesis and development of breast cancer, and presumably prostate and endometrium cancer as well (Davis *et al.*, 1998). Raloxifene, is a relative new SERM that acts as an estrogen agonist in bone and lipid metabolism but an estrogen antagonist on uterine endometrium and breast tissue (Scott *et al.*, 1999).

### Interaction of isoflavones with estrogen receptor

The chemical structure of isoflavones are strikingly similar to that of mammalian estrogens and many of them are estrogenic (Setchell and Adlercreutz, 1988). Isoflavones may either elicit weak estrogenic responses or block estrogenic actions in estrogen-responsive tissues. Whether they act as agonists or antagonists depend on many factors including receptor numbers, occupancy and concentrations and binding affinities of competing estrogens. Isoflavones have weakly estrogenic effect in the absence or low concentrations of estrogen but may exert an antagonistic effect when estrogen concentration is high (Messina *et al.*, 1994). Recent animal studies have shown that isoflavone may prevent bone loss in mice (Ishimi *et al.*, 1999) and rats (Yamaguchi and Gao, 1998) by its estrogenic and anabolic actions but exhibit anti-breast and endometrium cancer activities (Cline and Hughes, 1998). Similar to raloxifen, effects of isoflavones are also tissue specific (Dodge *et al.*, 1997). Detail structure–activity relationship studies should reveal the molecular basis for the agonistic and antagonistic actions of isoflavones (Brzozowski *et al.*, 1997) and provide important information for the design and synthesis of better, isoflavone based SERMs.

Recently, a novel member of ER family was cloned (Kuiper *et al.*, 1996), named ER$\beta$ to distinguish it from the "classical" ER$\alpha$ subtype. The two subtypes of ER may play different roles in gene regulation (Paech *et al.*, 1997). The tissue distribution (Table 11.4) and relative ligand binding affinities of ER$\beta$ and ER$\alpha$ are different (Setchell and Cassidy, 1999). As shown in Table 11.4, ER$\beta$ is found mostly in brain, bone, bladder, prostate and vascular epithelia (Kuiper *et al.*, 1997; Paech *et al.*, 1997, Tetsuka and Hillier, 1997), tissues that are responsive to classical hormone replacement therapy. Anti-estrogen such as Tamoxifen showed some agonistic activities with ER$\alpha$ but not ER$\beta$ (McInerney *et al.*, 1998). However, phytoestrogens including isoflavones exhibit significantly higher affinities for ER$\beta$ than for ER$\alpha$ (Kuiper *et al.*, 1997), suggesting that ER$\beta$ may be important to the action of nonsteroidal estrogens including isoflavones. The preferential binding of nonsteroidal estrogens to the ER$\beta$ receptor also suggests that they may exert their actions through distinct and separate pathways from those of classical steroidal estrogens. The absence of a specific lipophilic region in an isoflavone may affect its binding to different type of ER (Cunningham *et al.*, 1997). In view of that, isoflavones may exert antagonistic effect on breast, uterus and prostate tissue to prevent estrogen-dependent processes involved in cancer formation and growth.

*Table 11.4* Tissue distribution of human ER$\alpha$ and ER$\beta$

| ER$\alpha$ | ER$\beta$ | ER ($\alpha$ and $\beta$) |
| --- | --- | --- |
| Adrenal | Bladder | Breast |
| Kidney | Bone | Ovary |
| Testes | Brain | Uterus |
| | Lung | Vascular |
| | Prostate | |
| | Thymus | |

On the other hand, isoflavones can exert agonistic effect on bone formation when the estrogen level is relatively low in post-menopausal women.

### Inhibition of steroid metabolizing enzymes

Isoflavones or their secondary metabolites (e.g. daidzein-4'-0-sulfate) have been shown to inhibit many steroid metabolizing enzymes involved in the biosynthesis of estrogen. These include aromatase/estrogen synthetase (Adlercreutz *et al.*, 1993; Chen, 1998; Kao *et al.*, 1998), 17$\beta$-hydroxysteroid oxidoreductase Type I (Mäkelä *et al.*, 1995, 1998); sterol sulfatase (Wong and Keung, 1997), 3$\beta$-HSD/isomerase (Wong and Keung, 1999) and 5$\alpha$-reductase (Weber *et al.*, 1999).

Among these enzymes, aromatase catalyzes the conversion of androstenedione and testosterone to estrone and estradiol, respectively, and its expression is tissue specific. It was found to express at a higher level in human cancerous than normal breast tissues (Chen, 1998). Apart from aromatase pathway, in postmenopausal women, in whom breast cancer most frequently occurs, biologically active estrogenic steroids in mammary tissue are derived almost exclusively from their inactive sulfoconjugates – estrone sulfate and dehydroepiandrosterone sulfate – via the peripheral sterol sulfatase pathway (Yamamoto *et al.*, 1993). At target tissue, estrone sulfate is converted to estrone and subsequently reduced to active 17$\beta$-estradiol by estrone sulfatase (Yamamoto *et al.*, 1993) and 17$\beta$-hydroxysteroid oxidoreductase Type I, respectively (Mäkelä *et al.*, 1995). Since aromatase, estrone sulfatase, 17$\beta$-hydroxysteroid oxidoreductase Type I activities in breast, prostate and endometrial tissues could be significantly higher than in normal tissues (Yamamoto *et al.*, 1993; Mäkelä *et al.*, 1998), inhibition of these enzymes by isoflavones and isoflavone sulfate conjugates may suppress the formation of estrogenic steroids important for the genesis and development of estrogen-dependent cancer.

Human 3$\beta$-HSD/isomerase is the key enzyme for estrogen and neurosteroids metabolism (Guennoun *et al.*, 1995). Inhibition of this enzyme by isoflavones can indirectly reduce the concentrations of active estrogens and estrogenic steroids in target tissues (e.g. breast tissue) and hence decrease the risk of tumorigenesis. Since 5$\alpha$-reductase can convert testosterone into 5$\alpha$-dihydrotestosterone, which may be a primary factor in the development of prostate disease (Pollard *et al.*, 1989; Lamb *et al.*, 1992; Weber *et al.*, 1999), its inhibition by isoflavones, which has been found in young adult Japanese men, could be an attractive working hypothesis for evaluating the low prostate cancer risk found in this demographic group (Ross *et al.*, 1992; Moyad, 1999). In fact, three studies showed that isoflavones could indeed reduce tumorigenesis of the prostate in laboratory animals (Messina *et al.*, 1994; Barnes, 1995; Zhou *et al.*, 1999).

The exact concentrations of isoflavones after intake in human tissues are unknown. Animal studies showed that the isoflavone daidzin administered (i.p.) subchronically accumulates in the liver. Its concentration reaches 70 $\mu$M, more than 10 times higher than the maximal plasma concentration obtained in pharmacokinetic studies (Keung *et al.*, 1995, 1996). Therefore, it is likely that other isoflavones may also accumulate in the liver or other target organs (e.g. breast, prostate, uterus etc.) and reach concentrations higher than the enzyme inhibition constants measured in *in vitro* experiments. Dietary isoflavones may exert the chemopreventive effect of breast, prostate and endometrial cancer by diminishing the supply of biologically active estrogenic steroids to the target tissues and reducing their risk of becoming cancerous. In fact, animal studies have shown that isoflavones exhibit anti-carcinogenic effects on the breast and uterus

(Cline and Hughes, 1998). Genistein was shown to enhance mammary gland differentiation and, as a consequence, these animals have significantly less proliferative gland and is significantly less susceptible to mammary cancer (Lamartiniere *et al.*, 1998). The consumption of isoflavones was also found to reduce serum estrogen concentrations in pre-menopausal Japanese women (Nagata *et al.*, 1998).

### Effect of isoflavones on sex-hormone binding globulin

It has been suggested that phytoestrogens could exert their effects by increasing the levels of sex hormone binding globulin (SHBG) and hence reducing the concentration of free estrogens circulating in the plasma (Adlercreutz *et al.*, 1987, 1992, 1995). However, in controlled intervention studies, it was shown that soybean products have no effects on SHBG levels in both pre-menopausal (Cassidy *et al.*, 1994, 1995; Petrakis *et al.*, 1996) and post-menopausal women (Baird *et al.*, 1995; Petrakis *et al.*, 1996). Therefore, existing evidence do not support the hypothesis that the chemoprotective effect of isoflavones is mediated by elevating SHBG level.

### Anti-proliferating effect of isoflavones on tumor cells

Isoflavones can exert their anti-tumor effects at sites other than the ER. For instance, genistein induces apoptosis and inhibits proliferation of human breast cancer cell (Li *et al.*, 1999) and prostate carcinoma (Geller *et al.*, 1998; Onozawa *et al.*, 1998; Zhou *et al.*, 1999). While the detail cellular mechanisms are not well understood (Mathiasen *et al.*, 1999), it was suggested that genistein and daidzein could influence the BRCA1 protein levels in human breast cell lines (Bernard-Gallon *et al.*, 1998). Moreover, genistein can also inhibit the growth of cell lines without ER by inhibiting tyrosine kinases which play important role in cell division (Spinozzi *et al.*, 1994; Uckun *et al.*, 1995; Oude Weernink *et al.*, 1996). Genistein has also been shown to exhibit anti-tumor activity by inhibiting DNA topoisomerase (Kondo *et al.*, 1991). Apart from the direct anti-tumor effect, genistein has been shown to inhibit angiogenesis, the formation of new blood vessels, an essential process required for the growth and expansion of malignant tumor (Fotsis *et al.*, 1995). A major concern is that some of these *in vitro* anti-tumor effects of isoflavones were observed only at concentrations (>5 µg/ml) that are much higher than those observed in plasma of human on diets containing high isoflavone content (Barnes, 1995; Zava and Duwe, 1997). It would be of interest to see if the levels of isoflavones accumulated in target tissues would reach the effective concentration necessary for their pharmacological action.

## Potential adverse effects of isoflavones

Although the beneficial effects of phytoestrogens were observed in hormone dependent diseases, concerns about the long-term effects of phytoestrogens in infants and young children were also addressed. For example, coumestrol, a phytoestrogen found in lucerne, clovers, peas and beans, binds more strongly to ERs than the isoflavones found in *pueraria* plant or soybean. It exerted more estrogenic than anti-estrogenic properties and exhibited adverse effect on the reproductive system of male and female lactating rats (Whitten *et al.*, 1995). Human breast milk contains comparatively low levels of soybean phytoestrogens but soya milk formulas contain at least ten times this amount (Bluck and

Bingham, 1997). Infants fed on soya milk formulas have plasma isoflavone levels that are orders of magnitude greater than those of infants fed on human or cow milk (Setchell *et al.*, 1997). Therefore, caution is endorsed for the potential adverse effects of high levels of phytoestrogens in infancy.

## PERSPECTIVES

### Potential dietary supplement

Accumulating evidences, mostly from animal and some from human studies, have indicated that isoflavones have multiple, beneficial pharmacological activities, particularly in the prevention of osteoporosis and the genesis of estrogen-dependent cancer (Bingham *et al.*, 1999). However, safe and effective dose range of dietary phytoestrogens necessary to achieve a significant beneficial health effect without undesirable side effects remain to be established. To provide a guideline for dietary isoflavones intake, the pharmacokinetic properties (ADME) of various isoflavones from different types of foods must be understood. In the food industry, improvements in food technology and genetic modifications of the isoflavone bearing agricultural products to enhance and modify isoflavone constituents must also be achieved.

### Design of new SERM

With increasing interests in developing new SERM, natural occurring isoflavones will definitely draw much attention in pharmaceutical industry (Jordan, 1998). As a potential pharmacological agent for chemopreventive effect on cancer, more pharmacokinetic studies about isoflavones should be investigated. Moreover, studies about the route of administration, structure–function relationship, metabolic fates, bioavailability, half-life, timing and level of exposure, intrinsic estrogenic state, receptor binding properties, non-hormonal secondary mediated actions and interaction with steroid metabolizing enzymes of isoflavones should be investigated in the future. In addition, the intracellular signal transduction mechanism (e.g. protein kinases, transcription factors etc.) which may be involved in mediating the estrogenic and anti-tumor activity of isoflavones, also required further investigation at molecular and cell level.

### Clinical trials

Although many *in vitro* and animal studies suggest beneficial effects of dietary isoflavones in many disease states, evidences from human studies are limited (Bingham *et al.*, 1998). Since the true efficacy and safety of a pharmaceutical or neutriceutical can only be determined in humans, more clinical data supporting the claimed health benefits of isoflavones remains to be sought. Therefore, other factors like hormonal status, genetic polymorphisms etc. must combine with dietary factors in future clinical trials. Other prospective clinical studies include the assessment of hormone– and gene–nutrient interactions and biomarkers in blood and urine for isoflavones intake. Although hormonal strategies hold promise for the treatment of estrogen-dependent diseases, the profile of risks and benefits and the suitability for use in particular target populations must be addressed in the future disease preventive program.

## CONCLUSION

In the past decades, major stride was made in women's health by launching SERM, e.g. tamoxifen and raloxifene (Jordan, 1998). Their promising results on the chemoprevention of breast cancer in clinical studies shed light on the development of new SERM in pharmaceutical industry. Isoflavones have become prominent agents for estrogen-dependent diseases because they can protect women from osteoporosis and exhibit anti-estrogenic properties in the breast, uterus and prostate tissue to prevent cancer formation. They are natural compounds derived from edible plants that have been consumed by Asians, South Americans, and vegetarians for long periods apparently without any undesirable side effects (Olsen and Love, 1997). Clinical and experimental evaluation of those isoflavones could led to the discovery of more efficacious agents with better pharmacokinetic properties. Further, balanced diets containing proper isoflavone contents could be recommended for pre- and post-menopausal women and patients with localized prostate cancer. Serious scientific enquiries on phytoestrogens and phytoestrogen containing plants (Kudzu, Soy) represent just one kind of many important bioactive non-nutrients found in plants. Many other active ingredients in natural products with potential pharmacological activity required further investigation in the future.

## REFERENCES

Adlercreutz, H. (1995) Phytoestrogens: epidemiology and a possible role in cancer protection. *Envir. Health Perspect.*, **103**, Suppl. 7, 103–112.

Adlercreutz, H., Bannwart, C., Wähälä, K., Mäkelä, T., Brunow, G., Hase, T., Arosemena, P.J., Kellis, J.T., Jr. and Vickery, L.E. (1993) Inhibition of human aromatase by mammalian lignans and isoflavonoid phytoestrogens. *J. Steroid Biochem.*, **44**, 147–153.

Adlercreutz, H., Fotsis, T., Lampe, J. *et al.* (1993) Quantitative determination of lignans and isoflavones in plasma of omnivorous and vegetarian women by isotope-dilution gas chromatography-mass spectrometry. *Scand. J. Clin. Lab. Invest.*, **53**, 5–18.

Adlercreutz, H., Höckerstedt, K., Bannwart, C., Bloigu, S., Hämäläinen, E., Fotsis, T. and Ollus, A. (1987) Effect of dietary components, including lignans and phytoestrogens, on enterohepatic circulation and liver metabolism of estrogens and on sex hormone binding globulin (SHBG). *J. Steroid Biochem.*, **27**, 1135–1144.

Adlercreutz, H., Markkanen, H. and Watanabe, S. (1993) Plasma concentrations of phyto-oestrogens in Japanese men. *Lancet*, **342**, 1209–1210.

Adlercreutz, H., Mousavi, Y., Clark, J., Höckerstedt, K., Hämäläinen, E., Wähälä, K., Mäkelä, T. and Hase, T. (1992) Dietary phytoestrogens and cancer: *in vitro* and *in vivo* studies. *J. Steroid Biochem. Mol. Biol.*, **41**, 331–337.

Armstrong, B. and Doll, R. (1975) Environmental factors and cancer incidence and mortality in different countries, with special reference to dietary practices. *Int. J. Cancer*, **15**, 617–631.

Baird, D.D., Umbach, D.M., Lansdell, L., Hughes, C.L., Setchell, K.D.R., Weinberg, C.R., Haney, A.F., Wilcox, A.J. and McLachlan, J.A. (1995) Dietary intervention study to assess oestrogenicity of dietary soy among postmenopausal women. *J. Clin. Endocrinol. Metab.*, **80**, 1685–1690.

Barnes, S., Kirk, M. and Coward, L. (1994) Isoflavones and their conjugates in soy foods: extraction conditions and analysis by HPLC-mass spectrometry. *J. Agric. Food Chem.*, **42**, 2466–2474.

Barnes, S. (1995) Effect of genistein on *in vitro* and *in vivo* models of cancer. *J. Nutr.*, **125**, 773S–783S.

Barnes, D., Urban, W., Grizzel, L., Coward, L., Kirk, M., Weiss, H. and Irwin, W. (1996) A double blind clinical trial of the effects of soy protein on risk parameters for prostate cancer. *Proc. of 2nd Int. Symposium on the role of soy in preventing and treating chronic disease*, pp. 37.

Bingham, S.A., Atkinson, C., Liggins, J., Bluck, L. and Coward, A. (1998) Phyto-estrogens: where are we now? *Br. J. Nutri.*, 79, 393–406.

Bernstein, L., Deapen, D., Cerhan, J.R., Schwartz, S.M., Liff, J., McGann-Maloney, E., Perlman, J.A. and Ford, L. (1999) Tamoxifen therapy for breast cancer and endometrial cancer risk. *J. Natl. Cancer Inst.*, 91, 1654–1662.

Bernard-Gallon, D.J., Maurizis, J.C., Rio, P.G. and Bignon, Y.J. (1998) Influence of genistein and daidzein on BRCA1 protein levels in human breast cell lines. *J. Natl. Cancer Inst.*, 90, 862–863.

Berrino, F., Muti, P., Bolleli, G., Krogh, V., Sciajno, R., Pisani, P., Panico, S. and Secreto, G. (1996) Serum sex hormone levels after the menopause and subsequent breast cancer. *J. Natl. Cancer Inst.*, 88, 291–296.

Bluck, L.J.C. and Bingham, S.A. (1997) Isoflavone content of breast milk and formulas. *Clin. Chem.*, 43, 851–852.

Brzozowski, A.M., Pike, A.C.W., Dauter, Z., Hubbard, R.E., Bonn, T., Engstrom, O., Ohman, L., Greene, G.L., Gustafsson, J.A. and Carlquist, M. (1997) Molecular basis of agonism and antagonism in the oestrogen receptor. *Nature*, 389, 753–758.

Cassidy, A., Bingham, S.A. and Setchell, K.D.R. (1994) Biological effects of a diet of soy protein rich in isoflavones on the menstrual cycle of premenopausal women. *Amer. J. Clin. Nutri.*, 60, 333–340.

Cassidy, A., Bingham, S. and Setchell, K. (1995) Biological effects of isoflavones in young women: importance of the chemical composition of soyabean products. *Br. J. Nurt.*, 74, 587–601.

Chen, S. (1998) Aromatase and breast cancer. *Front. Biosci.*, 3, 922–933.

Cline, J.M. and Hughes, C.L. (1998) Phytochemicals for the prevention of breast and endometrial cancer. In J.M. Cline and C.L. Hughes (eds), *Biological and hormonal therapies of cancer*, Kluwer Academic Publishers, Boston, pp. 107–134.

Coleman, M.P., Esteve, J., Damieki, P., Arslan, A. and Renard, H. (1993) Trends in cancer incidence and mortality. *International agency for research on cancer scientific publication no. 121*, IARC Scientific publications, Lyon, France.

Constantinou, A. and Huberman, E. (1995) Genistein as an inducer of tumor cell differentiation: possible mechanisms of action. *Proc. Soc. Exp. Biol. Med.*, 208, 109–115.

Cunningham, A.R., Klopman, G. and Rosenkranz, H.S. (1997) A dichotomy in the lipophilicity of natural estrogens, xenoestrogens, and phytoestrogens. *Environ. Health Perspect.*, 105, 665–668.

Das, D.K. (1994) Naturally occurring flavonoids: structure, chemistry, and high-performance liquid chromatography methods for separation and characterization. *Methods Enzymol.*, 234, 410–420.

Davis, D.L., Axelrod, D., Bailey, L., Gaynor, M. and Sasco, A.J. (1998) Rethinking breast cancer risk and the environment: the case for the precautionary principle. *Envi. Health Persp.*, 106, 523–529.

Dodge, J.A., Jugar, C.W., Cho, S., Short, L.L., Sato, M., Yang, N.N., Spangle, L.A., Martin, M.J., Phillips, D.L., Glasebrook, A.L., Osborne, J.J., Frolik, C.A. and Bryant, H.U. (1997) Evaluation of the major metabolites of Raloxifene as modulators of tissue selectivity. *J. Steroid Biochem. Mol. Biol.*, 61, 97–106.

Dwyer, J.T., Goldin, B.R., Saul, N., Gualtieri, L., Barakat, S. and Adlercreutz, H. (1994) Tofu and soy drinks contain phytoestrogens. *J. Am. Diet Assoc.*, 94, 739–743.

Eldridge, A. and Kwolek, W.F. (1983) Soybean isoflavones: effect of environment and variety on composition. *J. Agri. Food Chem.*, 31, 394–496.

Elgammal, A.A. and Mansour, R.M. (1986) Antimicrobial activities of some flavonoid compounds. *Zentralbl Mikrobiol*, 141, 561–565.

Fan, L.L., O'Keefe, D.D. and Powell, W.W.J. (1985) Pharmacological studies on *Radix puerariae*. *Chin. Med. J.*, **98**, 821–932.

Finlay, G.J., Holdaway, K.M. and Bagulet, B.C. (1994) Comparison of the effects of genistein and amsacrine on leukemia cell proliferation. *Oncol. Res.*, **6**, 33–37.

Fotsis, T., Pepper, M., Adlercreutz, H., Hase, T., Montesano, R. and Schweigerer, L. (1995) Genistein, a dietary ingested isoflavonoids, inhibits cell proliferation and *in vivo* angiogenesis. *J. Nutr.*, **125**, 790S–797S.

Franke, A.A., Custer, L.J., Cerna, C.M. and Narala, K.K. (1994) Quantitation of phytoestrogens in legumes by HPLC. *J. Agric. Food. Chem.*, **42**, 1905–1913.

Geller J., Sionit, L., Partido, C., Li, L., Tan, X., Youngkin, T., Nachtsheim, D. and Hoffman, R.M. (1998) Genistein inhibits the growth of human-patient BPH and prostate cancer in histoculture. *Prostate*, **34**, 75–79.

Griffiths, K., Morton, M.S. and Denis, L. (1999) Certain aspects of molecular endocrinology that relate to the influence of dietary factors on the pathogenesis of prostate cancer. *Eur. Urol.*, **35**, 443–455.

Guennoun, R., Fiddes, R.J., Gouézou, M., Lombés, M. and Baulieu, E.E. (1995) A key enzyme in the biosynthesis of neurosteroids, 3ß-hydroxysteroid dehydrogenase/$\Delta^5$-$\Delta^4$-isomerase (3ß-HSD), is expressed in rat brain. *Mol. Brain Res.*, **30**, 287–300.

Head, K.A. (1999) Ipriflavone: an important bone-building isoflavone. *Altern. Med. Rev.*, **4**, 10–22.

Herman, C., Adlercreutz, T., Goldin, B.R., Gorbach, S.L., Hockerstedt, K.A.V., Watanabe, S., Hamalainen, E.K., Markkanen, M.H., Makela, T.H., Wahala, K.T., Hase, T.A. and Fotsis, T. (1995) Soybean phytoestrogen intake and cancer risk. *J. Nutr.*, **125**, 757S-770S.

Ho, S.M., Lee, K.F. and Lane, K. (1997) Neoplastic transformation of prostate. In K.N. Rajesh (ed.), *Prostate: basic and clinical aspects*, CRC Press, New York, pp.73–114.

Ingram, D., Sanders, K., Kolybaba, M. and Lopez, D. (1997) Case-control study of phyto-oestrogens and breast cancer. *Lancet*, **350**, 990–994.

Ishimi, Y., Miyaura, C., Ohmura, M., Onoe, Y., Sato, T., Uchiyama, Y., Ito, M., Suda, T. and Ikegami, S. (1999) Selective effects of genistein, a soybean isoflavone, on B-lymphopoiesis and bone loss caused by estrogen deficiency. *Endocrinol.*, **140**, 1893–1900.

Joannou, G.E., Kelly, G.E., Reeder, A.Y., Waring, M.A. and Nelson, C. (1995) A urinary profile study of dietary phytoestrogens. The identification and mode of metabolism of new isoflavones. *J. Steroid Biochem.*, **54**, 167–184.

Jones, A.E., Price, K.R. and Fenwick, G.R. (1989) Development of a high-performance liquid chromatographic method for the analysis of phytoestrogens. *J. Sci. Food Agric.*, **46**, 357–364.

Jordan, V.C. (1998) Designer estrogens. *Sci. Amer.*, **October**, 36–43.

Kao, Y.C., Zhou, C., Sherman, M., Laughton, C.A. and Chen, S. (1998) Molecular basis of the inhibition of human aromatase (estrogen synthetase) by flavone and isoflavone phytoestrogens: a site-directed mutagenesis study. *Environ. Health Perspect*, **106**, 85–92.

Keung, W.M., Lazo, O., Kunze, L. and Vallee, B.L. (1995) Daidzin suppresses ethanol consumption by Syrian golden hamsters without blocking acetaldehyde metabolism. *Proc. Natl. Acad. Sci. USA*, **92**, 8990–8993.

Keung, W.M., Lazo, O., Kunze, L. and Vallee, B.L. (1996) Potentiation of the bioavailability of daidzin by an extract of *Radix puerariae*. *Proc. Natl. Acad. Sci. USA*, **93**, 4284–4288.

Keung, W.M. and Vallee, B.L. (1993) Daidzin and daidzein suppress free-choice ethanol intake by Syrian Golden hamster. *Proc. Natl. Acad. Sci. USA*, **90**, 10008–10012.

King, R.A. and Bursill, D.B. (1998) Plasma and urinary kinetics of the isoflavones daidzein and genistein after a single soy meal in humans. *Am. J. Clin. Nutr.*, **67**, 867–872.

Kondo, K., Tsuneizumi, K., Watanabe, T. and Oishi, M. (1991) Induction of *in vitro* differentiation of mouse embryonal carcinoma (F9) cells by inhibitors of topoisomerases. *Cancer Res.*, **51**, 5398–5404.

Kuiper, G.G.J.M., Carlsson, B., Grandien, K., Enmark, E., Haggblad, J., Nilsson, S. and Gustafsson, J.A. (1997) Comparison of the ligand binding specificity and transcript tissue distribution of ERs alpha and beta. *Endocrinology*, **138**, 863–870.

Kuiper, G.G.J.M., Enmark, E., Peltohuikki, M., Nilsson, S. and Gustaffson, J.A. (1996) Cloning of a novel estrogen receptor expressed in rat prostate and ovary. *Proc. Natl. Acad. Sci. USA*, **93**, 5925–5930.

Lamartiniere, C.A., Zhang, J.X. and Cotroneo, M.S. (1998) Genistein studies in rats: potential for breast cancer prevention and reproductive and development toxicity. *Am. J. Clin. Nutr.*, **68**, 1400S–1405S.

Lamb, J.C., Levy, M.A., Johnson, R.K. and Isaacs, J.T. (1992) Response of rat and human prostatic cancers to the novel 5alpha-reductase inhibitor, SK&F 105657. *Prostate*, **21**, 15–34.

Leclerq, G. and Heuson, J.C. (1979) Physiological and pharmacological effects of estrogens in breast cancer. *Biochim. Biophys. Acta*, **560**, 427–455.

Lee, C.T. and Fair, W.R. (1999) The role of dietary manipulation in biochemical recurrence of prostate cancer after radical prostatectomy. *Sem. Urol. Oncol.*, **17**, 154–163.

Lippman, M.E. (1989) In L.I. DeGroot (ed.), *Endocrinology*, Saunders, Philadelphia, pp. 2706–2725.

Li, Y., Upadhyay, S., Bhuiyan, M. and Sarkar, F.H. (1999) Induction of apoptosis in breast cancer cells MDA-MB-231 by genistein. *Oncogene*, **18**, 3166–3172.

Lundh, T. (1995) Metabolism of estrogenic isoflavones in domestic animals. *Proc. Soc. Exp. Biol. Med.*, **208**, 33–39.

Mäkelä, S., Poutanen, M., Lehtimäki, N., Kostian, M.L., Santti, R. and Vihko, R. (1995) Estrogen-specific 17ß-hydroxysteroid oxidoreductase type I (E.C. 1.1.1.62) as a possible target for the action of phytoestrogens. *Proc. Soc. Exp. Biol. Med.*, **208**, 51–59.

Mäkelä, S., Poutanen, M., Kostian, M.L., Lehtimäki, N., Strauss, L., Santti, R. and Vihko, R. (1998) Inhibition of 17$\beta$-hydroxysteroid oxidoreductase by flavonoids in breast and prostate cancer cells. *Proc. Soc. Exp. Biol. Med.*, **217**, 310–316.

Mathiasen, I.S., Lademann, U. and Jaattela, M. (1999) Apoptosis induced by vitamin D compounds in breast cancer cells is inhibited by bcl-2 but does not involve known caspases or p53. *Cancer Res.*, **59**, 4848–4856.

Mazur, W., Fotsis, T., Wähala, K. *et al.* (1996) Isotope dilution gas chromatographic – mass spectrometric method for the determination of isoflavonoids, coumestrol, and lignans in food samples. *Anal. Biochem.*, **233**, 169–180.

McInerney, E.M., Weis, K.E., Sun, J., Mosselman, S. and Katzenellenbogen, B.S. (1998) Transcription activation by the human estrogen receptor subtype $\beta$ (ER$\beta$) studied with ER$\beta$ and ER$\alpha$ receptor chimeras. *Endocrinol.*, **139**, 4513–4522.

McPherson, K., Steel, C.M. and Dixon, J.M. (1994) Breast cancer-epidemiology, risk factors, and genetics. *Br. Med. J.*, **309**, 1003–1006.

Messina, M., Persky, V., Setchell, K.D.R. and Barnes, S. (1994) Soy intake and cancer risk: a review of the *in vitro* and *in vivo* data. *Nutr. Cancer*, **21**, 113–131.

Miksicek, R.J. (1994) Interaction of naturally occurring nonsteroidal estrogens with expressed recombinant human estrogen receptor. *J. Steroid Biochem. Mol. Biol.*, **49**, 153–160.

Miksicek, R.J. (1995) Estrogenic flavonoids: structural requirement for biological activity. *Proc. Soc. Exp. Biol. Med.*, **208**, 44–50.

Morton, M.S., Matos-Ferreira, A., Abranches-Monteiro, L., Correia, R., Blacklock, N., Chan, P.S.F., Cheng, C., Lioyd, S., Chiehping, W. and Griffith, K. (1997) Measurement and metabolism of isoflavonoids and lignans in the human male. *Cancer lett.*, **114**, 145–151.

Moyad, M.A. (1999) Soy, disease prevention, and prostate cancer. *Sem. Urol. Oncol.*, **17**, 97–102.

Murkies, A.L., Lombard, C., Strauss, B.J.G., Wilcox, G., Burger, H.G. and Morton, M.S. (1995) Dietary flour supplementation decreases postmenopausal hot flushes: effect of soy and wheat. *Maturitas*, **21**, 189–195.

Nagata, C., Takatsuka, N., Inaba, S., Kawakami, N. and Shimizu, H. (1998) Effect of soymilk consumption on serum estrogen concentrations in premenopausal Japanese women. *J. Natl. Cancer Inst.*, **90**, 1830–1835.

Nakamoto, H., Iwasaki, Y. and Kizu, H. (1977) The study of aqueous extract of *Puerariae radix*. IV. The isolation of daidzin from the active extract (MTF-101) and its antifebrile and spasmolytic effect (author's transl). *Yakugaku Zasshi*, **97**, 103–105.

Olsen, M.R. and Love, R.R. (1997) Hormonal strategies for the prevention of breast cancer. In K.A. Foon and H.B. Muss (eds), *Biol. hormonal ther. cancer*, Kluwer Academic Publishers, Boston, pp.133–157.

Onozawa, M., Fukuda, K., Ohtani, M., Akaza, H., Sugimura, T. and Wakabayashi, K. (1998) Effects of soybean isoflavones on cell growth and apoptosis of the human prostatic cancer cell line LNCaP. *Jpn. J. Clin. Oncol.*, **28**, 360–363.

Oude Weernink, P.A., Verheul, E., Kerkhof, E., van Veelen, C.W. and Rijksen, G. (1996) Inhibitors of protein tyrosine phosphorylation reduce the proliferation of two human glioma cell lines. *Neurosurgery*, **38**, 108–113.

Paech, K., Webb, P., Kuier, G.G.L., Nilsson, S., Gustafsson, J.A., Kushner, P.J. and Scanlan, T.S. (1997) Differential ligand activation of estrogen receptors ER$\alpha$ and ER$\beta$ at AP1 sites. *Sciences*, **277**, 1508–1510.

Parkin, D.M., Muir, C.S., Whelan, S.L., Gao, Y.T., Ferlay, J. and Powell, J. (eds) (1992) Cancer incidence in five continents, vol. VI. Genenva: World Health Organization.

Pelissero, C., Bennetau, B., Babin, P., Le Menin, F. and Dunogues, J. (1991) The estrogenic activity of certain phytoestrogens in the siberian sturgeon. *Acipenser baeri*, *J. Steroid Biochem.*, **38**, 293–299.

Persson, I., Weiderpass, E., Bergkvist, L., Bergstrom, R. and Schairer, C. (1999) Risks of breast and endometrial cancer after estrogen and estrogen-progestin replacement. *Cancer Causes Control*, **10**, 253–260.

Petrakis, N.L., Barnes, S., King, E.B., Lowenstein, J., Wiencke, J., Lee, M.M., Miike, R., Kirk, M. and Coward, L. (1996) Stimulatory influence of soy protein isolate on breast secreation in pre- and postmenopausal women. *Cancer Epidemiol. Biomarkers Prev.*, **5**, 785–794.

Pollard, M., Luckert, P.H. and Snyder, D.L. (1989) The promotional effect of testosterone on inducton of prostate cancer in MNU-sensitive L-W rats. *Cancer Lett.*, **45**, 209–212.

Potter, S.M., Baum, J.A., Teng, H., Stillman, R.J., Shay, N.F. and Erdman, J.W. (1998) Soy protein and isoflavones: their effects on blood lipids and bone density in postmenopausal women. *Am. J. Clin. Nutr.*, **68**, 1375S–1379S.

Ross, R.K., Bernstein, L., Lobo, R.A., Shimizu, H., Stanczyk, F.Z., Pike, M.C. and Henderson, B.E. (1992) 5-alpha-reductase activity and risk of prostate cancer among Japanese and US white and black males. *Lancet*, **339**, 887–889.

Scott, J.A., Da Camara, C.C. and Early, J.E. (1999) Raloxifene: a selective estrogen receptor modulator. *Am. Fam. Physician.*, **60**, 1131–1139.

Seo, A. and Morr, C.V. (1984) Improved high-performance liquid chromatographic analysis of phenolic acids and isoflavonoids from soybean protein products. *J. Agri. Food Chem.*, **32**, 530–533.

Setchell, K.D.R. (1998) Phytoestrogens: biochemistry, physiology, and implications for human health of soy isoflavones. *Am. J. Clin. Nutr.*, **68**, 1333S–1346S.

Setchell, K.D.R. and Adlercreutz, H. (1988) Mammalian lignans and phyto-oestrogens. Recent studies on their formation, metabolism and biological role in health and disease. In I.R. Rowland (ed.), *Role of the gut flora in toxicity and cancer*, Academic Press, London, UK, pp. 315–345.

Setchell, K.D.R., Borriello, S.P., Hulme, P., Kirk, D.N. and Axelson, M. (1984) Non-steroidal oestrogens of dietary origin: possible roles in hormone dependent disease. *Am. J. Clin. Nutr.*, **40**, 569–578.

Setchell, K.D.R. and Cassidy, A. (1999) Dietary isoflavones: biological effects and relevance to human health. *J. Nutr.*, **129**, 758S–767S.

Setchell, K.D.R., Nechemias-Zimmer, L.Z., Cai, J. and Heubi, J.E. (1997) Exposure of infants to phytoestrogens from soy infant formulas. *Lancet*, **350**, 23–27.

Setchell, K.D.R., Welsh, M.B. and Lim, C.K. (1987) HPLC analysis of phytoestrogens in soy protein preparations with ultraviolet, electrochemical, and thermospray mass spectrometric detection. *J. Chromatogram.*, **385**, 267–274.

Shutt, D.A. and Cox, R.I. (1972) Steroid and phytoestrogen binding to sheep uterine receptors *in vivo*. *J. Endocrinol.*, **52**, 299–310.

Spinozzi, F., Pagliacci, M.C., Migliorati, G., Moraca, R., Grignani, F., Riccardi, C. and Nicoletti, I. (1994) The natural tyrosine kinase inhibitor genistein produces cell cycle arrest and apoptosis in Jurkat T-leukemia cells. *Leukemia Res.*, **18**, 431–439.

Strom, S.S., Yamamura, Y., Duphorne, C.M., Spitz, M.R., Babaian, R.J., Pillow, P.C. and Hursting, S.D. (1999) Phytoestrogen intake and prostate cancer: a case-control study using a new database. *Nutr. Cancer*, **33**, 20–25.

Tesarik, J., Garrigosa, L. and Mendoza, C. (1999) Estradiol modulates breast cancer cell apoptosis: a novel nongenomic steroid action relevant to carcinogenesis. *Steroids*, **64**, 22–27.

Tetsuka, M. and Hillier, S.G. (1997) Androgen receptor gene expression in rat granulosa cells: the role of follicle-stimulating harmone and steroid harmones. *Endocrinology*, **137**, 4392–4397.

Uckun, F.M., Evans, W.E., Forsyth, C.J., Waddick, K.G., Ahlgren, L.T., Chelstrom, L.M., Burkhardt, A., Bolen, J. and Myers, D.E. (1995) Biotherapy of B-cell precursor leukemia by targeting genistein to CD19-associated tyrosine kinases. *Science*, **267**, 886–891.

Wang, H.J. and Murphy, P.A. (1994) Isoflavone content in commercial soybean foods. *J. Agric. Food Chem.*, **42**, 1666–1673.

Watanabe, T., Kondo, K. and Oishi, M. (1991) Induction of *in vitro* differentiation of mouse erythroleukemia cells by genistein, an inhibitor of tyrosine protein kinases. *Cancer Res.*, **51**, 764–768.

Weber, K.S., Jacobson, N.A., Setchell, K.D. and Lephart, E.D. (1999) Brain aromatase and 5alpha-reductase, regulatory behaviors and testosterone levels in adult rats on phytoestrogen diets. *Proc. Soc. Exp. Biol. Med.*, **221**, 131–135.

Wei, H., Wei, L., Frenkel, K., Bowen, R. and Barnes, S. (1993) Inhibition of tumor pro-moter-induced hydrogen peroxide formation *in vitro* and *in vivo* by genistein. *Nutr. Cancer*, **20**, 1–12.

Whitten, P.L., Lewis, C., Russel, E. and Naftolin, F. (1995) Potential adverse effects of phytoestrogens. *J. Nutri.*, **125**, 771S–776S.

WHO (1994) Assessment of fracture risk and its application for screening for postmenopausal osteoporosis. Technical report series no. 834. Geneva: WHO.

WHO (1997) Conquering Suffering-Enriching Humanity. The World Health Report, Geneva: World Health Organization.

Winter, M.L., Bosland, M.C., Wade, D.R., Falvo, R.E., Nagamani, M. and Liehr, J.G. (1995) Induction of benign prostatic hyperplasia in intact dogs by near-physiological levels of 5$\alpha$-dihydrotestosterone and 17$\beta$-estradiol. *Prostate*, **26**, 325–333.

Wong, C.K. and Keung, W.M. (1997) Daidzein sulfoconjugates are potent inhibitors of sterol sulfatase (E.C. 3.1.6.2). *Biochem. Biophys. Res. Comm.*, **233**, 579–583.

Wong, C.K. and Keung, W.M. (1999) Bovine adrenal 3$\beta$-hydroxysteroid dehydrogenase (E.C. 1.1.1.145)/5-ene-4-ene isomerase (E.C. 5.3.3.1): characterization and its inhibition by isoflavones. *J. Steroid Biochem. Mol. Biol.*, **71**, 191–202.

Xie, C.I., Lin, R.C., Antony, V., Lumeng, L., Li, T.K., Mai, K., Liu, C., Wang, Q.D., Zhao, Z.H. and Wang, G.F. (1994) Daidzin, an antioxidant isoflavonoid, decreases blood alcohol levels and shortens sleep time induced by ethanol intoxication. *Alcohol. Clin. Exp. Res.*, **18**, 1443–1447.

Xu, X., Harris, K.S., Wang, H., Murphy, P. and Hendrich, S. (1995) Bioavailability of soybean isoflavones depends on gut microflora in women. *J. Nutri.*, **125**, 2307–2315.

Yamaguchi, M. and Gao, Y.H. (1998) Anabolic effect of genistein and genistin on bone metabolism in the femoral-metaphyseal tissues of elderly rats: the genistein effect is enhanced. *Mol. Cell. Biochem.*, **178**, 377–382.

Yamamoto, T., Kitawaki, J., Urabe, M., Honjo, H., Tamura, T., Noguchi, T., Okada, H., Sasaki, H., Tada, A., Terashima, Y. *et al.* (1993) Estrogen productivity of endometrium and endometrial cancer tissue; influence of aromatase on proliferation of endometrial cancer cells. *J. Steroid Biochem. Mol. Biol.*, **44**, 463–468.

Yu, H., Diamandis, E.P. and Hoffman, B. (1999) Elevated estradiol and testosterone levels and risk for breast cancer. *Ann. Intern. Med.*, **131**, 715.

Zava, D.T., Dollbaum, C.M. and Blen, M. (1998) Estrogen and progestin bioactivity of foods, herbs, and spices. *Proc. Soc. Exp. Biol. Med.*, **217**, 369–378.

Zava, D.T. and Duwe, G. (1997) Estrogenic and antiproliferative properties of genistein and other flavonoids in human breast cancer cells *in vitro*. *Nutr. Cancer*, **27**, 31–40.

Zhou, J.R., Gugger, E.T., Tanaka, T., Guo, Y., Blackburn, G.L. and Clinton, S.K. (1999) Soybean phytochemicals inhibit the growth of transplantable human prostate carcinoma and tumor angiogenesis in mice. *J. Nutr.*, **129**, 1628–1635.

# 12 Chemistry and hepatoprotective effect of *Pueraria* saponins

*Junei Kinjo and Toshihiro Nohara*

## INTRODUCTION

*Flos puerariae*, the flowers of *P. lobata*, is a crude drug used to counteract the over-consumption of alcohol in traditional Japanese and Chinese therapeutic systems (Kinjo *et al.*, 1988). We have found that the total saponin fraction in this crude drug was also effective for alcohol intoxication (Niiho *et al.*, 1989). The saponin fraction was also shown effective in an experimental *in vivo* model of hepatic injury (Niiho *et al.*, 1990): the total saponin fraction decreased serum alanine aminotransferase (ALT) level in high fatty food and $CCl_4$-induced experimental liver injury. In an effort to reveal the chemical constituents in the flowers of *P. lobata*, we have isolated and identified three saponins together with several isoflavones (Kinjo *et al.*, 1988).

*Radix puerariae*, the roots of *P. lobata*, is an important oriental crude drug used as a perspiration-inducing, antipyretic and antispasmodic agent. This drug, known in Japan as Kakkon, is also the major ingredient of Kakkon-To, a composite formulation used for treatment of the common cold since ancient times in China and Japan. Among the pharmacological actions of *Radix puerariae*, antispasmodic activity has been attributed to isoflavones, such as daidzein (Shibata *et al.*, 1959). However, the active principles of perspiration-inducing and antipyretic activity have not been pursued in a routine manner to date. Very recently, Kurokawa *et al.* (1996) disclosed that Kakkon-To significantly suppressed rise of the interleukin (IL)-1 level in mouse serum. Since IL-1 is a mediator of the inflammation responsible for fever induction, the unsolved actions (perspiration and antipyretic) of this drug are assumed to be due to similar anti-inflammatory action. We have also clarified that the oleanene-type triterpene saponins from *Herba Abri* (the whole plants of *Abrus cantoniensis*) are effective in the treatment of *in vivo* experimental hepatic injury (Takeshita *et al.*, 1990). On the other hand, Higuchi *et al.* (1992), using *in vitro* immunological liver injury, confirmed the hepatoprotective effects of oleanene-type triterpene saponins from soybeans. They used an antiserum against rat plasma membranes as a lesion model of liver cells. The mechanism of this immunological liver injury is thought to be caused by complement-mediated cell damage (Shiki *et al.*, 1984). Since the antibody-complement system plays an important role in humoral immunity, we have tried to clarify the actions of oleanene-type triterpene saponins in *Radix puerariae* on *in vitro* immunological liver injury of rat primary hepatocyte cultures. Herein, we describe the structures of saponins and their hepatoprotective activities and discuss the structure–activity relationships.

## SAPONINS IN ROOTS

### Extraction, isolation and structural determination

A methanolic extract of the fresh roots of *P. lobata* was partitioned between 1-BuOH and water. Both layers were concentrated and subjected to polystyrene gel column chromatography. Each total saponin fraction was subjected to normal and reversed-phase chromatography to yield saponins 1–13 (Arao *et al.*, 1995, 1997a). Further, similar experiments with *P. thomsonii* gave saponins 1 and 14–17 (Arao *et al.*, 1996).

The structures of saponins were determined by both chemical and physicochemical methods. Since we had already elucidated the structures of sapogenols (Kinjo *et al.*, 1985), the aglycone part of the isolated saponins was easily identified. Further, since the $^{13}$C-NMR data of a representative saponin, Soyasaponin I (1) has been assigned using HMBC spectra (Miyao *et al.*, 1996), all carbon signals could be assigned based on the $^{13}$C-NMR spectral data of each saponin. The absolute configurations of the sugar moieties were determined according to the procedure developed by Hara *et al.* (1987). The structures of the obtained saponins are shown in Figure 12.1.

No saponin spot was found in the roots of *P. tuberosa* known as Indian Kudzu by TLC.

### HPLC profile analysis of total saponin fractions

Since the structures of saponins obtained from *P. lobata* and *P. thomsonii* were considerably different, a qualitative analysis seemed to be important. In order to confirm the constitution of saponins, HPLC profile analyses of the total saponin fractions of some *Pueraria* roots produced worldwide were performed. The test samples were six domestic and foreign *P. lobata* roots and two foreign *P. thomsonii* roots. HPLC conditions were referred to the previous paper (Kinjo *et al.*, 1994). Figures 12.2–12.4 show the total saponin fractions from domestic *P. lobata* roots, foreign *P. lobata* roots and foreign *P. thomsonii* roots, respectively.

Although the producing countries were different, all total saponin fractions originated from *P. lobata* gave similar complicated profiles. In contrast, the HPLC profiles of the total saponin fractions originated from *P. thomsonii* differed from those of *P. lobata*. The peaks derived from the more polar saponin were not observed in *P. thomsonii*. Not only HPLC profiles but also yields of the total saponin fractions were different. The total saponin fraction in *P. lobata* roots was most abundant (Arao *et al.*, 1995, 1996).

### Hepatoprotective effect

#### Assay method

We devised a similar assay method using *in vitro* immunological liver injury according to Higuchi *et al.* (1992). The detailed procedure was described in a previous paper (Arao *et al.*, 1997b). One day after the isolated rat hepatocytes were plated, the cultured cells were exposed to a medium (300 µl) containing antiserum against rat plasma membranes (40 µl/ml) and 4 µl of the tested compounds in DMSO (final concentration 0 (reference), 10, 30, 90, 200, 500 µM). Forty minutes after the cells were treated with the antiserum, the medium was taken for determination of ALT activity. Hepatoprotective

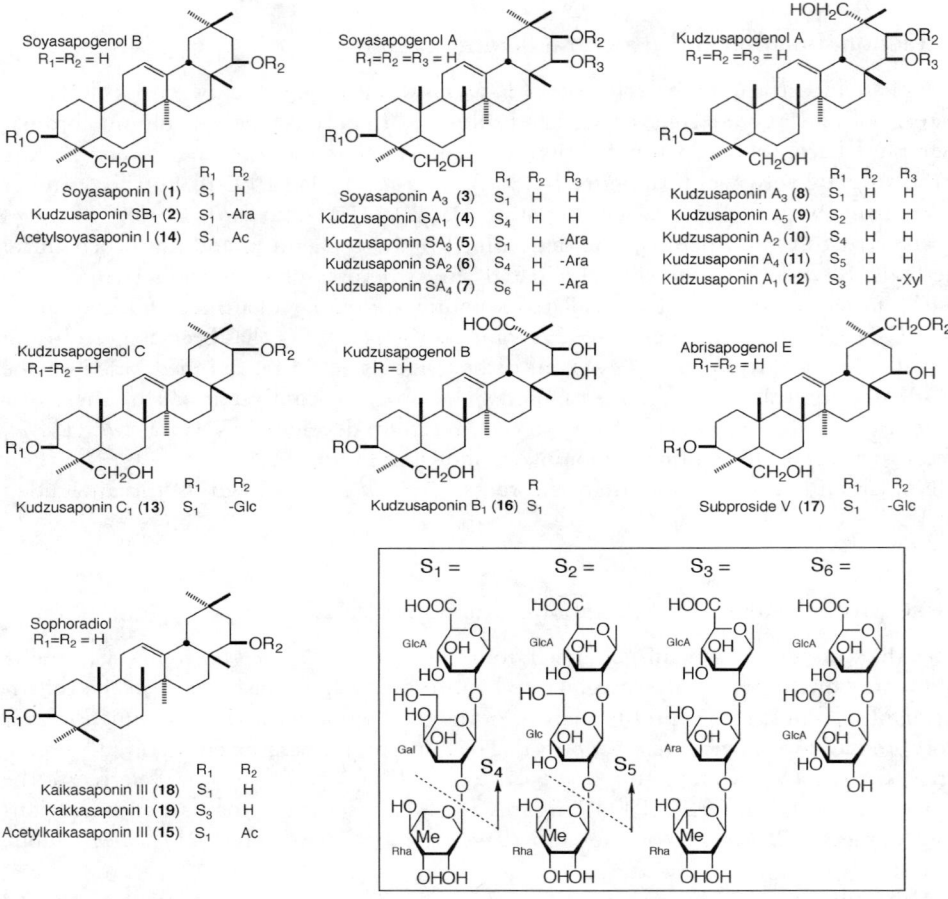

*Figure 12.1* The structures of saponins obtained from *Pueraria* plants.

activity is expressed as percent protection, calculated as $\{1-(\text{Substance}-\text{Control})/(\text{Reference}-\text{Control})\}\times 100$. Control is the value of hepatocytes which were not treated with the antiserum. Reference value was derived from cells treated with the antiserum but not with the tested samples. ALT activity was assayed by auto-analyzer COBAS MIRA (Roche) using commercial kits based on the ALT assay method. Data are the mean $\pm$SD ($n=4$). After analysis of variances, Sheffe's test was employed to determine the significance of differences between reference and experimental samples.

### Hepatoprotective activity

At first, hepatoprotective activities of the crude drugs originated from *Radix puerariae lobatae* (the roots of *P. lobata* produced in Kumamoto Prefecture, Japan) and *Radix puerariae thomsonii* (the roots of *P. thomsonii* produced in Yunnan Province, China) were determined and compared. The total saponin fraction of *Radix puerariae lobatae* provided

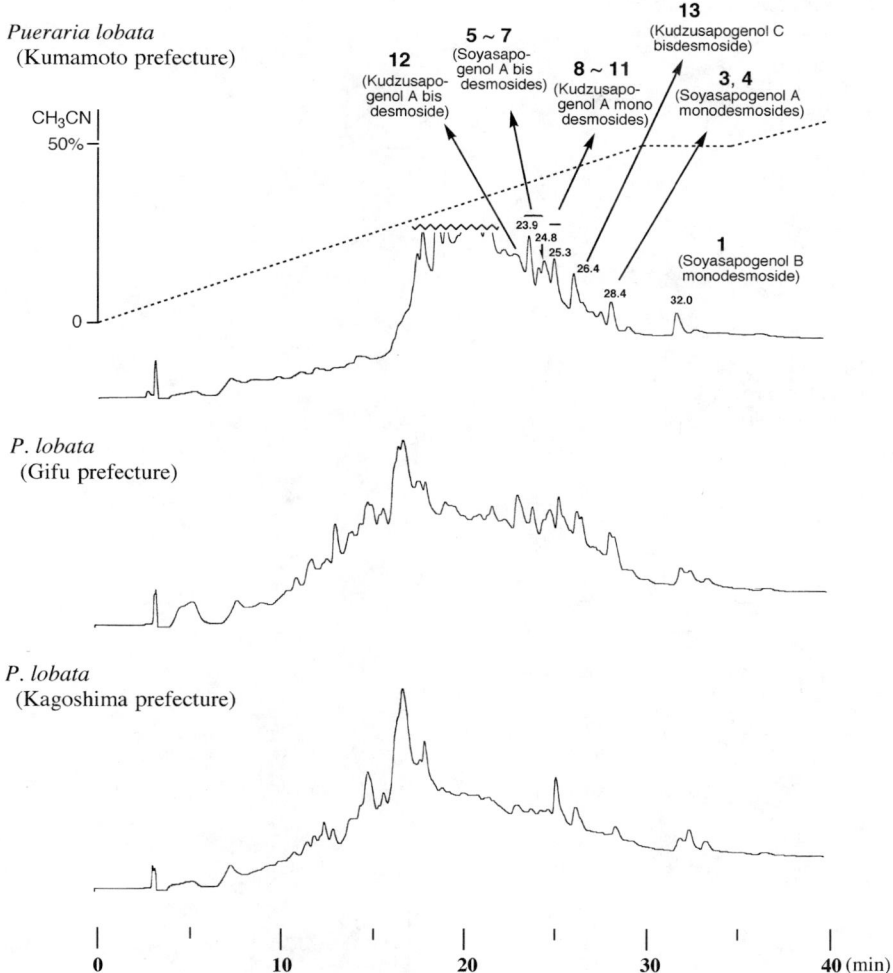

*Figure 12.2* HPLC profiles for total saponin fractions obtained from domestic *Pueraria lobata* roots. HPLC conditions were as follows; column, Nova-Pak C18 (4 μm, 8×100 mm) with Radial-Pak RCM×10 module, solvent A, $H_2O$:TFA=100:0.05 (v/v); solvent B, $CH_3CN$:$H_2O$:TFA=60:40:0.05 (v/v). Elution was done with the following process: 0→83 per cent solvent B (5 min), 83 per cent solvent B (5 min), 83 → 100 per cent solvent B (5 min), 100 per cent solvent B (5 min). Flow rate was 1 ml/min. Detection was done by UV at 205 nm. Sample of 50 μl of total saponin (1 mg/ml in MeOH) were injected.

65 per cent protection at 200 μg/ml, 84 per cent at 500 μg/ml, whereas that of *Radix puerariae thomsonii* provided 13 per cent at 200 μg/ml, 34 per cent at 500 μg/ml (Kinjo and Nohara, 1998). As described in the previous section (Figures 12.2–12.4), *Radix puerariae lobatae* was a rich source of more polar saponins, such as soyasapogenol A bisdesmosides (5–7). Since, as mentioned later, 6 and 7 were more hepatoprotective (Arao *et al.*, 1998) than 1, the total saponin of *Radix puerariae lobatae* was thought to be more potent than that of *Radix puerariae thomsonii*.

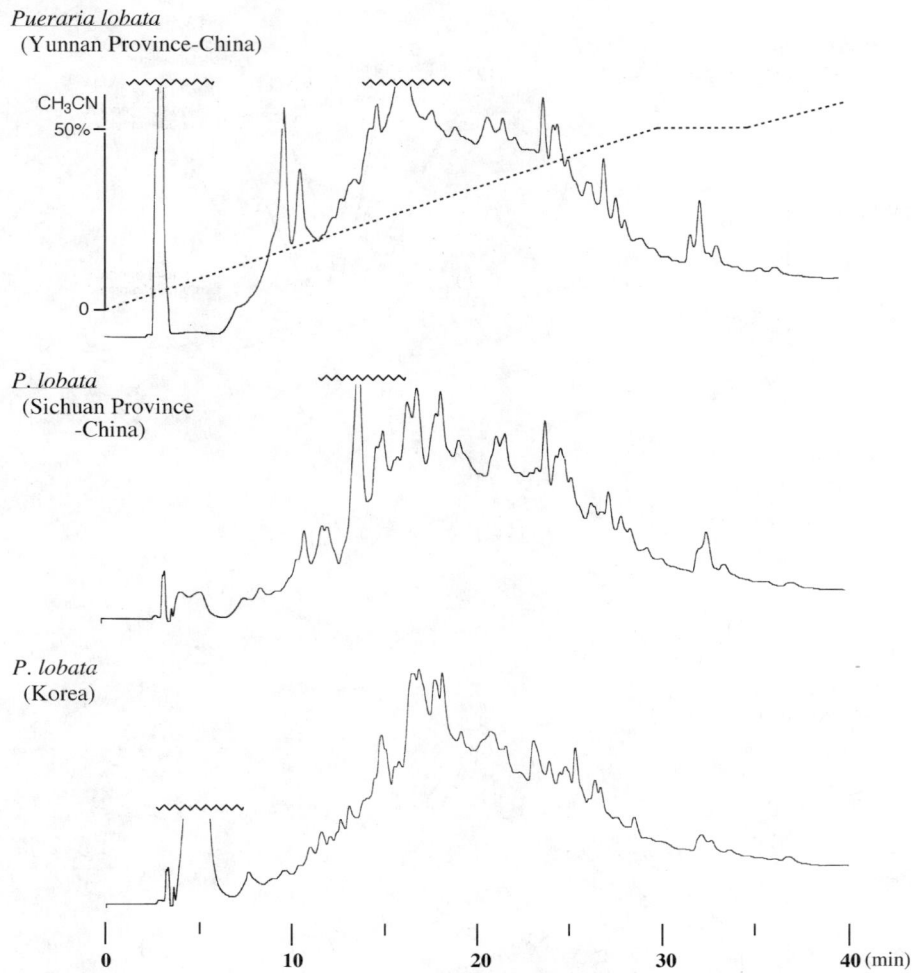

*Figure 12.3* HPLC profiles for total saponin fractions obtained from foreign *Pueraria lobata* roots. HPLC analysis was done as described in Figure 12.2.

Next, we compared the hepatoprotective activity of isolated saponins (Table 12.1). These tested saponins were classified into three groups, namely, soyasapogenol A glycosides (**3, 6** and **7**), kudzusapogenol A glycosides (**8–12**) and kudzusapogenol C glycoside (**13**). On the other hand, these saponins were also divided into six groups, the derivatives of $S_1$ (**3, 8** and **13**), $S_2$ (**9**), $S_3$ (**12**), $S_4$ (**10** and **6**), $S_5$ (**11**) and $S_6$ (**7**). The relationships between the hepatoprotective activity and their structures were investigated and discussed based on the above mentioned classifications. Within the $S_1$ derivatives group, **3** is significantly effective at $90\,\mu M$, while the effect of **8** was very weak, even at $500\,\mu M$. Moreover, all kudzusapogenol A glycosides except for **11** were less effective than glycyrrhizin, although soyasapogenol A glycosides were the most effective. Therefore, the

*Pueraria thomsonii*
(Yunnan Province-China)

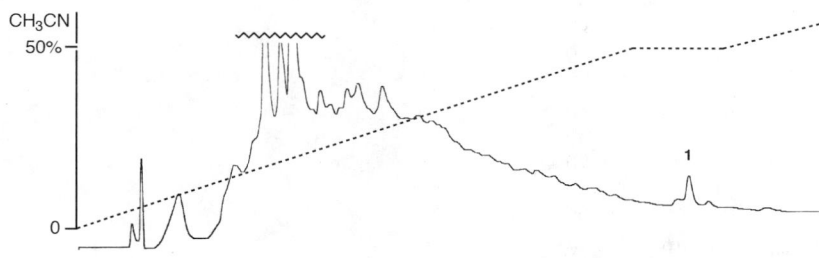

*P. thomsonii*
(Guangxi Province-China)

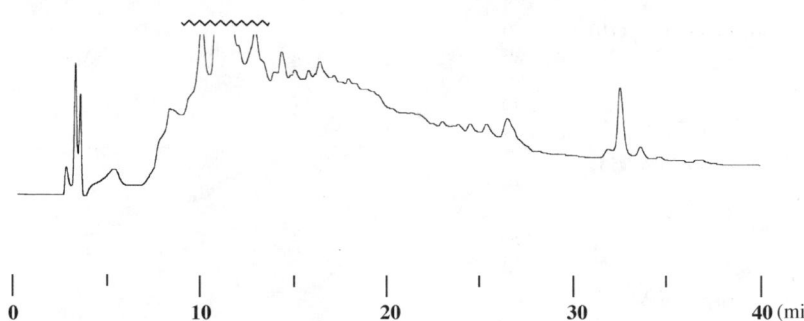

*Figure 12.4* HPLC profiles for total saponin fractions obtained from foreign *Pueraria thomsonii* roots. HPLC analysis was done as described in Figure 12.2.

*Table 12.1* Hepatoprotective activity of *Pueraria* saponins

| Substances | Dose (µM) | Protection (%) |
|---|---|---|
| Soyasaponin I (1) | 10 | −24 |
| | 30 | −23 |
| | 90 | −9 |
| | 200 | 9 |
| | 500 | 56* |
| Soyasaponin A₃ (3) | 10 | 7 |
| | 30 | 3 |
| | 90 | 40*** |
| | 200 | 87*** |
| | 500 | 100*** |
| Kudzusaponin SA₂ (6) | 10 | 7 |
| | 30 | 1 |
| | 90 | 17 |
| | 200 | 82*** |
| | 500 | 94*** |

*Table 12.1* (Continued)

| Substances | Dose (μM) | Protection (%) |
|---|---|---|
| Kudzusaponin SA$_4$ (7) | 10 | 8 |
| | 30 | 4 |
| | 90 | 7 |
| | 200 | 24*** |
| | 500 | 94*** |
| Kudzusaponin A$_3$ (8) | 10 | 4 |
| | 30 | 5 |
| | 90 | 5 |
| | 200 | 12** |
| | 500 | 24*** |
| Kudzusaponin A$_5$ (9) | 10 | 5 |
| | 30 | 8 |
| | 90 | 10 |
| | 200 | 10 |
| | 500 | 21*** |
| Kudzusaponin A$_2$ (10) | 10 | 5 |
| | 30 | 5 |
| | 90 | 3 |
| | 200 | 7 |
| | 500 | 23*** |
| Kudzusaponin A$_4$ (11) | 10 | 7 |
| | 30 | 7 |
| | 90 | 13** |
| | 200 | 16** |
| | 500 | 40*** |
| Kudzusaponin A$_1$ (12) | 10 | 3 |
| | 30 | 4 |
| | 90 | 4 |
| | 200 | 9*** |
| | 500 | 25*** |
| Kudzusaponin C$_1$ (13) | 10 | 6 |
| | 30 | 7 |
| | 90 | 11 |
| | 200 | 19* |
| | 500 | 78*** |
| Kaikasaponin III (18) | 10 | −10 |
| | 30 | −8 |
| | 90 | 26* |
| | 200 | 37*** |
| | 500 | 76*** |
| Kakkasaponin I (19) | 10 | −12 |
| | 30 | −6 |
| | 90 | 6 |
| | 200 | 30* |
| | 500 | 66*** |

Notes
Hepatoprotective actions of Soyasaponins I, A$_3$, Kudzusaponins SA$_2$, SA$_4$, A$_3$, A$_5$, A$_2$, A$_4$, A$_1$, C$_1$, Kaikasaponin III and Kakkasaponin I toward *in vitro* immunological liver injury on primary cultured rat hepatocytes. Significantly different from Ref., effective
* $p < .05$, ** $p < .01$, *** $p < .001$.

hydroxy group at C-29 would reduce the hepatoprotective activity of these types of saponins. On the other hand, the preventive effect of 3 is at least two times higher than that of 1. Hence, the hydroxy group at C-21 appears essential for a higher hepato-protective activity. Among kudzusapogenol A glycosides, 8 and 10 have a galactosyl unit, whereas 9 and 11 have a glucosyl unit. When the hepatoprotective activities of 8 and 9 were compared, 8 was slightly more effective than 9. In contrast, when the hepatoprotective activities of 10 and 11 were compared, 11 was more effective than 10. Therefore, the configuration of the hydroxy group at C-4″ in the hexosyl unit could be less important. Similarly, as with the soyasapogenol A glycosides, the hepatoprotective activity of 6 with a terminal galactosyl unit was almost identical to that of 7 which has a terminal glucuronyl unit. Since the saponin having a hexosyl unit shows greater action than that of the pentosyl unit (Kinjo *et al.*, 1998), the oxygen-bearing group at C-5″ seems to be the factor to enhance the hepatoprotective activity.

### Structure–hepatoprotective relationship

Some structure–hepatoprotective relationships, obtained through our experiments are summarized in Figure 12.5: (i) the $\beta$-hydroxy group on C-21 was important for hepato-protective activity whereas the hydroxy group on C-29 contributed negatively to the activity, (ii) in the sugar unit bound at C-3, an oxygen-bearing group at C-5″ seems to enhance the hepatoprotective activity, and (iii) the terminal rhamnopyranosyl group does not seem to contribute to the activity.

The cause of the immunological liver injury is thought to be complement-mediated (Shiki *et al.*, 1984; Kiso *et al.*, 1987). Therefore, kudzusaponins seem to prevent the injury from occurring by attack of the complement system toward hepatocyte membrane. At that time, kudzusaponins could have strong affinity for hydrophobic areas and hydro-philic areas on the hepatocyte membrane.

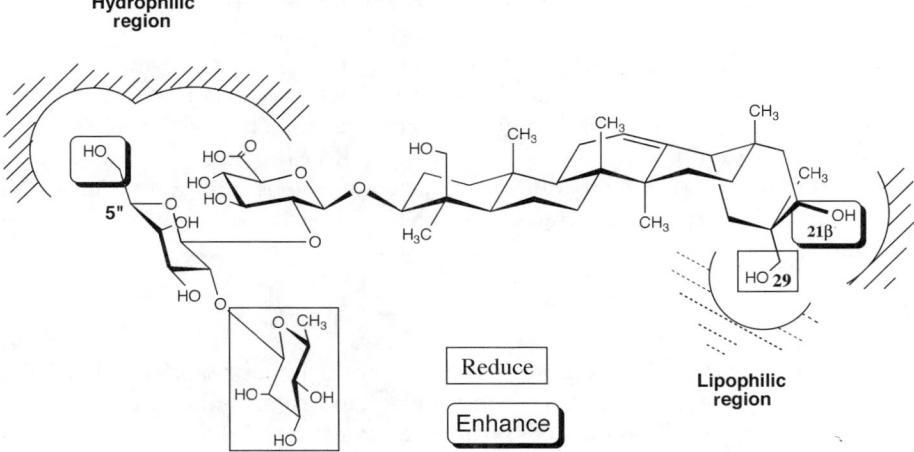

*Figure 12.5* Structure–hepatoprotective relationships of *Pueraria* roots saponins.

## SAPONINS IN FLOWERS

### Extraction, isolation and structures

A methanolic extract of *Flos puerariae thomsonii* (the flowers of *P. thomsonii*) was separated by Sephadex LH-20 column chromatography to obtain the total saponin fraction. Similarly, the total saponin fraction from *Flos puerariae lobatae* was prepared. After repeated silica gel chromatography of the total saponin fraction from *Flos puerariae thomsonii*, saponins 1, 18 and 19 were obtained (Kinjo *et al.*, 1999b) and were identified as Soyasaponin I, Kaikasaponin III, and Kakkasaponin I, respectively (Figure 12.1). Kaikasaponin III (18) was also found to be the major saponins in the leaves of *P. lobata* (Kinjo *et al.*, 1988).

### HPLC profile analysis of total saponin fraction

In order to confirm the constitution of saponins, HPLC profile analyses of the total saponin fractions of both *Flos puerariae* were performed (Figure 12.6). Although the HPLC profile for the total saponin fraction in the roots of *P. lobata* differed from that of *P. thomsonii* (Figures 12.2–12.4), no great difference in the HPLC profiles with respect to saponins in the flowers was readily apparent. Since the isolated saponins (1, 18 and 19) from *Flos puerariae thomsonii* had already been isolated from *Flos puerariae lobatae* (Kinjo *et al.*, 1988), the constitution of saponins in both crude drugs was concluded to be almost identical.

### Hepatoprotective effect

#### In vivo *efficacy*

In alcohol metabolism, a reduction effect of blood alcohol and the acetaldehyde level was observed after oral administration of the methanol extract for *Flos puerariae lobatae* (Niiho *et al.*, 1989). However, such reduction effect was not recognized by treatment with the total saponin fraction. The concentrations of blood alcohol and acetaldehyde decreased further after treatment with the total isoflavone fraction. Furthermore, the methanol extract and isoflavone fraction suppressed the increment of spontaneous movement induced by alcohol administration. The saponin fraction did not prevent such decrease in the increment caused by alcohol. The methanol extract and saponin fraction inhibited the increase of triglyceride and the blood urea nitrogen level induced by alcohol, although the saponin fraction did not affect the blood glucose level (Niiho *et al.*, 1990). Moreover, the saponin and isoflavone fractions significantly inhibited the increase in the aspartate aminotransferase (AST) and ALT levels induced by high fatty food and $CCl_4$ in control animals.

#### In vitro *efficacy*

Next, we compared the hepatoprotective actions of the isolated saponins (1, 18 and 19) using *in vitro* immunological liver injury model and the results are shown in Table 12.1. While all tested compounds exhibited protective activity, the values varied significantly. It may be speculated that hepatoprotective activity depends on some structural features. The tested saponins were classified into two groups, namely,

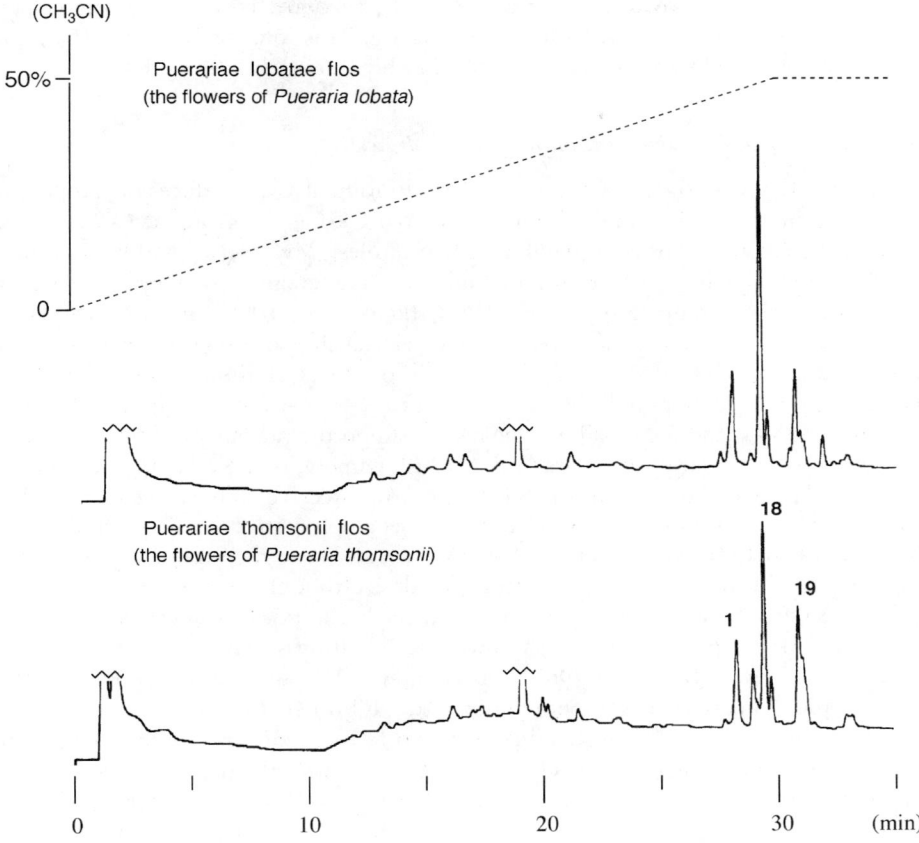

*Figure 12.6* HPLC profile for total saponin fractions originated from the flowers of *Pueraria lobata* and *P. thomsonii*. HPLC analysis was done as described in Figure 12.2 except for the column. The column used is Mightysil RP-18 GP 250–4.6 (5 μm) from Kanto Chemical Co., Inc.

Soyasapogenol B glycoside (**1**) and Sophoradiol glycosides (**18** and **19**). These saponins were also divided into two groups which have different sugar moieties linked at C-3. These were $\alpha$-L-rhamnopyranosyl-(1 → 2)-$\beta$-D-galactopyranosyl-(1 → 2)-$\beta$-D-glucurono-pyranosyl derivatives ($S_1$) (**1** and **18**) and an $\alpha$-L-rhamnopyranosyl-(1 → 2)-$\alpha$-L-arabino-pyranosyl-(1 → 2)-$\beta$-D-glucuronopyranosyl derivative ($S_3$) (**19**). When the actions of **1** and **18** were compared, saponin **18** was apparently more effective. Significant pro-tective activity was observed for **18** at a concentration 100 μM, whereas **1** was only effective at the highest dose (500 μM). This indicates that the hydroxy group at C-24 contributed negatively to the hepatoprotective activity. This information supported previously obtained structure–hepatoprotective relationship data which was measured with a different model (Miyao *et al.*, 1998). Since we reported a similar effect for the hydroxyl group at C-23 (Kinjo *et al.*, 1999a), the presence of a hydroxymethyl group at C-4 seems to reduce the hepatoprotective action, regardless of configuration. From a comparative study in the $S_1$ and $S_2$ groups, the saponin having a galactosyl unit (**18**) in the central sugar moiety shows greater action at any dose than that of the one with an arabinosyl unit (19). Hence, the presence of a hydroxymethyl group on the galactosyl unit would enhance the hepatoprotective activity. This information also substantiated previously obtained structure–activity relationship data (Kinjo *et al.*, 1998).

### Preparation of sophoradiol glucuronides and their activity

As described above, the presence of a hydroxyl group at C-24 reduces hepatoprotective activity, whereas Sophoradiol glucuronides having a methyl group at C-24 was much more effective than Soyasapogenol B glucuronides (Miyao *et al.*, 1998; Kinjo *et al.*, 1999b). Since the disaccharide and monoglucuronide saponins show greater action than the trisaccharide group (Kinjo *et al.*, 1998; Ikeda *et al.*, 1998), in order to clarify in more detail the structure–hepatoprotective relationship of saponins, we investigated the hepatoprotective effects of the hydrolytic products of Kaikasaponin III (18), i.e. Kaikasaponin I (Sophoradiol glycoside having the $S_4$ sugar unit), sophoradiol monoglucuronide (SoMG) and Sophoradiol. The results supported previously obtained structure–hepatoprotective relationship data (Table 12.2), namely, that Kaikasaponin I was more effective than 18. Kaikasaponin I showed hepatoprotective activity even at $30 \mu M$. On the other hand, not only did SoMG exhibit no hepatoprotective activity, but also become hepatotoxic at high concentration ($500 \mu M$). In contrast, the sapogenol (Sophoradiol) showed hepatoprotective activity at the same dose although less potent.

Since SoMG showed strong cyotoxicity at the highest dose, its cytotoxicity, together with that of soyasapogenol B analogs, toward liver cells was also examined in the absence of an antiserum (Table 12.2). Only SoMG showed hepatotoxicity at doses of $200 \mu M$ and $500 \mu M$. This is the first saponin known to exhibit hepatotoxic effect although some glucuronides of oleanolic acid have been shown to be hepatotoxic (Kinjo *et al.*, 1999a). Since SoMG showed hepatoprotective activity at $200 \mu M$, the hepatoprotective activity of saponin could represent a balance between hepatoprotective action and hepatotoxicity.

*Table 12.2* Hepatoprotective and hepatotoxic activity for hydrolytic products of kaikasaponin III and soyasaponin I

| Substances | Dose (µM) | Protection (%)[a] | Cytotoxicity (%)[b] |
|---|---|---|---|
| Kaikasaponin III (18) | 10 | 5 | 95 |
| | 30 | 3 | 90 |
| | 90 | 10 | 80 |
| | 200 | 34** | 95 |
| | 500 | 81** | 75 |
| Kaikasaponin I | 10 | 3 | 113 |
| | 30 | 17* | 144 |
| | 90 | 76** | 113 |
| | 200 | 93** | 138 |
| | 500 | 94** | 156 |
| Sophoradiol monoglucuronide (SoMG) | 10 | 4 | 115 |
| | 30 | 6 | 105 |
| | 90 | 0 | 115 |
| | 200 | 18* | 350† |
| | 500 | −49† | 660† |

*Table 12.2* (Continued)

| Substances | Dose (μM) | Protection (%)[a] | Cytotoxicity (%)[b] |
|---|---|---|---|
| Sophoradiol | 10 | 0 | 82 |
| | 30 | 3 | 82 |
| | 90 | 2 | 91 |
| | 200 | 0 | 95 |
| | 500 | 13* | 100 |
| Soyasaponin I (1) | 10 | | 121 |
| | 30 | | 95 |
| | 90 | | 95 |
| | 200 | | 105 |
| | 500 | | 89 |
| Soyasaponin III | 10 | | 82 |
| | 30 | | 100 |
| | 90 | | 106 |
| | 200 | | 100 |
| | 500 | | 141 |
| Soyasapogenol B monoglucuronide (SBMG) | 10 | | 90 |
| | 30 | | 85 |
| | 90 | | 80 |
| | 200 | | 85 |
| | 500 | | 110 |
| Soyasapogenol B | 10 | | 118 |
| | 30 | | 82 |
| | 90 | | 106 |
| | 200 | | 112 |
| | 500 | | 100 |

Notes

a Hepatoprotective activity toward *in vitro* immunological liver injury in primary cultured rat hepatocytes. Significantly different from reference, effective * $p < .01$, ** $p < .001$, toxic † $p < .001$.

b Hepatotoxicity in primary cultured rat hepatocytes. The percent of cytotoxicity is calculated as (sample/reference) × 100. Reference is the value of hepatocytes which were not treated with the tested samples. Significantly different from reference, toxic † $p < .001$.

### Structure–hepatotoxic relationship of SoMG

The structure–activity relationship of saponins obtained from the results of Sophoradiol derivatives was divided into two models. One model is for **18** and Kaikasaponin I

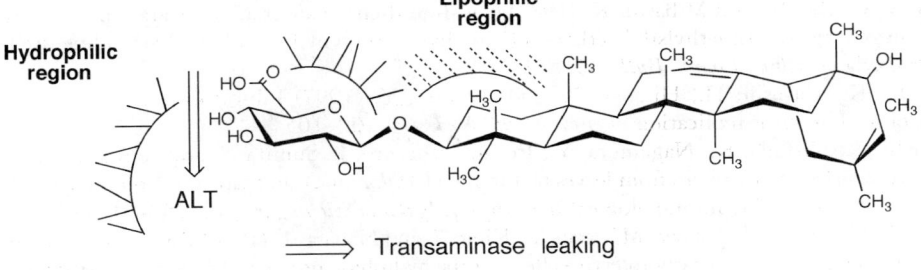

*Figure 12.7* Structure–hepatotoxic relationship for SoMG.

which accorded with structure–hepatoprotective relationships obtained from the previous results (Kinjo and Nohara, 1999). They seemed to show hepatoprotective action due to the suitable distance they keep from hepatocyte membranes. The other model for SoMG is completely different (Figure 12.7). Because of its strong affinity toward hepatocyte membranes, SoMG appeared to be too close to the membranes for protection. Hence, it showed hepatotoxicity, i.e. injured hepatocytes, although it also showed protective action against an antiserum. Consequently, the transaminase (ALT) leaked out of the hepatocytes. A similar observation has been obtained from the experiments with oleanolic acid-type glucuronides (Kinjo *et al.*, 1999a).

In the meantime, Shiraki reported that glycyrrhizin inhibited the replication of hepatitis B virus after penetrating hepatocyte membranes (Shiraki, 1995). Since soyasaponins I and II also showed antiviral effects against several viruses (Hayashi *et al.*, 1997), saponin might show not only protective action on hepatocyte membranes but also antiviral action after penetrating the membranes.

## ACKNOWLEDGMENTS

We express our appreciation to Mr K. Takeuchi of Yamada Yakken Co. Ltd. for supply of various *Radix puerariae*. We also express our appreciation to Mr Y. Niiho, Mr T. Yamazaki, Dr Y. Nakajima and H. Itoh of Ohta's Isan Co. Ltd. for *in vivo* experiments.

## REFERENCES

Arao, T., Kinjo, J., Nohara, T. and Isobe, R. (1995) Oleanene-type triterpene glycosides from Puerariae Radix. II. Isolation of saponins and the application of tandem mass spectrometry to their structure determination. *Chem. Pharm. Bull.*, 43, 1176–1179.

Arao, T., Idzu, T., Kinjo, J., Nohara, T. and Isobe, R. (1996) Oleanene-type triterpene glycosides from Puerariae Radix. III. Three new saponins from *Pueraria thomsonii*. *Chem. Pharm. Bull.*, 43, 1970–1972.

Arao, T., Kinjo, J., Nohara, T. and Isobe, R. (1997a) Oleanene-type triterpene glycosides from Puerariae Radix. IV. Six new saponins from *Pueraria lobata*. *Chem. Pharm. Bull.*, 45, 362–366.

Arao, T., Udayama, M., Kinjo, J., Funakoshi, T., Kojima, S. and Nohara, T. (1997b) Preventive effects of saponins from Puerariae Radix (the root of *Pueraria lobata*) on *in vitro* immunological injury of rat primary hepatocyte cultures. *Biol. Pharm. Bull.*, 20, 988–991.

Arao, T., Udayama, M., Kinjo, J. and Nohara, T. (1998) Preventive effects of saponins from the *Pueraria lobata* root on *in vitro* immunological liver injury of rat primary hepatocyte cultures. *Planta Med.*, 64, 413–416.

Hara, S., Okabe, H. and Mihashi, K. (1987) Gas–liquid chromatographic separation of aldose enantiomers as trimethylsilyl ethers of methyl 2-(polyhydroxyalkyl)-thiazolidine-4(*R*)-carboxylates. *Chem. Pharm. Bull.*, 35, 501–506.

Hayashi, K., Hayashi, H., Hiraoka, N. and Ikeshiro, Y. (1997) Inhibitory activity of soyasaponin II on virus replication *in vitro*. *Planta Med.*, 63, 102–105.

Higuchi, T., Nishida, K., Nagamura, Y., Ito, M., Ishiguro, I., Sumita, S. and Saito, S. (1992) Preventive effect of saponin from leaves of Taranoki (*Alalia elata*), and saponins from hypocotyls of soybean on *in vitro* immunological liver injury. *Igaku to Seibutugaku*, 124, 57–61.

Ikeda, T., Udayama, M., Okawa, M., Arao, T., Kinjo, J. and Nohara, T. (1998) Partial hydrolysis of soyasaponin I and the hepatoprotective effects of the hydrolytic products. Study of the structure–hepatoprotective relationship of soyasapogenol B analogs. *Chem. Pharm. Bull.*, 46, 359–361.

Kinjo, J., Miyamoto, I., Murakami, K., Kida, K., Tomimatsu, T., Yamasaki, M. and Nohara, T. (1985) Oleanene-sapogenols from *Puerariae radix*. *Chem. Pharm. Bull.*, 33, 1293–1296.

Kinjo, J., Takeshita, T., Abe, Y., Terada, N., Yamashita, H., Yamasaki, M., Takeuchi, K., Murakami, K., Tomimatsu, T. and Nohara, T. (1988) Studies on the constituents of *Pueraria lobata*. IV. Chemical constituents in the flowers and the leaves. *Chem. Pharm. Bull.*, 36, 1174–1179.

Kinjo, J., Kishida, F., Watanabe, K., Hashimoto, F. and Nohara, T. (1994) Five new triterpene glycosides from Russel lupine. *Chem. Pharm. Bull.*, 42, 1874–1878.

Kinjo, J., Imagire, M., Udayama, M., Arao, T. and Nohara, T. (1998) Structure–hepatoprotective relationships study of soysaponins I ~ IV having soyasapogenol B as aglycone. *Planta Med.*, 64, 233–236.

Kinjo, J. and Nohara, T. (1998) Hepatoprotective oleanene glucuronides in Fabaceae. H. Ageta. In N. Aimi, Y. Ebizuka, T. Fujita and G. Honda (eds), *Towards Natural Medicine Research in the 21st Century*, Elsevier, Tokyo, pp. 237–248.

Kinjo, J., Okawa, M., Udayama, M., Sohno, Y., Hirakawa, T., Shii, Y. and Nohara, T. (1999a) Hepatoprotective and hepatotoxic actions of oleanolic acid-type triterpenoidal glucuronides on rat primary hepatocyte cultures. *Chem. Pharm. Bull.*, 47, 290–292.

Kinjo, J., Aoki, K., Okawa, M., Shii, Y., Hirakawa, T., Nohara, T., Nakajima, Y., Yamazaki, T., Hosono, T., Someya, M., Niiho, Y. and Kurashige, T. (1999b) HPLC profile analysis of hepatoprotective oleanene-glucuronides in *Puerariae flos. Chem. Pharm. Bull.*, 47, 708–710.

Kiso, Y., Kawakami, Y., Kikuchi, K. and Hikino, H. (1987) Assay method for antihepatotoxic activity using complement-medicated cytotoxicity in primary cultured hepatocytes. *Planta Med.*, 57, 241–247.

Kurokawa, M., Iwakita, M., Kumeda, C., Yukawa, T. and Shiraki, K. (1996) Kakkon-to suppressed interleukin-1 alpha production responsive to interferon and alleviated influenza infection in mice. *J. Tradit. Med.*, 13, 201–209.

Miyao, H., Sakai, Y., Takeshita, T., Kinjo, J. and Nohara, T. (1996) Triterpene saponins from Abrus cantoniensis. I. Isolion and characterization of four new saponins and a new sapogenol. *Chem. Pharm. Bull.*, 44, 1222–1227.

Miyao, H., Arao, T., Udayama, M., Kinjo, J. and Nohara, T. (1998) Kaikasaponin III and soyasaponin I, major triterpene saponins of *Abrus cantoniensis*, act on GOT and GPT: Influence on transaminase elevation of rat liver cells concomitantly exposed to CCl$_4$ for one hour. *Planta Med.*, 64, 5–7.

Niiho, Y., Yamasaki, T., Nakajima, Y., Itoh, H., Takeshita, T., Kinjo, J. and Nohara, T. (1989) Pharmacological studies on Puerariae Flos. I. The effects of Puerariae Flos on alcoholic metabolism and spontaneous movement in mice. *Yakugaku Zasshi*, 109, 424–431.

Niiho, Y., Yamasaki, T., Nakajima, Y., Itoh, H., Takeshita, T., Kinjo, J. and Nohara, T. (1990) Pharmacological studies on Puerariae Flos. II. The effects of Puerariae Flos on alcohol-induced unusual metabolism and experimental liver injury in mice. *Yakugaku Zasshi*, 110, 604–611.

Shiki, Y., Shirai, K., Saito, Y., Yoshida, S., Wakashin, M. and Kumagai, A. (1984) Effect of glycyrrhizin on the release of transaminases from isolated rat hepatocytes. *Wakan-Iyaku Gakkaishi*, 1, 11–14.

Shiraki, K. (1995) Effect of glycyrrhizin on hepatitis B surface antigen. *Wakan-Iyakugaku Zasshi*, 12, 24–28.

Shibata, S., Harada, M. and Murakami, T. (1959) Constituents of Japanese and Chinese crude drugs. II. Antispasmodic action of the constituents of *Pueraria* root. *Yakugaku Zasshi*, 79, 863–868.

Takeshita, T., Ito, Y., Sakai, Y., Nohara, T., Yasuhara, M., Saito, H., Kitagawa, I., Ariga, T., Irino, N. and Takaoka, T. (1990) Studies on the constituents of Abri Herba effective on experimental liver injuries. *J. Pharmacobio-Dynamics*, 13, s-54.

# 13 Mammalian metabolism of *Pueraria* isoflavonoids

*Keisuke Ohsawa and Takaaki Yasuda*

## INTRODUCTION

*Radix puerariae* (RP) is the root of *Pueraria lobata* Ohwi, a perennial plant of the Genus *Pueraria*. It is known as "Kakkon" in Japan and "Ge Gen" in China. In these countries, RP is an important herbal medicine used for antipyretic and antispasmodic purposes. Its major pharmacologically active components are isoflavonoid compounds such as daidzin, daidzein, puerarin, and genistein. Some pharmacokinetic data have been reported for daidzein (Su and Zhu, 1979), puerarin (Zhu *et al.*, 1979), and genistein (Cayen *et al.*, 1964); however, to date their metabolism, including the chemical structures of their metabolites, has not been studied in detail. Based on the study using genistein, Griffiths and Smith (1972) proposed a general mechanism for the *in vivo* biotransformation of isoflavonoids (Figure 13.1). However, the methods used so far have been largely paper chromatography and TLC which may not have sufficient sensitivity to detect all metabolites, especially the minor ones. Besides, only metabolites of isoflavone aglycones obtained from enzymatic hydrolysis were studied.

In Japan and China, many Kakkon containing composite formulations are used in the practice of traditional herbal medicine. While isoflavonoids are believed to be the major pharmacologically active prinicples in these formulations, the pharmacokinetics of this class of compounds have never been studied in detail. Pharmacokinetic studies are essential to the elucidation of the mechanism of action of a drug and provide crucial information based on which the potency, efficacy, and safety of the drug can be improved. In addition, a better understanding of the pharmacokinetic properties will provide a scientific explanation for the efficacy of the composite formulations of traditional herbal medicines that have been used based only on empirical clinical experiences.

To study isoflavone metabolism in greater detail, we used a three dimensional (3D-) HPLC equipped with a photodiode array detector as a new tool in the detection and structural identification of the metabolites. Samples of urine, bile, and feces from rats

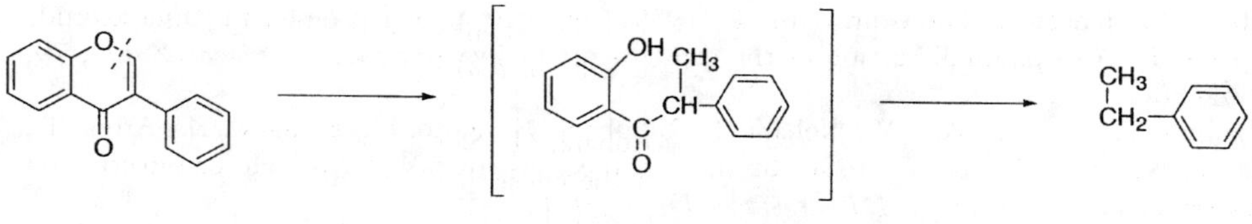

*Figure 13.1* Postulated scission of isoflavonoid molecules *in vivo*.

received oral dose(s) of isoflavone(s) were collected and analyzed to reveal the biological fate of these isoflavones in this animal species.

In this chapter, we intended to provide a brief account on the current research on the metabolism of the major isoflavones found in RP – daidzin, daidzein, puerarin, genistein, and genistin – in rats and humans.

## METABOLISM OF DAIDZIN AND DAIDZEIN IN RATS

Urine samples from rats received oral dose of daidzin were collected and analyzed for daidzin metabolites using 3D-HPLC. Four peaks **M1**, **M2**, **M3**, and **M4** were detected. The same peaks were also detected in the urine samples of rats that had received oral doses of daidzein, the aglycone of daidzin. Structural analysis of **M1**, **M2**, **M3** and **M4** using IR, MS and NMR identified these metabolites as daidzein-4′,7-di-O-sulfate, daidzein-7-O-β-D-glucuronide, daidzein-4-O-sulfate, and daidzein, respectively (Figure 13.2) (Yasuda *et al.*, 1994). The total amounts of these metabolites in 48-h urine samples were $4.80 \pm 0.08$ per cent (mean±S.E., $n=5$) and $4.57 \pm 0.25$ per cent (mean±S.E.,

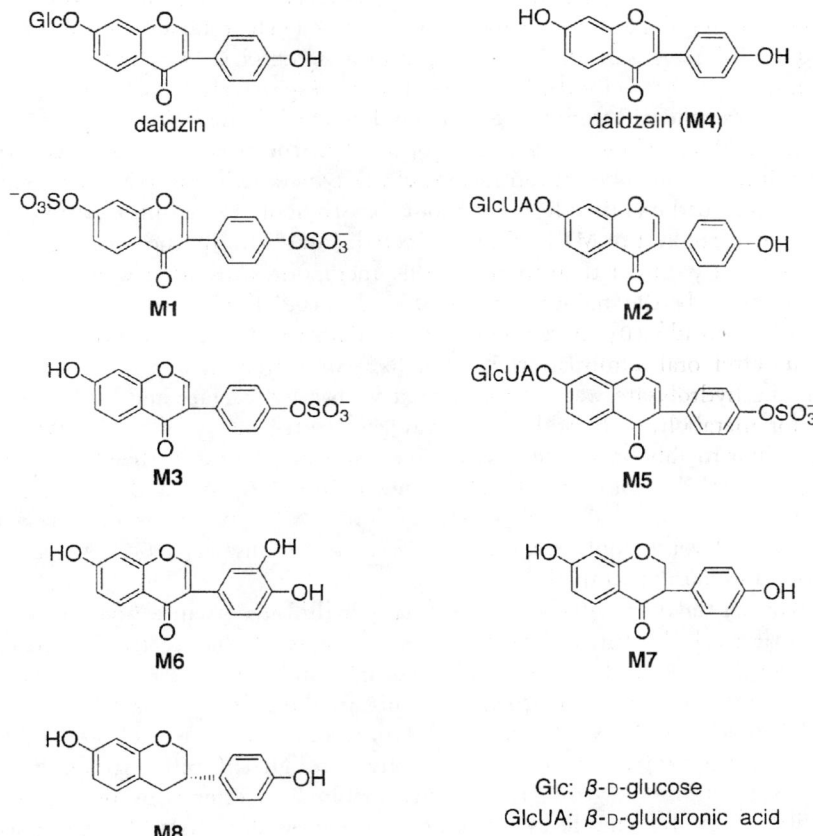

Glc: β-D-glucose
GlcUA: β-D-glucuronic acid

*Figure 13.2* Metabolites of daidzin and daidzein in rats.

$n=5$) for daidzin and daidzein, respectively. These results indicated that approximately 5 per cent of the given doses of daidzin and daidzein was excreted through the urine and that there is no significant difference in the composition of metabolites found in the urine of rats administered with the glycoside daidzin or its aglycone daidzein.

The same metabolites, **M1–M4**, were also detected in the 24-h bile samples from rats received oral administration of daidzin or daidzein. In addition, a new major peak **M5** was found in the bile sample and considered to be one of the major biliary metabolites. Structural analyses using IR, MS, and NMR identified **M5** as daidzein-4'-0-sulfate-7-0-$\beta$-D-glucuronide (Yasuda *et al.*, 1994). The total amounts of excretion of these metabolites in the 36-h bile samples were $10.13 \pm 2.59$ per cent (mean$\pm$S.E., $n=3$) and $8.31 \pm 1.89$ per cent (mean$\pm$S.E., $n=3$) for daidzin and daidzein, respectively. In other words, about 10 per cent of the given dose was excreted through the bile (Yasuda *et al.*, 1994). Like the urine samples, there is no significant difference between metabolites found in the biliary excretions of rats receiced the glycoside or aglycone. Further, in both rats, biliary excretion of these metabolites continued 36h after daidzin and daidzein administration.

Daidzein (Figure 13.2), the corresponding aglycone of daidzin, was detected along with a trace amount of the unchanged compound in 48-h feces from rats given oral dose of daidzin (Yasuda *et al.*, 1995). In the feces of rats receiving daidzein, however, only the unchanged compound but no metabolite was detected. The total amounts of excreted metabolites in 48-h feces was $10.42 \pm 2.15$ per cent (mean$\pm$S.E., $n=3$) and $17.03 \pm 3.06$ per cent (mean$\pm$S.E., $n=3$) for daidzin and daidzein, respectively.

Plasma samples collected from rats given oral dose of daidzin also contained **M1–M4** (Yasuda *et al.*, 1994b). These metabolites appeared in the plasma as early as 30 min after oral administration. The concentrations of these metabolites in plasma continued to increase and reached maximal levels at about the 8th hour after administration. After 24 h, only a small amount of **M1** could be detected. Interestingly, the concentration of **M1** was significantly higher than those of other metabolites. In other words, daidzein di-sulfate appears to be the major metabolite found in the plasma.

To search for and identify more metabolites of daidzin, the urine samples collected from the rats after oral administration of daidzin were first treated with enzymatic hydrolysis. The hydrolysate was then extracted with ethyl acetate and analyzed with 3D-HPLC for metabolites. In addition to daidzein, three new peaks (**M6–M8**) were detected as minor metabolites. These same peaks were also found in the urine samples from rats given oral dose of daidzein. These metabolites **M6**, **M7**, and **M8** were then isolated and identified as 3',4',7-trihydroxyisoflavone, 4,7-dihydroxyisoflavanone, and 4',7-dihydroxyisoflavan (equol), respectively (Yasuda and Ohsawa, 1998). **M7** and **M8** were also found in plasma samples.

It is known that flavonoid glycosides are usually hydrolyzed by intestinal flora in the gastrointestinal tract (Griffiths, 1982). Our study showed that orally administered daidzin was not detected in the plasma and the urinary and biliary excretions of the rats. We also showed that daidzin is hydrolyzed by intestinal flora and absorbed as aglycone like other flavonoid glycosides. The conjugated metabolites are excreted from the blood stream partially through the urine and partially through the bile. Although these metabolites were eliminated from the plasma within 24h after administration, the excretion through the bile and feces tend to continue even after 36h. This suggests an enterohepatic circulation as well as an efficient first-pass metabolism for the compound. The amounts of daidzin and daidzein metabolites in the biliary excretion are about two times higher than those in the urine. Therefore, like other flavonoids, the fecal excretion through the bile is the major excretion route for the isoflavones as well. **M5** was not detected in urine but its level in bile is higher than those of **M1–M4**. This may be attributed to the facts that it is relatively larger in size and higher in polarity, as these could be important factors in facilitating biliary excretion.

**M1–M4** were detected 30 min after oral administration of daidzin, suggesting that once the glycoside is hydrolyzed, the resulted aglycone is able to enter the plasma relatively rapidly. The fact that these metabolites were not detected in the plasma 24 h after administration suggests that they were also eliminated very rapidly. Su and Zhu (1979) reported that radioactivity was detected in the plasma collected from rats 30 min after oral administration of $^{14}$C-daidzein and reached maximal level 6–8 h later. Within 24 h, 64.6 per cent of the radioactivity was absorbed by the digestive tract. When administered intravenously, radioactivity in the plasma underwent a two-phase decay with $t_{1/2}$ of 13 min and 42 min. Highest amount of radioactivity was found in the kidney and liver, followed by the plasma, lung, heart, skeleton muscle, spleen, testis, and brain. In terms of excretion, 71.2 per cent of the radioactivity was eliminated through the urine and 17.4 per cent was eliminated through the feces within 24 h after administration through intravenous injection. However, when administered orally, 34.3 per cent was eliminated through the urine and 33.1 per cent was eliminated through the feces. In the 24-h biliary excretions, recovery of radioactivity was 47.4 per cent when the $^{14}$C-daidzein was administered intravenously and 39.1 per cent if it was given orally. Only a small amount of unchanged daidzein was detected in the urine of rats after intravenous injection and oral administration, suggesting a rapid *in vivo* metabolism of the compound. Moreover, 42.2 per cent of the compound are protein-bound and can be recovered by extracting with 95 per cent alcohol. Abe *et al.* (1993) studied the plasma levels in rats after oral administration of naringin and naringenin, which are flavanone derivatives. Based on a higher $t_{max}$ value of naringin compared to naringenin, they proposed that naringin is absorbed as its aglycone after hydrolysis. However, the fact that daidzin and daidzein had almost the same $t_{max}$ value and were also similar in the excretion patterns suggests that aglycone and glycoside have similar behaviors in absorption, distribution, and excretion. Interestingly, in our study, the major metabolite found in the plasma was daidzein-di-sulfate which is consistent with the results obtained by Abe *et al.* (1993) showing naringenin-di-sulfate as the major metabolite in the plasma.

**M6** (a 3′-hydroxylated derivative of daidzein), **M7** (an isoflavanone derivative), and equol of the isoflavan derivative were detected as the minor metabolites of daidzin. Equol, an estrogenic compound, has been reported as the metabolite of daidzein generated in various animal species (Marrian and Haslewood, 1932; Klyne and Wright, 1957, 1959; MacRae *et al.*, 1960; Common and Ainsworth, 1961; Braden *et al.*, 1967; Axelson and Setchell, 1981). In our studies, we identified **M6** and the isoflavanone derivative **M7** as the potential intermediates of the biotransformation from daidzein to equol. These results may provide useful information for the understanding of a new metabolic pathway of isoflavonoids. In Figure 13.3, the metabolic pathway of daidzin and daidzein is summarized based on the knowledge accumulated through these studies. Furthermore, this pathway also included metabolite **M7** found in the plasma. This information will be very useful for the study of the biological effects of these metabolites, including the conjugated metabolites **M1–M3**, of drug efficacy and mechanism of action. Important results are anticipated to emerge from these studies.

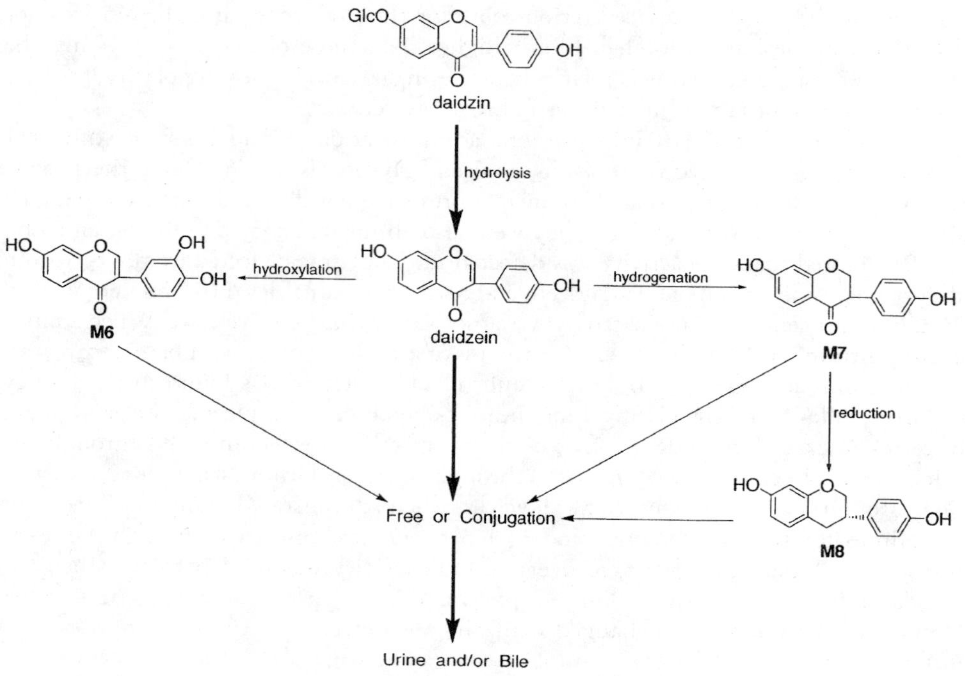

*Figure 13.3* Proposed metabolic pathways of daidzin and daidzein in rats. Thickness of arrows indicate relative importance of the pathways.

## METABOLISM OF PUERARIN IN RATS

Urine samples collected from rats given oral dose of puerarin were analyzed for metabolites using 3D-HPLC. As shown in Figure 13.4, **M1–M4** as well as an unchanged puerarin peak were detected. These metabolites were isolated and their chemical structures were determined by spectral (MS, NMR, IR) analyses (Yasuda *et al.*, 1995). Urine samples, collected up to 48 h after oral administration, showed that the total amount of metabolites was $3.63 \pm 0.05$ per cent (mean $\pm$ S.E., $n = 5$) of the administered dose and is slightly lower than those determined for daidzin and daidzein. The major species found in the urine was the original compound puerarin.

The unchanged compound as well as the major metabolites **M9** and **M10** (Figure 13.4) were also detected in the 24-h biliary excretions from rats after oral administration of puerarin. These biliary metabolites were also isolated and their chemical structures identified by spectral analyses as puerarin-4'-*O*-sulfate and puerarin-7-*O*-β-D-glucuronide, respectively (Yasuda *et al.*, 1995). The total cumulative amounts of puerarin and its metabolites in the biliary excretion collected up to 36 h after oral administration was $15.04 \pm 0.78$ per cent (mean $\pm$ S.E., $n = 3$), much higher than those of daidzin and daidzein (Yasuda *et al.*, 1994). Moreover, biliary excretion of puerarin and its metabolites virtually completed in 36 h after oral administration. Like **M5**, the major biliary metabolites **M9** and **M10** are relatively larger in size and more polar than puerarin. These may have

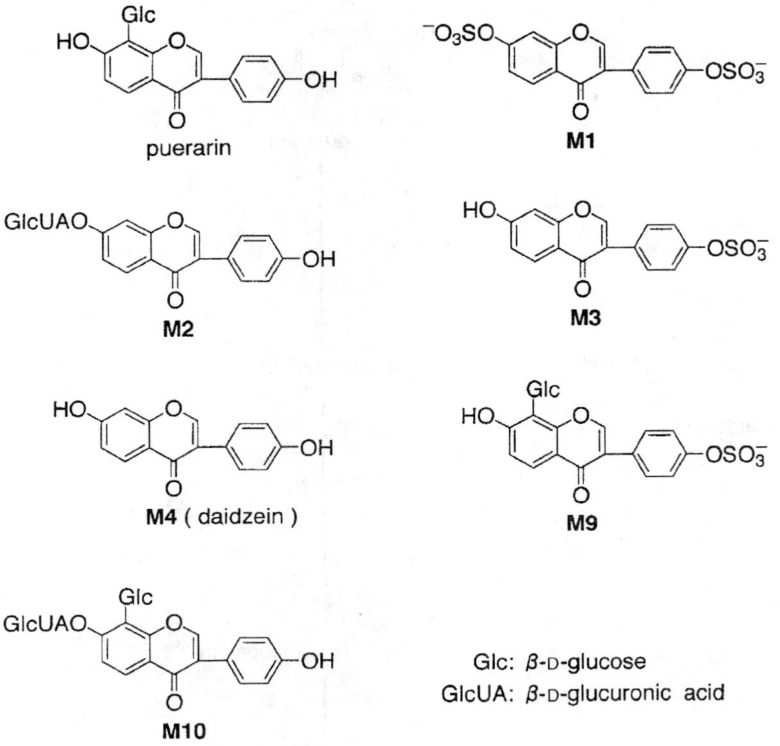

*Figure 13.4* Metabolites of puerarin in rats.

facilitated their excretion through bile. To summarize, puerarin is first converted to these metabolites through sulfation at the C-4 position and glucuronidation at the C-7 position and then excreted through the bile.

Only the unchanged compound and no metabolites were detected in the 48-h feces of rats received oral dose of puerarin. The total recovery of the unchanged compound found in the 48-h fecal samples was $34.49 \pm 2.52$ per cent (mean$\pm$S.E., $n = 3$), about two to three times higher than the numbers obtained from the studies on daidzin and daidzein (Yasuda *et al.*, 1995).

Plasma samples collected from rats received oral puerarin contained only the unchanged compound but not **M1**–**M4**. Zhu *et al.* (1979) reported that puerarin, when administered intravenously, underwent a two-phase decay in the plasma ($t_{1/2}$: 3 min and 18 min). Distribution of puerarin was the highest in kidney, followed by the plasma, liver, and lung. The spleen and brain accumulated little orally administered puerarin. In the first 24 h after oral administration, puerarin was recovered mostly from the feces and the digestive tract contents (35.7 per cent), whereas only 1.85 per cent was recovered from the urine. When humans (1 male and 2 females) were given 500 mg of puerarin orally, only 0.78 per cent of the administered dose was recovered in 36-h urine and 73.3 per cent was recovered in the 72-h feces. The amount of plasma protein bound puerarin was 24.6 per cent.

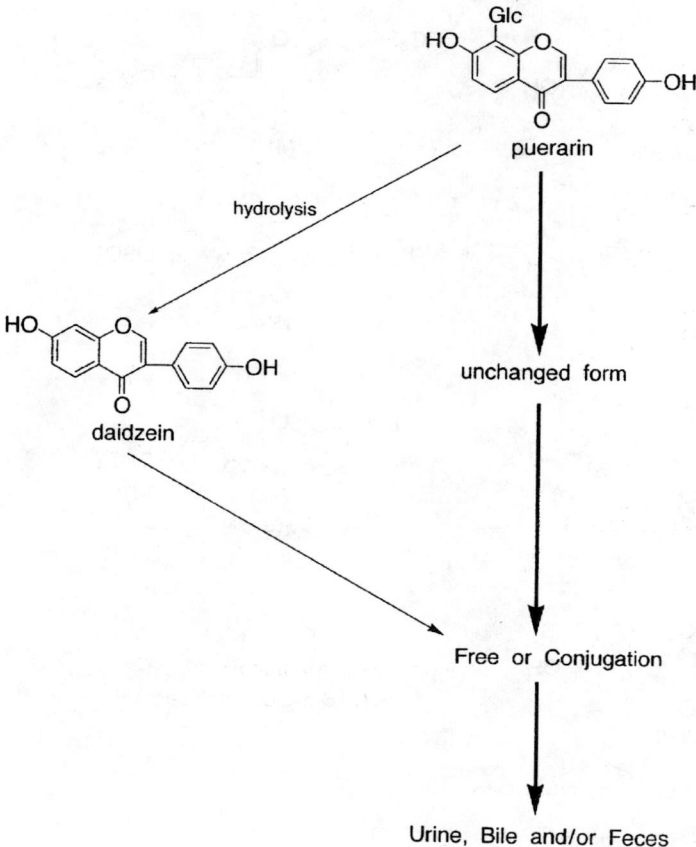

*Figure 13.5* Proposed metabolic pathways of puerarin in rats. Thickness of arrows indicate relative importance of the pathways.

In summary, as an *O*-glycoside, daidzin is mainly hydrolyzed, and then absorbed and excreted as its aglycone. On the other hand, as a *C*-glycoside, puerarin shows up in the plasma, urine, and feces mainly as the unchanged compound. This could be attributed to the fact that C–C linkage between the sugar and isoflavone moiety in puerarin is more resistant than the glycosidic linkage in daidzin to the hydrolytic actions of enzymes found in the intestinal flora in drug-metabolizing organs, such as liver, etc. To date, the metabolism of flavonoid glycosides in mammals has been studied mainly for *O*-glycosides. This study provided very interesting results on the metabolism of *C*-glycoside of an isoflavone. A proposed pathway for puerarin metabolism is shown in Figure 13.5.

## METABOLISM OF GENISTIN AND GENISTEIN IN RATS

3D-HPLC analysis of the urine samples collected from rats after an oral dose of genistin detected four major peaks **M11–M14** (Figure 13.6). The same peaks were also detected

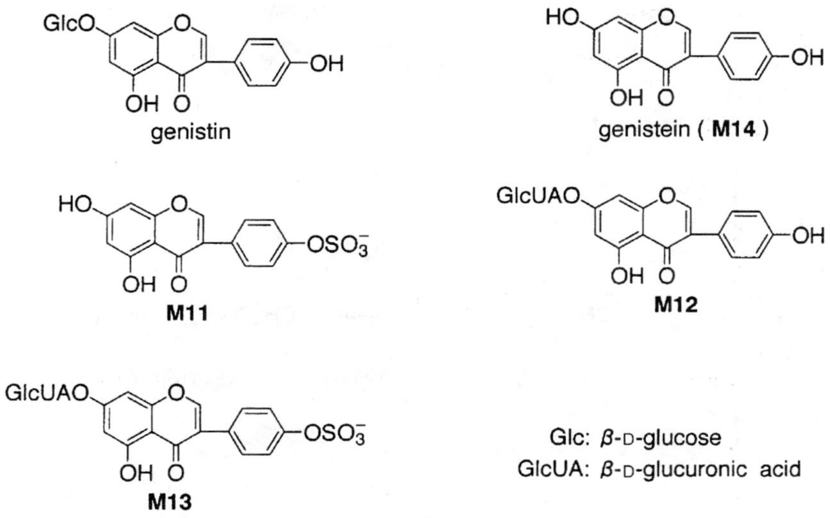

*Figure 13.6* Metabolites of genistin and genistein in rats.

in the urine samples of rats receiving the aglycone genistein. Structural analyses (IR, MS, and NMR) identified **M11**, **M12**, **M13**, and **M14** as genistein-4'-O-sulfate, genistein-7-O-β-D-glucuronide, genistein-4'-O-sulfate-7-O-β-D-glucuronide, and genistein, respectively (Yasuda and Ohsawa, 1996). The total amount of these metabolites in the 48-h urine of rats given oral genistein was 5.67±0.97 per cent (mean±S.E., $n=5$), slightly lower than those of daidzin and daidzein. The most abundant metabolite of oral genistein found in the urine was **M11**.

**M12** and **M13** were detected in the biliary excretions collected from rats up to 24 h after oral administration of genistein. The total amount of **M13** found in 36-h biliary excretion was 16.02±3.66 per cent (mean±S.E., $n=3$), higher than that of daidzin but almost the same as that of daidzein (Yasuda *et al.*, 1996). Moreover, the biliary excretion of these metabolites continued for 36 h after oral administration. As compared to the major urinary metabolite **M11**, the major biliary metabolite **M13** is relatively large and polar, which may facilitate its excretion via the bile. It appears that orally administered genistein is first converted to these metabolites through sulfation at the C-4 position and glucuronidation at the C-7 position and then excreted as **M5** through the bile.

Only genistein, the corresponding aglycone and a trace amount of the unchanged compound were detected in the 48-h feces collected from rats received genistin orally (Yasuda *et al.*, 1996). Moreover, when genistein was orally administered, only the unchanged compound was detected but no metabolites. The total amounts of metabolites excreted in the 48-h feces was 6.50±2.56 per cent (mean±S.E., $n=3$) and 21.08±8.88 per cent (mean±S.E., $n=3$) for genistin and genistein, respectively.

Batterham *et al.* (1965) and Griffiths and Smith (1972) carried out metabolic studies using sheep and rat microflora and identified *p*-ethylphenol as the major metabolite of genistein. They also reported that this metabolite was derived mostly from the B ring, some was originated from the C ring of genistein, but none from the A ring. Cayen *et al.*

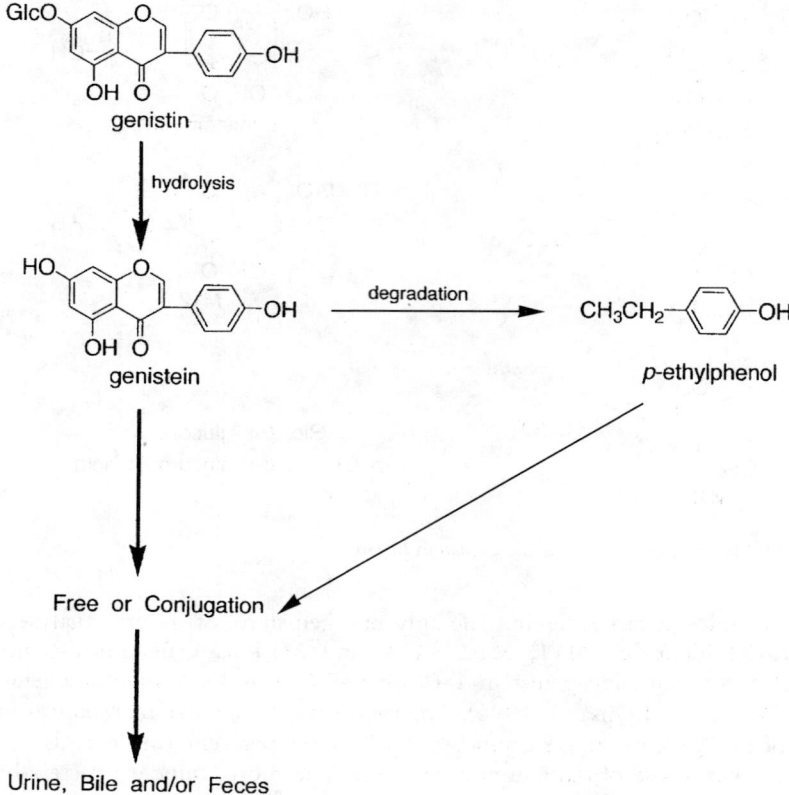

*Figure 13.7* Proposed metabolic pathways of genistin and genistein in rats. Thickness of arrows indicate relative importance of the pathways.

(1964) conducted a study in which $^3$H-genistein was given to hens through intramuscular and found equol as a metabolite in the urine. However, our studies detected neither *p*-ethylphenol nor equol in rats. A proposed pathway for genistin and genistein metabolism in rats is shown in Figure 13.7.

Genistein is known to have estrogenic activity (Farmakalidis and Murphy, 1984) and tyrosine- (Akiyama *et al.*, 1987) and histidine-kinase inhibitory activity (Huang *et al.*, 1992). Therefore, studies on the effects of these metabolites on the drug efficacy and mechanism of action will be very interesting, and will certainly provide useful information.

## METABOLITES OF ORALLY ADMINISTERED KAKKON-TO IN RAT AND HUMAN URINE

Kakkon-to is a composite formulation of Chinese herbal medicine which contains Kakkon and has been used mainly as a traditional treatment in the early stage of a common cold. The composite formulation consists of seven dried herbal medicines. They are pueraria root [*Pueraria lobata* Ohwi (Leguminosae)], ephedra herb [*Ephedra sinca* Stapf (Ephedraceae)], ginger rhizome [*Zingiber officinale* Roscoe (Zingiberaceae)], jujube fruit [*Zizyphus jujuba* Miller var. *inermis* Rehder (Rhamnaceae)], cinnamon bark [*Cinnamomum cassia* Blume (Lauraceae)], peony root [*Paeonia lactiflora* Pallas (Paeoniaceae)], and glycyrrhiza root [*Glycyrrhiza uralensis* Fisher (Leguminosae)].

First, in the urine samples collected from rats after oral administration of a hot water extract of Kakkon, which is the main drug component of Kakkon-to, the presence of **M1–M4** as well as puerarin was confirmed by 3D-HPLC (Yasuda *et al.*, 1995). Moreover, **M1–M4** as well as puerarin were also detected from urine samples collected from rats after oral administration of Kakkon-to. These results suggest that, similar to the purified single components daidzin, daidzein, and puerarin, the effective components of a crude water extract of Kakkon-to, are also absorbed through the digestive tract.

Since drug metabolism may differ significantly in different species, a study was conducted in humans administered with Kakkon-to. Urine samples collected from healthy

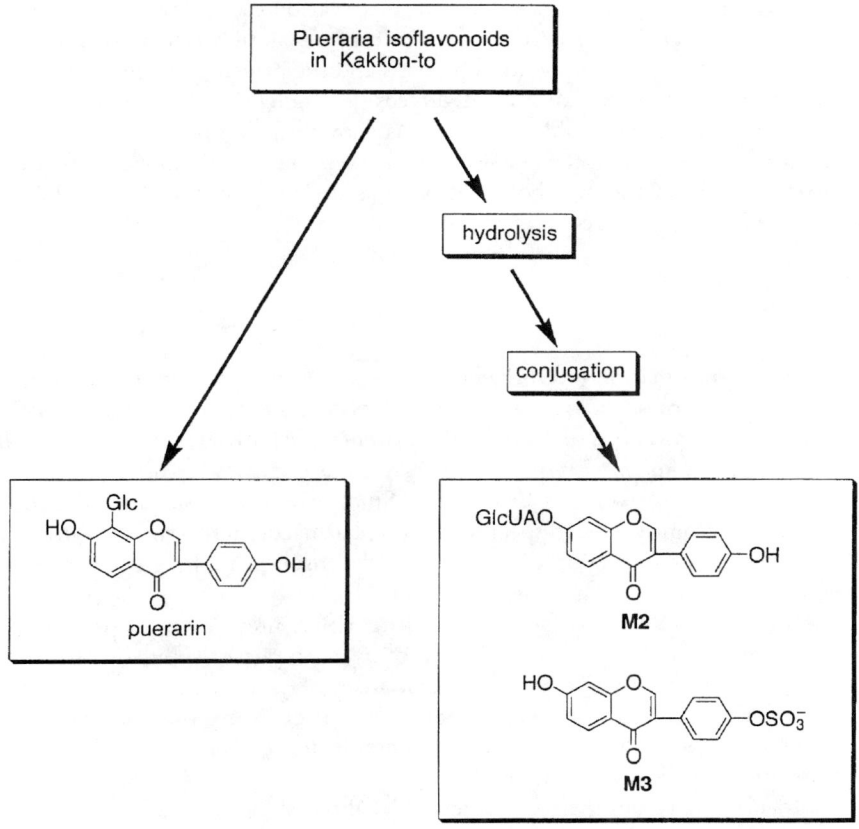

*Figure 13.8* Proposed metabolic pathways of pueraria isoflavonoids in humans.

male adults (22 and 26 years old) who have received daily dose (oral) of Kakkon-to formulation showed the peaks corresponding to **M2** and **M3**, but no peaks for **M1**, **M4** or any other metabolites. Moreover, equol could be also detected in the urine sample after enzymatic hydrolysis of conjugated metabolites. These results suggest that daidzin and daidzein in Kakkon-to are absorbed by the human body and excreted through the urine mainly as their mono-glucuronide and mono-sulfate of daidzein, and some in the form of the conjugate(s) of equol. The metabolites of genistein and genistin were not detected in urine samples of rats and humans. This could be due to fact that Kakkon-to contains very little genistin and genistein.

In the previous section, we showed that puerarin was excreted intact in the urine of rats and concluded that the *C*-glycoside (puerarin) is more resistant to hydrolysis in rats compared to *O*-glycoside (daidzin). According to the quantitative analyses performed previously in our laboratory, the dose of Kakkon-to given is equivalent to a dose of about 25 mg of daidzin, 3 mg of daidzein, and 130 mg of puerarin (Yasuda *et al.*, 1993). Interestingly, although the formulation has a high content of puerarin, a puerarin peak was not detected from any of the human samples. Zhu *et al.* (1979) reported that when humans (1 male and 2 females) were given 500 mg of puerarin, 0.78 per cent of the unchanged compound was detected in the 36-h urine samples. Moreover, through an *in vitro* study using human intestinal flora, Hattori *et al.* (1995) reported the hydrolysis of homoorientin, a *C*-glycoside. Therefore, the observation that puerarin was not found in human urine collected after oral Kakkon-to may be attributed to the low urinary excretion of the unchanged compound in humans and the presence of a human intestinal flora that is capable of hydrolyzing the *C*-glycoside efficiently. A proposed metabolic pathway for pueraria isoflavonoids in human is shown in Figure 13.8. In the future, studies on the metabolism of these isoflavone compounds in humans should also include plasma levels of the metabolites as well as detailed information on the effects from differences in sex and age, and corresponding diseases.

## CONCLUSIONS

Pharmacokinetic information (ADME) of a drug is crucial to the understanding of its efficacy and mechanism of biological action. It is also essential to the evaluation of the drug's safety and drug–drug interactions. Currently, ADME results obtained from various animals and some *in vitro* studies using various enzymes are used to predict the pharmacokinetics of isoflavonoids in humans. Since aspects of drug metabolism may differ significantly from species to species, *in vitro* pharmacokinetic studies using human organs or human enzymes produced through genetic engineering have been attracting a great deal of attention in recent years.

In addition to the attentions focusing on drug–drug interactions in multiple-drug therapy, drug–metabolite(s) interactions should be noted in pharmacokinetic mass-balance studies, especially in cases where phase II metabolism plays a central role in the metabolism and regeneration of a drug. Precise information regarding the pharmacokinetics of the conjugated metabolite itself is indispensable for a kinetic investigation of a drug–conjugated metabolite interaction study.

The results obtained from the studies described above show that isoflavones taken by rats and humans are mainly absorbed as their aglycones and undergo phase II metabolism forming various conjugated compounds. Sulfation of the poorly water soluble isoflavones significantly increase their polarity and as a result accelerate excretion. However, many sulfuric esters tend to bind with proteins or form complexes with phospholipids. Due to these properties, the permeative diffusion can be facilitated through biomembranes. Therefore, sulfation not only accelerates excretion but also improves other drug distribution and elimination processes. For instance, the conjugated compounds of morphine have a higher pharmacological activity than mother compound. The major product of glucuronidation of morphine is 3-glucuronide. However, the 6-glucuronide of morphine shows a higher analgesic effect than mother compound. Therefore, it is very important to gain better understanding on the pharmacological activities of the conjugated metabolites.

Adlercreutz *et al.* (1991) proposed an interesting hypothesis that the low incidence in hormone (estrogen)-dependent cancers observed in Japan is due to the fact that soybean products are major components of typical Japanese diets. Soy products contain high contents of isoflavones and are thought to be useful in preventing estrogen-dependent cancers owing to their antiestrogenic effect. Better understanding of the ADME of the pharmacologically active compounds we consume from our diet will provide crucial information for the development of better and more efficacious agents or food supplements for chemoprevention of estrogen-dependent diseases.

# REFERENCES

Abe, K., Katayama, H., Suzuki, A. and Yumioka, E. (1993) Biological fate of orally administered naringin and naringenin in rats. *Shoyakugaku Zasshi*, 47, 402–407.

Adlercreutz, H., Honjo, H., Higashi, A., Fotsis, T., Hamalainen, E., Hasegawa, T. and Okada, H. (1991) Urinary excretion of lignans and isoflavonoid phytoestrogens in Japanese men and women consuming a traditional Japanese diet. *Am. J. Clin. Nutr.*, 54, 1093–1100.

Akiyama, T., Ishida, J., Nakagawa, S., Ogawara, H., Watanabe, S., Itoh, N., Shibuya, M. and Fukami, Y. (1987) Genistein, a specific inhibitor of tyrosine-specific protein kinases. *J. Biol. Chem.*, 262, 5592–5595.

Axelson, M. and Setchell, K.D.R. (1981) The excretion of lignans in rats – evidence for an intestinal bacterial source for this new group of compounds. *FEBS Lett.*, 123, 337–342.

Batterham, T.J., Hart, N.K., Lamberton, J.A. and Braden, A.W.H. (1965) Metabolism of oestrogenic isoflavones in the sheep. *Nature*, 206, 509.

Braden, A.W.H., Hart, N.K. and Lamberton, J.A. (1967) The oestrogenic activity and metabolism of certain isoflavones in sheep. *Aust. J. Agric. Res.*, 18, 335–348.

Cayen, M.N., Carter, A.L. and Common, R.H. (1964) The conversion of genistein to equol in the fowl. *Biochim. Biophys. Acta*, 86, 56–64.

Common, R.H. and Ainsworth, L. (1961) Identification of equol in the urine of the domestic fowl. *Biochim. Biophys. Acta*, 53, 403–404.

Farmakalidis, E. and Murphy, P.A. (1984) Oestrogenic response of the CD-1 mouse to the soyabean isoflavones genistein, genistin and daidzin. *Food Chem. Toxicol.*, 22, 237–239.

Griffiths, L.A. and Smith, G.E. (1972) Metabolism of apigenin and related compounds in the rat. *Biochem. J.*, 128, 901–911.

Griffiths, L.A. (1982) In J.B. Harbone and T.J. Mabry (eds), *The Flavonoids: Advance in Research*, Chapman and Hall, London, pp. 692–694.

Hattori, M., Shu, Y., El-Sedwy, A. and Namba, T. (1995) Metabolism of homoorientin by human intestinal bacteria. *J. Nat. Prod.*, 51, 874–878.

Huang, J., Nasr, M., Kim, Y. and Matthews, H.R. (1992) Genistein inhibits protein histidine kinase. *J. Biol. Chem.*, 267, 15511–15515.

Klyne, W. and Wright, A.A. (1957) Steroids and other lipids of pregnant goat's urine. *Biochem. J.*, 66, 92–101.

Klyne, W. and Wright, A.A. (1959) Steroids and other lipids of pregnant cows' urine. *J. Endocr.*, 18, 32–45.

MacRae, H.F., Dale, D.G. and Common, R.H. (1960) Formation *in vivo* of 16-epiestriol and 16-ketoestradiol-17$\beta$ from estriol by the laying hen and occurrence of equol in hens' urine and faeces. *Can. J. Biochem. Physiol.*, 38, 523–532.

Marrian, G.F. and Haslewood, G.A.D. (1932) Equol, a new inactive phenol isolated from the ketohydroxyoestrin fraction of mares' urine. *Biochem. J.*, 26, 1227–1232.

Su, G. and Zhu, X. (1979) The metabolic fate of the effective components of *Radix puerariae*. *Acta Pharm. Sin.*, 14, 129–134.

Yasuda, T., Momma, N. and Ohsawa, K. (1993) A simultaneous determination of daidzin and puerarin and determination of daidzein in oriental pharmaceutical decoctions containing *Puerariae Radix* by ion-pair high-performance liquid chromatography. *Yakugaku Zasshi*, 113, 881–889.

Yasuda, T., Kano, Y., Saito, K. and Ohsawa, K. (1994a) Urinary and biliary metabolites of daidzin and daidzein in rats. *Biol. Pharm. Bull.*, 17, 1369–1374.

Yasuda, T., Nakazawa, T., Sekiguchi, M., Kano, Y., Saito, K. and Ohsawa, K. (1994b) Biological fate of orally administered major component of *Puerariae Radix* in rats. *Tohoku Yakka daigaku Kenkyu Nempo.*, 41, 161–166.

Yasuda, T., Ohtsubo, S., Kano, J. and Ohsawa, K. (1994c) Fecal metabolites after oral administration of major components of *Puerariae Radix* in rats. *Tohoku Yakka daigaku Kenkyu Nempo.*, 42, 191–198.

Yasuda, T., Kano, Y., Saito, K. and Ohsawa, K. (1995) Urinary and biliary metabolites of puerarin in rats. *Biol. Pharm. Bull.*, 18, 300–303.

Yasuda, T., Kano, Y., Saito, K. and Ohsawa, K. (1995) Urinary compounds after oral administration of Ge-Gen-Tang in rats and human. *J. Trad. Med.*, 12, 66–70.

Yasuda, T., Mizunuma, S., Kano, Y., Saito, K. and Ohsawa, K. (1996) Urinary and biliary metabolites of genistein in rats. *Biol. Pharm. Bull.*, 19, 413–417.

Yasuda, T. and Ohsawa, K. (1996) Fecal metabolites after oral administration of genistein and genistin in rats. *Tohoku Yakka daigaku Kenkyu Nempo.*, 43, 159–164.

Yasuda, T. and Ohsawa, K. (1998) Urinary metabolites of daidzin orally administered in rats. *Biol. Pharm. Bull.*, 21, 953–957.

Zhu, X., Su, G., Li, Z., Yue, T., Yan, X. and Wei, H. (1979) The metabolic fate of the effective components of *puerariae*. *Acta Pharm. Sin.*, 14, 349–355.

# 14 Synthesis of naturally occurring isoflavones and their analogs

*Anwar Jardine*

## INTRODUCTION

The development of synthetic routes to naturally occurring isoflavonoids received considerable attention since their isolation and characterization in the early 1900s. The multitude of substitution possibilities on the 3-phenylchroman skeleton (**A**) of isoflavonoids naturally lends itself to vast structural and/or molecular diversification. Hitherto, hundreds of isoflavonoids have been isolated and therefore it constitutes the largest group of naturally occurring *O*-heterocycles found mainly in the subfamily *Papilionoideae* of the *Leguminosae*.

**A**

The renaissance of chromatographic and spectroscopic methods during the 1960s and their continued development until today paralleled the explosion of structural data and characterization of naturally occurring isoflavonoids. The development of improved disease models, high throughput screening and lead optimization through combinatorial chemistry, added a new dimension to the research around this class of compounds. Contemporary developments in the area of isoflavonoid chemistry has been thoroughly reviewed (Dewick, 1993; Donelly and Boland, 1995, 1998). Supporting chapters in this book will address the many interesting biological aspects of isoflavones from the *Pueraria* genus. The scope of this chapter will be focussed around the synthesis of naturally occurring isoflavones and analogs particularly of the *Pueraria lobata* (Kudzu) plant species.

### Structure–activity relationships

Traditional medicines derived from natural sources are increasingly being used as the starting point for the development of new drugs. Drug discovery programs have broaden their screening from purely synthetic focused chemical libraries to natural resource libraries. Other than pure compounds, natural product libraries now include the well documented crude extracts as well. The process of finding the active component(s) in a whole natural resource has obtained an elevated status once again. After all, the majority of drugs on the market today are of natural origin. The goal is to shorten the time-line from drug discovery to the market, thus meeting the demand and cost for that particular drug. Meanwhile, the marketing of the whole natural resource or extracts thereof has attracted some attention. New names have been coined for the latter group of products, i.e. nutriceuticals, functional foods and herbal supplements. However, for these products to gain legitimacy in conventional medicine will require rigorous identification and standardization of active ingredients, plus FDA approved clinical testing. Central to this discussion is the role of phytoestrogens in health and disease. Amongst other classes of compounds, phyto-estrogens include isoflavones that have been isolated from fruits, vegetables and whole grains commonly consumed by humans. Soybeans and numerous other seeds and plant roots (Kudzu) are the most significant dietary sources of isoflavones.

*In vitro* and *in vivo* studies suggest that dietary isoflavones play an important role in prevention of menopausal symptoms (Nestel *et al.*, 1999), osteoporosis (Arjmandi *et al.*, 2000), cancers (Messina, 1999) and heart diseases (Greaves *et al.*, 1999). Proposed mechanisms include estrogenic and anti-estrogenic effects, induction of cancer cell differentiation, inhibition of tyrosine kinase and DNA topoisomerase, anti-angiogenesis and antioxidant effects etc. Some of the major isoflavone biological activities are summarized in Table 14.1. The mode of action of these isoflavones are currently under investigation.

The isoflavone skeleton provides a platform for diverse biological activities. These multitude of activities are established through a combination of antioxidant properties and the spatial orientation of hydroxyl groups with the propensity to form hydrogen bonds with proteins and metals. Principles determining the structure–activity relationships (SAR) of isoflavones with respect to antioxidant properties and modes of receptor binding are not well defined yet. However, the structural requirements for effective antioxidant and radical scavenging ability for flavones include, the C-4 carbonyl conjugated to the C-2, C-3 double bond and polyhydroxylation at ring A and B (Cook and Samman, 1996). Similarly, isoflavone antioxidant properties can be rationalized. Inflammatory oxidants such as hypochlorous acid (HOCl) and peroxynitrate ($ONOO^-$), have been shown to react selectively with the isoflavones genistein, daidzein and biochanin A (Boersma *et al.*, 1999). Antioxidant abilities of daidzein and genistein have been improved by 8-hydroxylation (Esaki *et al.*, 1998). Furthermore, antioxidant abilities (and radical scavenging potential) of isoflavones in aqueous media were determined to be in the order genistein > daidzein = genistin ≅ biochanin A = daidzin > formononetin ≅ ononin (no activity) (Ruizlarrea *et al.*, 1997). The presence of a 4'-OH group proved to be essential for activity. This correlates well with the 4-OH group of tyrosine, which is essential for reactivity and may share a common function in comparison to isoflavones. Based on the above mentioned antioxidant properties, it is no surprise that genistein, but not daidzein inhibits the oxidation of LDL, which plays a potential role in atheroscelerosis (Kapiotis *et al.*, 1997).

Isoflavones solicit much attention for their anticancer properties. Genistein, a protein tyrosine kinase (PTK) inhibitor, is probably the most studied of this class of compounds. The exact mode of action is not known, but it is likely that a combination of antioxidant and antiangiogenic properties is operative. Genistein is a specific inhibitor of tyrosine autophosphorylation of the epidermal growth factor (EGF) receptor (Kim *et al.*, 1998).

*Table 14.1* Structure and function of naturally ocurring isoflavones and their analogs

| Structure | Activity | References |
|---|---|---|
| $R^7=R^5=R^{4'}$=OH (Genistein) | Increases Metallothionein expression in human intestinal cells, Caco-2. <br> Protein tyrosine kinase and topoisomerase II inhibitor. Anticancer Anti-atherosclerosis <br> Topical treatment, prevention of skin cancer and photo-damage | (Kuo and Leavitt, 1999) <br> (Jing and Waxman, 1995) <br> (Kapiotis *et al.*, 1997) |
| $R^7=R^5$=OH, $R^{4'}$=Cl (4'-Chlorogenistein) | Potent EGFR tyrosine kinase inhibitor 10× better than genistein | (Traxler *et al.*, 1998) |
| $R^7=R^{4'}$=OH,(Daidzein) <br> $R^7$=OGlc, $R^{4'}$=OH (Daidzin) <br> $R^7$=O(CH$_2$)$_n$CO$_2$R, $R^{4'}$=OH | Alcohol dehydrogenase inhibitors <br> Antidipsotropic properties 4–10× better than daidzin | (Keung and Vallee, 1993) <br> (Rooke *et al.*, 2000) |
| $R^8$=C- Glc, $R^7=R^{4'}$=OH (Puerarin) | Anti-angina, Increases Choroidal bloodflow during treatment of ischemic retinopathy and age-related muscular degeneration | (Xuan *et al.*, 1999) |
| $R^7=R^8=R^{4'}$=OH (8-Hydroxydaidzein) <br> $R^7=R^8=R^5$=OsH (8-Hydroxygenistein) | Potent anti-oxidant properties | (Esaki *et al.*, 1998) |
| $R^7=R^3=R^{4'}$=OMe (Cabreuvin) | Anti-helicobacter Pylori | (Osaki *et al.*, 1999) |
| $R^7$=isopropyl (Ipriflavone) | Stimulates bone formation. Anti-osteoporosis Anti-inflammatory, Antihypersensitive | (Cheng *et al.*, 1994) <br> (Wu *et al.*, 1992) |
| $R^7$=OMe, $R^5=R^{4'}$=OH (Prunetin) | Zoospore attracting activity | (Sekizaki *et al.*, 1988) |
| $R^7=R^{4'}$=OH, $R^6$=OMe (Tectorigenin), <br> $R^7$=OGlc, $R^{4'}$=OH, $R^6$=OMe (Tectoridin) | COX-2 inhibitor, Anti-inflammatory | (Kim *et al.*, 1999) |

On the basis of a pharmacophore model for competitive inhibitors of the EGFR-ATP-binding site and X-ray crystal structure data of other PTK inhibitors, a putative binding mode for genistein was proposed. Based on the latter data, a potent PTK inhibitor, 3'-chlorogenistein, was designed (Traxler *et al.*, 1999).

A recent study demonstrated that genistein can induce the expression of metal-lothionein (a zinc-binding protein with antioxidant property) (Kuo *et al.*, 1998; Kuo and Leavitt, 1999).

*O*-Alkylation, *C*- and *O*-glycosylation of the isoflavone skeleton gives rise to another class of interesting biological activities. Daidzin (the 7′-*O*-Glc derivative of daidzein) was identified as the major active principle of RP, an ancient Chinese treatment for "alcohol addiction" (Keung and Vallee, 1993). *Radix pueraria* is prepared from the root of *Pueraria lobata*, which serves as a remedy for numerous other common ailments as well. *In vitro* and *in vivo* studies showed that daidzin reversibly and selectively inhibits ALDH-2, whereas daidzein inhibits ADH more potently. These enzymes could play important roles in ethanol and biogenic aldehyde metabolism and therefore could serve as potential drug targets for the treatment of alcohol abuse (Rooke *et al.*, 2000). Furthermore, 7-*O*-ω-(carboxyalkyl) derivatives of daidzein have been synthesized and shown to be more potent than daidzin toward ALDH-2 inhibition *in vitro* and alcohol intake suppression *in vivo* (Keung *et al.*, 1997). Similarly, 7,4′-*O*-α-bis (carboxyalkyl) derivatives of daidzein improved its antihypoxia activity significantly (Shao *et al.*, 1980).

A selective anti-Helicobacter pylori isoflavone, cabreuvin (7,3′4′-trimethoxy-isoflavone) has been isolated from a Brazilian medicinal plant, *Myroxylon perniferum*. Structure–activity studies indicated that methylation and specifically the 3′-*O*-methylation is essential for activity (Ohsaki *et al.*, 1999).

The 8-*C*-glycoside Puerarin, an abundant isoflavone found in *Pueraria lobata*, displays anti-angina properties (Yang *et al.*, 1990). Derivatives of puerarin also proved to be useful for the improvement of ocular blood flow and retinal functions and therefore could be used for the treatment of ischemic retinopathy as well as age-related muscular degeneration (AMD) (Xuan *et al.*, 1999).

Leguminous plants defend themselves against microbial pathogens and herbivores through tissue accumulation of isoflavones before, during, and after attack. Some isoflavones can also function as chemoattractants. Naturally occurring prunetin (7-*O*-methylated genistein), displayed fungal zoospore attracting activity (Sekizaki *et al.*, 1993). The SAR of isoflavones in relation to *Aphanomyces euteiches* zoospore attraction has been investigated. However, R&D in antifungal agents for medicinal and agricultural purpose have not been forthcoming in this class of compounds. It was shown that C-5 hydroxylation was necessary for strong attraction and an additional hydroxyl group at C-7 and C-4′ strengthened the attracting ability, as did methylation at C-7.

The 6-*O*-methylated isoflavones tectorigenin and tectoridin (the 7-*O*-Glc-derivative) were recently isolated as the major anti-inflammatory isoflavones from the rhizomes of the Korean *Belamcarda chinensis* (Iridaceae), a Chinese traditional medicine for the treatment of inflammation (Kim *et al.*, 1999). Further investigation showed that the anti-inflammatory activity of these compounds is due to their ability to inhibit prostaglandin $E_2$ production and COX-2 induction.

The synthetic 7-*O*-isopropyloxy isoflavone, ipriflavone is reputed for its ability to inhibit bone resorption (anti-osteoporosis) and enhances bone formation without exerting estrogenic activity (Arjmandi *et al.*, 2000).

It is obvious that the isoflavones exhibit a wide spectrum of biological activities. Some of the many activities may share the same or a similar molecular mechanism. But most of them stem from the vast structural or molecular diversifications. Such biological activities can be manipulated and diversified through synthetic means. The search of better, safer and more efficacious lead compounds, delineating the molecular mechanism of each of the many activities, is an intellectual challenge in waiting. Thus, the vehicle that will expedite the drug discovery process will be an efficient synthetic route to isoflavones.

## SYNTHETIC ROUTES TO ISOFLAVONE CORE STRUCTURES

Increased commercial interest in isoflavones revived the need for efficient synthetic routes towards this class of compounds and analogs thereof. Furthermore, the continued research to unravel isoflavone mechanism of action and metabolism in mammalian systems necessitates efficient synthetic routes that would permit isotopic labeling.

### Natural products and analogs

Since 1925, variations of at least three synthetic approaches have been developed. The early syntheses of isoflavones were based on the formylation of a substituted deoxybenzoin (Baker and Robinson, 1928) (Scheme 14.1).

Deoxybenzoins undergo α-keto formylation, intramolecular acetal formation, followed by a facile dehydration to give the pyrone ring C and hence, the isoflavones. Deoxybenzoins have been synthesized utilizing the following strategies (Chapman and Stephen, 1923): (1) reacting phenyl acetic acid (X=CO$_2$H) with the appropriately substituted phenol in the presence of ZnCl$_2$, (2) reacting the acid chloride (X=COCl) with the appropriately substituted phenol in the presence of AlCl$_3$, and (3) Hoesch's method (Hoesch, 1915) of condensing the nitrile (X=CN) with the phenol to give an intermediate ketimine hydrochloride, which can be hydrolyzed to the deoxybenzoin with ease and in high overall yield (92 per cent) (Shriner and Hull, 1945). Selective protection of

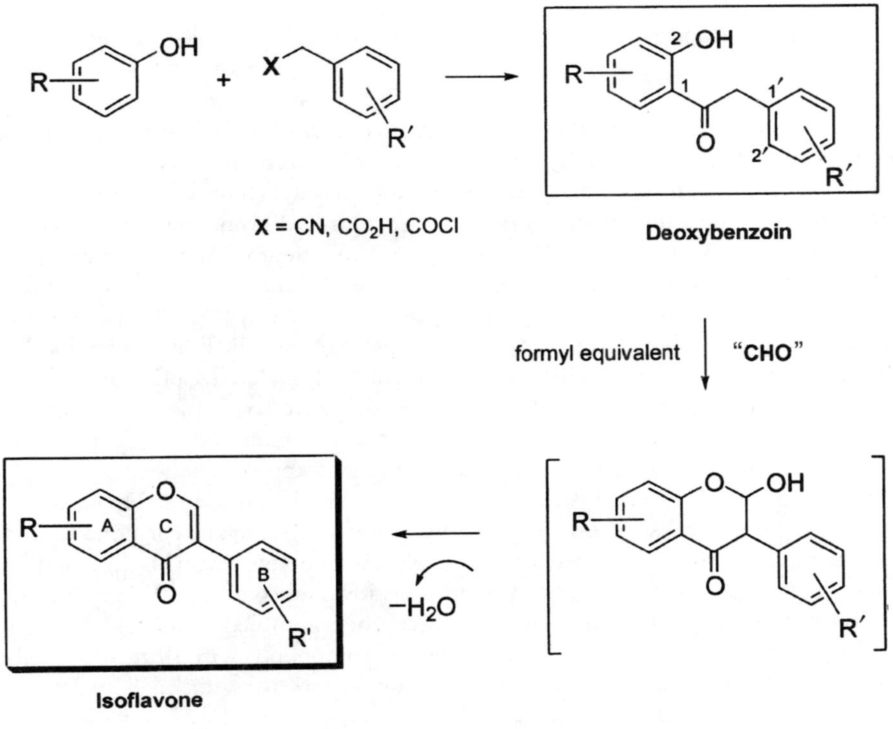

X = CN, CO$_2$H, COCl

**Deoxybenzoin**

formyl equivalent | "CHO"

−H$_2$O

**Isoflavone**

*Scheme 14.1*

all phenolic oxygens but the 2-hydroxy group is a prerequisite in order to prevent ring formylation and subsequent polymerization. Following the latter strategy, using ethyl formate (as formyl equivalent) in the presence of sodium and benzyl protected deoxybenzoins, Venkataraman *et al.* synthesized formononetin, daidzein and ψ-baptigenin in good yield (Joshi and Venkataraman, 1934; Mahal *et al.*, 1934a, 1934b).

**Formononetin**

**Daidzein**

*ψ* **-Baptigenin**

Other formyl equivalents that have been used includes; *N,N*-dimethylformamide dimethylacetal (Pelter and Foot, 1976), triethyl orthoformate (Hishmat *et al.*, 1988), *N,N*-dimethylformamide (Parmar *et al.*, 1988) and ethoxalyl chloride (Mohan *et al.*, 1987).

Due to commercial unavailability of deoxybenzoins, isoflavones have been synthesized by a linear synthetic strategy, hampered by protection–deprotection sequences. In order to shorten the synthesis, some one-pot reactions have been attempted. Boron trifluoride promoted Friedel-Crafts acylation of unprotected (or selectively *O*-methylated) resorcinols with *O*-methylated phenylacetic acids in the presence of DMF and methanesulfonyl chloride (or $POCl_3$) gave excellent yields of polyhydroxylated isoflavones (Luk *et al.*, 1983; Wahala and Hase, 1991). The latter approach gave access to a greater number of naturally occurring isoflavones in a more convergent fashion and in higher yield. The aromatic ring activating and directing capabilities of appropriately substituted polyhydroxylated phenols promotes the formation of natural isoflavones. This procedure is favored for the general synthesis of naturally occurring isoflavones or unnatural isoflavone analogs providing the substituted phenols has favorable directing functionality.

In order to expedite drug discovery, it is often advantageous to be able to incorporate maximum diversity in building blocks that can be assembled in a highly convergent manner resulting in a molecular diverse library of compounds with drug-like attributes. Although isoflavones display many medicinal properties (as outlined in supporting chapters), its potential as a candidate for combinatorial diversification has not yet been fully realized. From this point of view, the synthesis of isoflavones via the oxidative

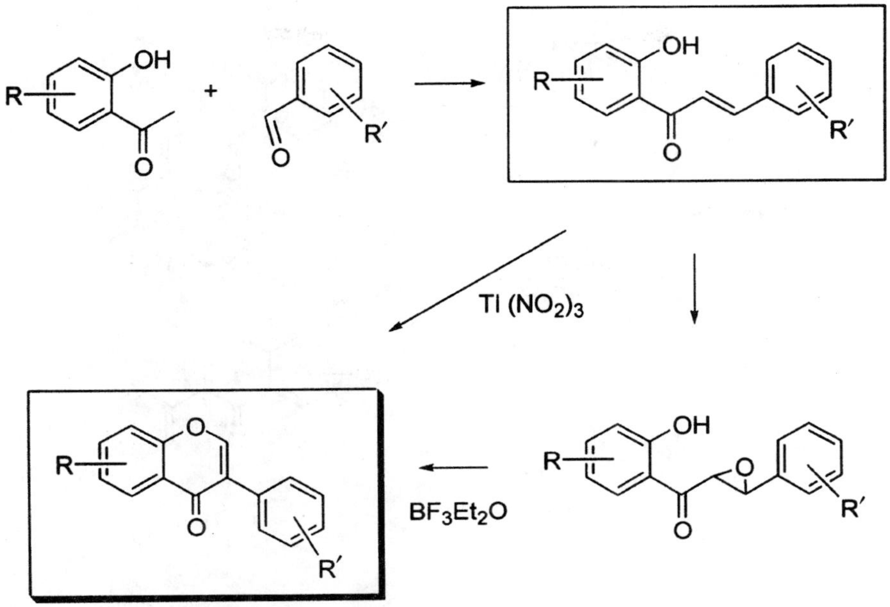

*Scheme 14.2*

rearrangement of chalcones becomes more appealing. Chalcones have been readily synthesized by condensation of substituted acetophenones and benzaldehydes (Powers *et al.*, 1998) (Scheme 14.2). The commercial availability of these building blocks are enormous, and hence would indirectly give access to a huge library of molecular diversity based upon the isoflavone skeleton.

Isoflavones have been synthesized via epoxidation of the intermediate 2'-hydroxy-chalcone, followed by boron trifluoride mediated rearrangement and cyclization (Rani, 1987). While the latter reaction gave poor yields, an improvement was achieved when thallium (III) nitrate was used in the presence of methanol (Nabaei-Bidhendi and Bannerjee, 1990) or trimethyl orthoformate (Parmar *et al.*, 1987) instead.

In plants, the enzyme 2-hydroxyisoflavone synthetase converts flavones into isoflavones. The biosynthesis proceeds via a cytochrome P450 dependent 2-hydroxylation of the flavanone, followed by an aryl ring migration and dehydration (Steele *et al.*, 1999). An expressed, labile 2-hydroxyisoflavone synthetase has been used to demonstrate the conversion of 4',7-dihydroxyflavanone (liquiritigenin) to 4',7-dihydroxyisoflavone (daidzein) and 4',5,7-trihydroxyflavanone (naringenin) to genistein [Koghs and Grisebach, 1986] (Scheme 14.3).

The biomimetic transformation of flavanones into isoflavones has been achieved in near quantitative yield with thallium (III) *p*-toluenesulfonate in propionitrile, while thallium (III) acetate gave 5–45 per cent of flavone in acetonitrile. Electron withdrawing groups on the C-2 aryl ring deactivate this reaction. However, thallium (III) nitrate in the presence of methanol–chloroform and 70 per cent perchloric acid or thallium (III) perchlorate in acetonitrile or DME gave better results (Kinoshita and Sankawa, 1990;

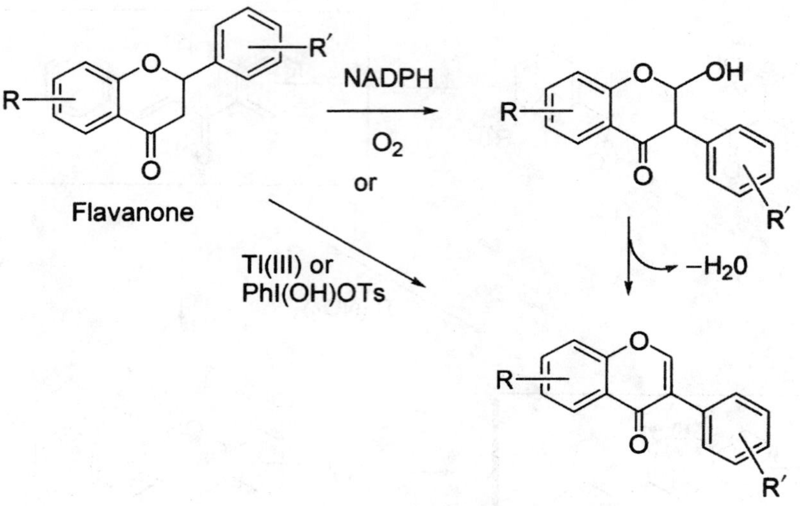

*Scheme 14.3*

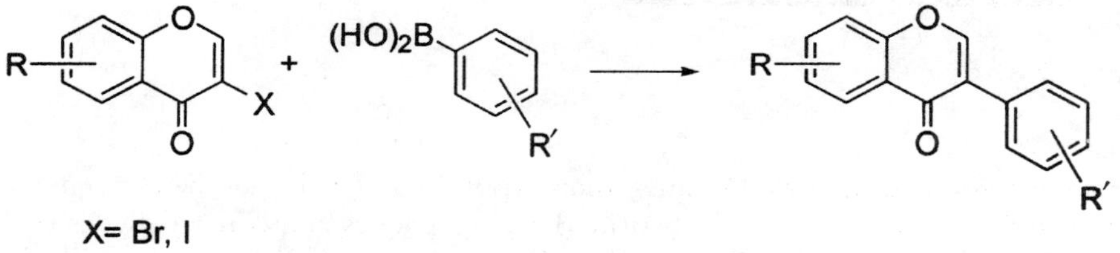

X= Br, I

*Scheme 14.4*

Singh and Kapil, 1993). Furthermore, [hydroxy(tosyloxy)iodo]benzene in acetonitrile also gave the desired isoflavone (Prakash *et al.*, 1990).

Palladium catalyzed cross-coupling (Suzuki-Coupling) between arylboronic acids and arylhalides found widespread applications in medicinal chemistry (Suzuki, 1999). This methodology has been well studied in recent years and displayed the versatility of both solid phase as well as large-scale solution phase synthesis. Commercial availability and diversity of substituted arylboronic acids are steadily increasing. This methodology was successfully applied to the synthesis of isoflavones from 3-halochroman-4-ones (Scheme 14.4) (Yokoe *et al.*, 1989; Howshino *et al.*, 1988).

Prenylated isoflavones are common amongst the *Leguminosae*, which in turn gives access to biotransformed pyrano or furano rings on the parent skeleton (Donelly and Boland, 1995, 1998). Under certain chemical or physical stimuli, some plants synthesize "stress metabolites." Using cupric chloride for this purpose, the stems of *Pueraria lobata* produced 8-prenyldaidzein (Hakamatsuka *et al.*, 1991). The ease of selective halogenation of the isoflavone skeleton manifests itself in the isolation of 8-chloro- and 6,8-dichloro-genistein from a *Streptomyces griseus* culture grown on soybean meal in the presence of sodium chloride (Anyanwutaku, 1992). Isoflavones react with perchlorate and nitrate

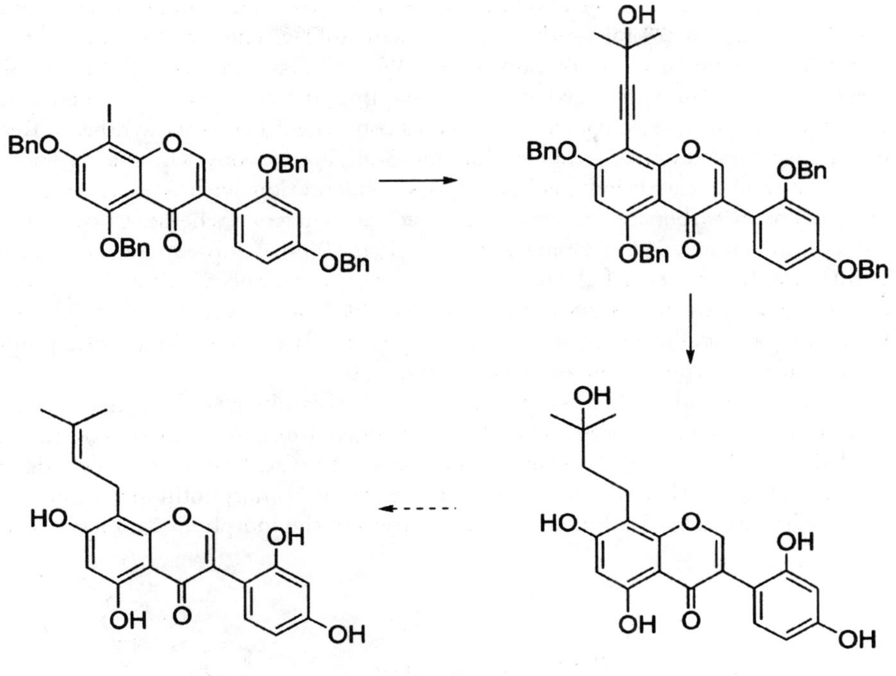

*Scheme 14.5*

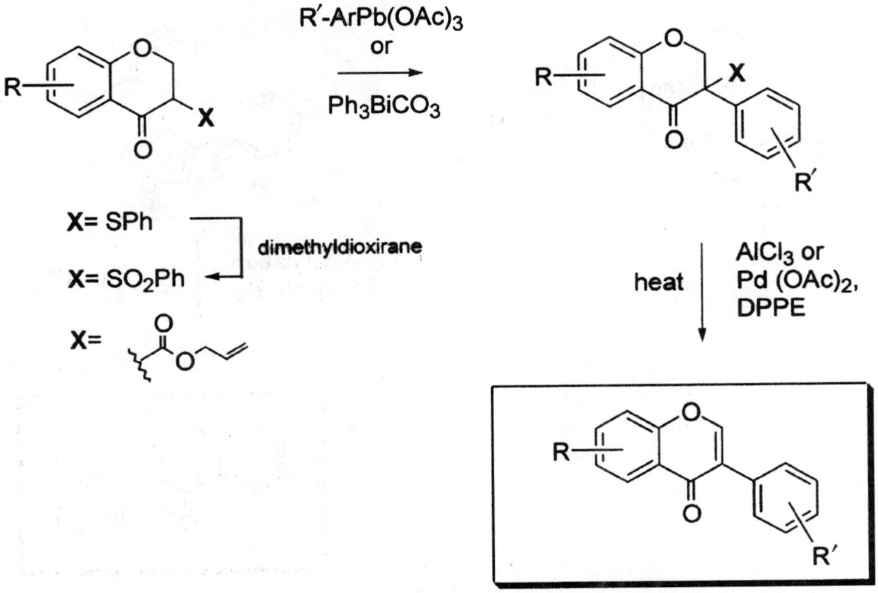

*Scheme 14.6*

and displayed remarkable selectivity in chlorination and nitration between genistein, biochanin-A and daidzein (Boersma *et al.*, 1999). Access to 8-iodo-isoflavone were, in part, responsible for the successful synthesis of 2,3-dehydrokievitone via Palladium catalyzed prenylation (Scheme 14.5) (Tsukayama *et al.*, 1994; Tsukayama *et al.*, 1998a, 1998b).

Relative to palladium catalyzed aryl cross-coupling, direct arylation of α-keto carbanions of chroman-4-ones also contributed toward convergent isoflavone synthesis. Readily accessible 3-phenylsulfonyl-, 3-phenylthio and 3-allyloxycarbonyl chroman-4-ones provided a common 3-carbanion that undergoes condensation with an aryl group from a triarylbismuth carbonate or aryllead (IV) acetate reagent (Scheme 14.6) (Santhosh and Balasubramanian, 1992; Donelly *et al.*, 1993a, 1993b). Subsequent, Lewis acid or palladium catalyzed thermal elimination of the respective phenylsulfonyl- or allyloxycarbonyl groups gave the isoflavones in good yield. The unavailability of substituted arylbismuth reagents limited its use, whereas aryllead (IV) acetates with more appropriate aryl-substitution patterns were more easily obtainable.

Isoflavones have also been synthesized via 3,4-bis-phenyl-substituted isoxazoles (Thomsen and Torssell, 1988). The aforementioned isoxazoles can be considered as masked α-formylated deoxybenzoins (Hoesch, 1915). Oxidation of salicylaldoximes gives a nitrile oxide, that which in turn reacts with an N-morpholino-enamine to give a 4,5-dehydroisoxazole. Acid mediated elimination of the morpholine moiety, gave the

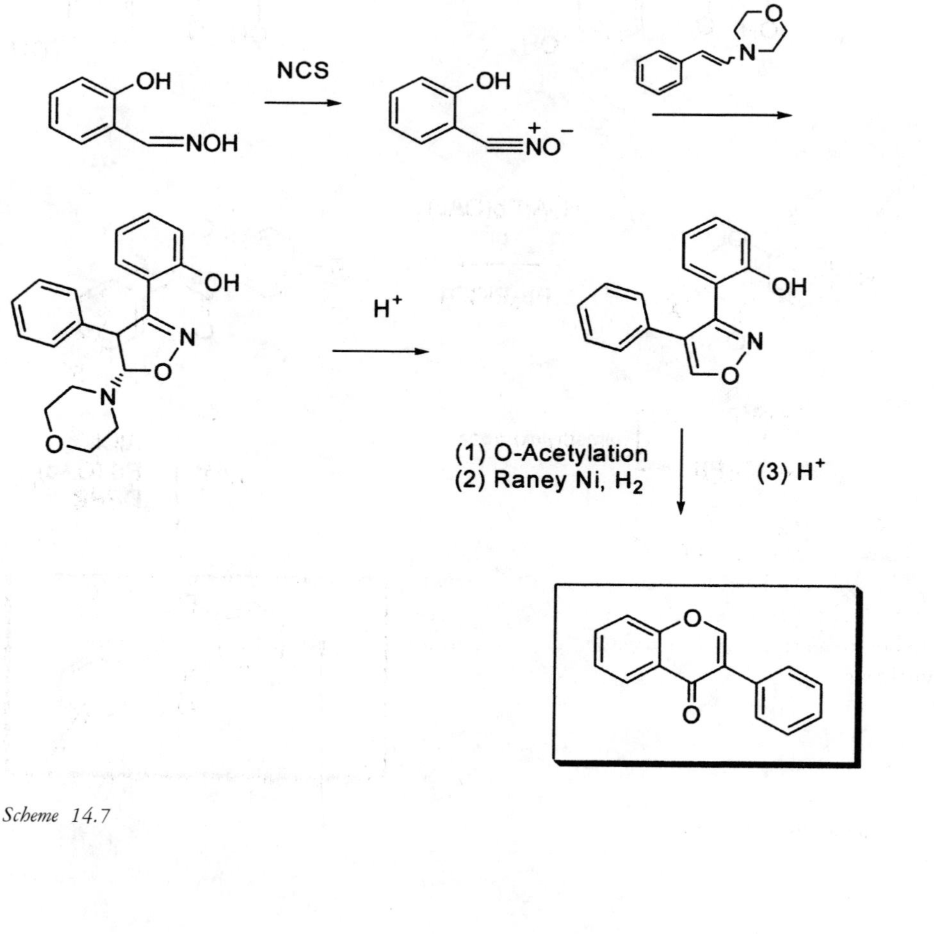

(1) O-Acetylation
(2) Raney Ni, H₂            (3) H⁺

*Scheme 14.7*

3,4-bis-phenylisoxazole, which was acetylated, reduced and cyclized to give isoflavones (Scheme 14.7).

## Glyco-conjugates

Naturally occurring isoflavone-*O*- and *C*-aryl glycosides have been shown to possess interesting medicinal properties (Ingham *et al.*, 1986; Shibuya, 1991; Messina *et al.*, 1997; Rong *et al.*, 1998; Park *et al.*, 1999). Commercial production of these compounds (e.g. daidzin, puerarin) has been primarily from natural origin. A continued increase in the demand for biologically active isoflavone-glycosides or derivatives thereof necessitates the development of viable and regulated synthetic procedure. Synthetic routes toward isoflavone-*O*-glycosides would also accelerate research toward a better understanding of its biological activity.

Hitherto, standard glucosylation of isoflavones only gave poor to moderate yields. Early synthetic procedures utilized aqueous potassium hydroxide and glycosyl halide for this purpose. However, low yields (9–22 per cent) were obtained largely due to C-ring cleavage and competing hydrolysis of the glycosyl halide. Subsequently, a small improvement was made when the glycosylation step preceded the final cyclization step of the isoflavone assembly (Zemplén and Farkas, 1943; Zemplén *et al.*, 1944; Farkas *et al.*, 1969; Bogna and Levai, 1973; Pivovarenko and Kihlya, 1993). Recently, the use of a biphasic reaction (liquid/liquid) in the presence of a phase transfer catalyst (tetrabuty-lammonium bromide) resulted in moderate yields (40–52 per cent) (Lewis *et al.*, 1998). Treatment of unprotected daidzein or genistein resulted in the regio- and stereoselective 7-*O*-glucosylation to give daidzin (40 per cent) and genistin (42 per cent) respectively. Subsequent methylation of the peracetylated sugars of daidzin and genistin gave, after deacetylation, biochanin A and ononin respectively (Scheme 14.8).

Furthermore, regioselective control in glycosylation could be achieved by controlling the molar equivalents of base (potassium *t*-butoxide in acetonitrile) in the presence of 18-crown-6 as phase transfer catalyst (Lewis *et al.*, 1998). In the presence of excess base, the unprotected isoflavone di- or tri-phenoxide precipitates. After the addition of 18-crown-6, the mixture becomes solubilized and presents the more nucleophilic 4′-phenoxide for β-selective glucosylation. For the first time, the naturally occurring daidzein- and genistein-4′-*O*-β-glucosides was obtained in 37 per cent and 41 per cent yields, respectively. The reverse selectivity was obtained by using one equivalent of base and 18-crown-6 in order to present the 7-phenoxide for β-selective glucosylation, thus giving daidzin and genistin in 50 per cent and 54 per cent yields, respectively. The difference in acidity of the 7- and 4′-hydroxyl groups has been exploited in the direct *O*-alkylation of daidzein and genistein to give isoflavone analogs with increased potency (Shao *et al.*, 1980; Lapcik *et al.*, 1998; Rooke *et al.*, 2000).

The enzymatic glycosylation of isoflavones have been achieved *in vitro*, using rabbit liver glycosyl transferase in a microsomal fraction (Labow and Layne, 1972). However, further investigation is necessary in order to assess the viability of a chemoenzymatic synthesis.

Puerarin, a *C*-aryl isoflavone-glycoside prevalent amongst the *Pueraria*, is renowned for its medicinal properties (Ingham *et al.*, 1986; Rong *et al.*, 1998). The total synthesis of this medically important isoflavone has not yet been published. However, the *de novo* synthetic approaches toward *C*-aryl glycosides of other natural products have been reviewed (Parker, 1994; Hart *et al.*, 1996).

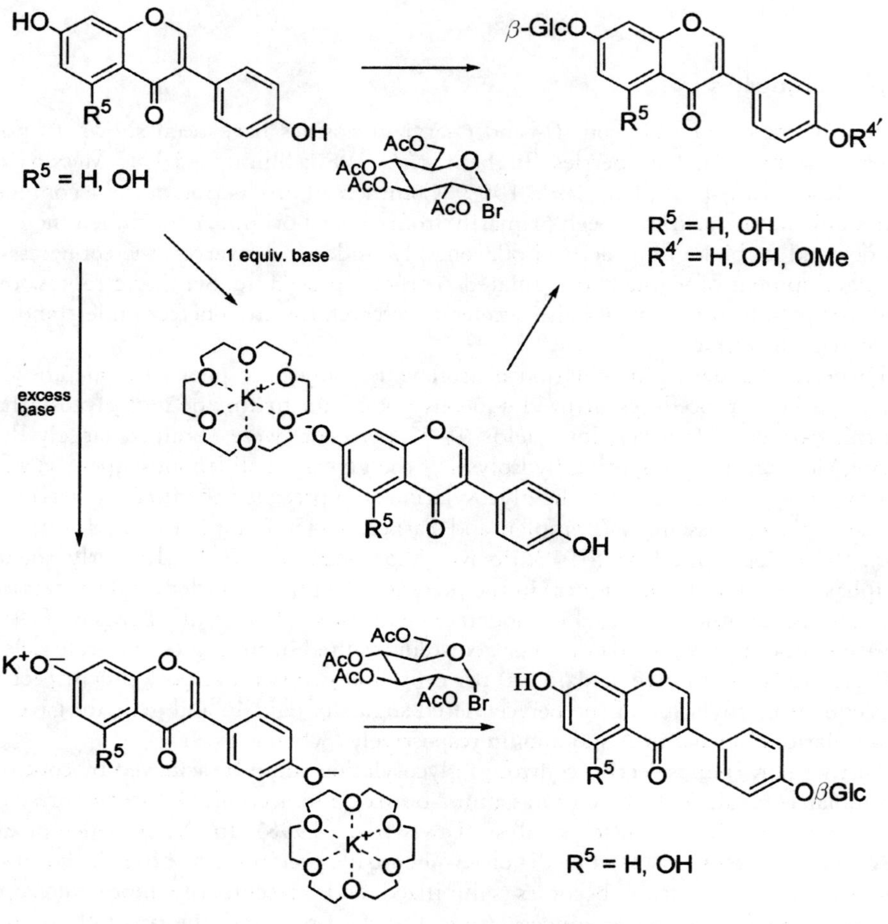

*Scheme 14.8*

## Radiolabelled isoflavones

Isotopically labelled biologically active compounds are invaluable tools for biological studies, pharmacokinetics and structure elucidation. The exchangeability of protons, in acidic media, *ortho* and *para* to the phenolic hydroxyl groups of isoflavones was exploited for the introduction of deuterium. Thus, electrophilic attack of the deuterium cation on daidzein, genistein, formononetin and biochanin A, could be achieved in deuterated trifluoroacetic or phosphoric acid (in the presence of a Lewis acid catalyst) (Scheme 14.9) (Wahala *et al.*, 1986; Rasku and Wahala, 1997; Rasku *et al.*, 1999). Deuteration at C-2′ and C-6′ is assisted by the presence of a 5-hydroxyl (genistein). The most labile deuteriums are at C-6 and C-8 and can easily be exchanged again.

Recently, the Hoesch method of isoflavone synthesis was considered as ideal for specific $^{13}$C-4 isotopic labeling of daidzein, genistein, formononetin and biochanin A

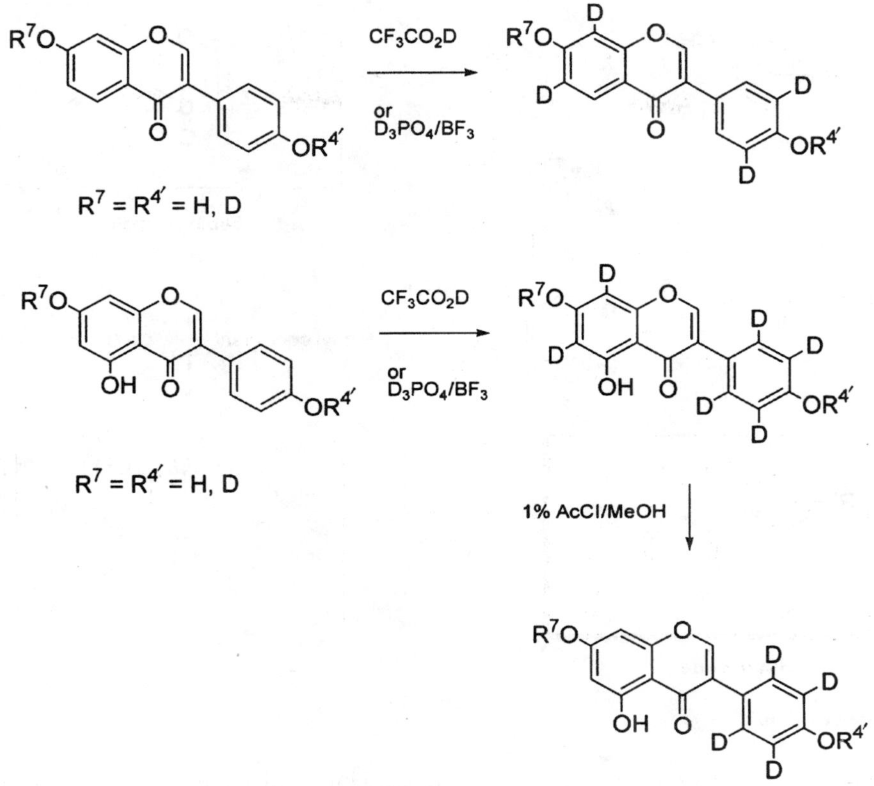

*Scheme 14.9*

(Whalley *et al.*, 2000). The Hoesch method (Hoesch, 1915) involves the condensation of a benzylnitrile with the appropriately substituted phenol to give an intermediate ketimine hydrochloride, which can be hydrolyzed to the deoxybenzoin. Subsequent addition of a formyl equivalent followed by cyclization provides the isoflavone in good yield. Carbon-13 labelled 4′-alkyloxybenzylnitrile was obtained from 4′-alkyloxybenzyl-alcohol in two steps via $K^{13}CN$ treatment of an intermediate 4′-alkyloxybenzylbromide (Scheme 14.10).

## CONCLUSIONS

Synthetic methodology to aid the preparation of naturally occurring isoflavones has been well developed for this purpose. Some difficulties related to insolubility do exist with glycosylation of the isoflavone skeleton, but can be overcome by selective protection–deprotection strategies. A more convergent approach toward the complete isoflavone glycoside structure has not yet been fully addressed. In addition to the natural biodiversity of isoflavones, there is still a need for synthetic methodology that would allow an explosion of diversity around the core isoflavone skeleton. Combinatorial

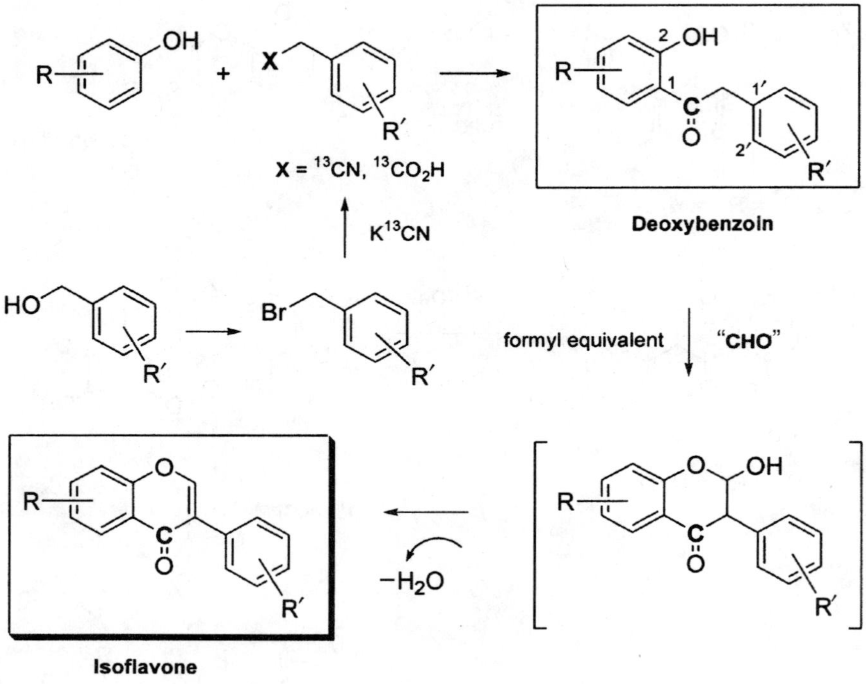

Scheme 14.10

synthesis is most convenient, when a readily available multi-component one-pot synthesis yields a single crystalline product in high yield. Most of the synthetic strategies described herein involve a multistep reaction, which complicates solution phase strategies. The modern trend toward solid supported reagents may alleviate this problem. A solid-phase strategy may suffice using the synthetic methodology at hand, but this would limit diversity and scale. Further research is necessary to develop an ideal synthetic method from those available. From the point of diversification, the synthesis of isoflavones via chalchones is favored since a fair amount of acetophenones and particularly aldehydes are commercially available (Powers *et al.*, 1998). Furthermore, a simplified synthesis toward a halo-substituted isoflavone skeleton would facilitate further diversification through the use of arylhalide displacement.

## REFERENCES

Arjmandi, B.H., Birnbaum, R.S., Juma, S., Barengolts, E. and Kukreja, S.C. (2000) The synthetic phytoestrogen, ipriflavone, and estrogen prevent bone loss by different mechanisms. *Calcified Tissue Int.*, 66, 61–65.

Baker, W. and Robinson, R. (1928) Synthetical experiments in the isoflavone group. Part III. A synthesis of genistein. *J. Chem. Soc.*, 3115–3118.

Boersma, B.J., Patel, R.P., Kirk, M., Jackson, P.L., Muccio, D. and Darley-Usmar, V.M. (1999) Chlorination and nitration of soy isoflavones. *Archiv. Biochem. Biophys.*, 368, 265–275.

Bogna, R. and Levai, A., (1973) Synthesis of 4-β-D-glucosyloxydeoxybenzoins and their conversion into 7-β-glucosyloxyisoflavones. *Acta Chim. Acad. Sci. Hung.*, 77, 435–442.

Chapman, E. and Stephen, H. (1923) Di- and trihydroxydeoxybenzoins. *J. Chem. Soc.*, 404–409.

Cheng, S.L., Zhang, S.F., Nelson, T.L., Warlow, P.M. and Civitelli, R. (1994) Stimulation of human osteoblast differentiation and function by ipriflavone and its metabolites. *Calcif. Tissue Int.*, 55, 365–372.

Cook, N.C. and Samman, S. (1996) Flavonoids-chemistry, metabolism, cardioprotective effects, and dietary sources. *J. Nutr. Biochem.*, 7, 66–76.

Dewick, P.M. (1993) Isoflavonoids. In J.B. Harbourne (ed.), *The Flavonoids: Advances in Research Since 1986*, Chapman and Hall, London.

Donelly, D.M.X. and Boland, G.M. (1995) Isoflavonoids and neoflavonoids: Naturally occuring 0-heterocycles. *Nat. Prod. Res.*, 321–338.

Donelly, D.M.X. and Boland, G.M. (1998) Isoflavonoids and related compounds. *Nat. Prod. Res.*, 241–260.

Donelly, D.M.X., Fitzpatrick, B.M., O'Reilly, B.A. and Finet, J. (1993a) Aryllead mediated synthesis of isoflavanone and isoflavone derivatives. *Tetrahedron*, 49, 7967–7976.

Donelly, D.M.X., Finet, J. and Rettigan, B.A. (1993b) Organolead-mediated arylation of allyl β-ketoesters: A selective synthesis of isoflavanones and isoflavones. *J. Chem. Soc. Perkin Trans.*, 1, 1729–1735.

Esaki, H., Onozaki, H., Morimitsu, Y., Kawakishi, S. and Osawa, T. (1998) Potent antioxidative isoflavones isolated from soybeans fermented with *aspergillus saitoi*. *Biosci. Biotechnol. Biochem.*, 62, 740–746.

Farkas, L., Mezey-Vandor, G., Nogradi, M. and Gottsege, A. (1969) Transacylation reactions in flavonoid series. 4. New syntheses of 5-methylgenistein, prunetin, biochanin-A and sisotrin. *Acta. Chim. Acad. Sci. Hung.*, 60, 293–300.

Greaves, K.A., Parks, J.S., Williams, J.K. and Wagner, J.D. (1999) Intact dietary soy protein, but not adding an isoflavone-rich soy extract to casein, improves plasma lipids in ovariectomized cynomolgus monkeys. *J. Nutr.*, 129, 1585–1592.

Hakamatsuka, T., Ebizuka, Y. and Sankawa, U. (1991) Induced isoflavonoids from copper chloride-treated stems of *Pueraria lobata*. *Phytochemistry*, 30, 1481–1482.

Hart, D.J., Merriman, G.H. and Young, D.G.J. (1996) Synthesis of C-aryl glycosides related to the Chrysomycins. *Tetrahedron*, 52, 14437–14458.

Hishmat, O.H., Abd El Rahman, A.H., EL-Diwani, H.I. and Abu-Bakr, S.H.M. (1988) Synthesis of substituted furobenzopyrones. *Rev. Roum. Chim.*, 33, 741–745.

Hoesch, K. (1915) Eine neue synthese aromatischer ketone. I. Darstellung einiger phenol-ketone. *Ber.*, 48, 1122–1133.

Howshino, Y., Miyaura, N. and Suzuki, A. (1988) Novel synthesis of isoflavones by the palladium cross-coupling reaction of 3-bromochromones with arylboronic acids or its esters. *Bull. Chem. Soc. Jpn.*, 61, 3008–3010.

Ingham, J.L., Markham, K.R., Dzeidzic, S.Z. and Pope, G.S. (1986) Puerarin 6″-0-β-apiofuranoside, a C-glycosylisoflavone 0-glycoside from *Pueraria mirifica*. *Phytochemistry*, 25, 1772–1775.

Jing, Y.K. and Waxman, S. (1995) Structural requirements for differentiation – induction and growth – inhibition of mouse erythroleukemia cells by isoflavones. *Anticancer Res.*, 15, 1147–1152.

Joshi P.C. and Venkataraman, K. (1934) Synthetical experiments in the chromone group. Part XI. Synthesis of isoflavone. *J. Chem. Soc.*, 113, 513–514.

Kapiotis, S., Hermann, M., Held, I., Seelos, C., Ehringer, H. and Gmeiner, B.M.K. (1997) Genistein, the dietary-derived angiogenesis inhibitor, prevents LDL oxidation and protects endothelial cells from damage by atherogenic LDL. *Arteriosclerosis, Thromb. Vasc. Biol.*, 17, 2868–2874.

Keung, W.M. and Vallee, B.L. (1993) Daidzin and daidzein suppress free-choice ethanol intake by Syrian Golden hamsters. *Proc. Natl. Acad. Sci. USA*, 90, 10008–10012.

Kim, H., Peterson, T.G. and Barnes, S. (1998) Mechanisms of action of the soy isoflavone genistein: emerging role for its effects via transforming growth factor beta signaling pathways. *Am. J. Clin. Nutr.*, 68, 1418S–1425S.

Kim, Y.P., Yamada, M., Lim, S.S., Lee, S.H., Ryu, N., Shin, K.H. and Ohuchi, K. (1999) Inhibition by tectorigenin and tectoridin of prostaglandin E2 production and cyclooxygenase-2 induction in rat peritoneal macrophages. *Biochim. Biophys. Acta*, 1438, 399–407.

Kinoshita, T. K. and Sankawa, U. (1990) One-step conversion of flavanones into isoflavones: A new facile biomimetic synthesis of isoflavones. *Tetrahedron Lett.*, 31, 7355–7356.

Kochs, G. and Grisebach, H. (1986) Enzymic synthesis of isoflavones. *Eur. J. Biochem.*, 155, 311–318.

Kuo, S.M., Leavitt, P.S. and Lin, C.P. (1998) Dietary flavonoids interact with trace metals and affect metallothionein level in human intestinal cells. *Biol. Trace Elem. Res.*, 62, 135–153.

Kuo, S.M. and Leavitt, P.S. (1999) Genistein increases metallothionein expression in human intestinal cells, Caco-2. *Biochem. Cell Biol.*, 77, 79–88.

Labow, R.S. and Layne, D.S. (1972) The formation of glucosides of isoflavones and of some other phenols by rabbit liver microsomal fractions. *Biochem. J.*, 128, 491–497.

Lapcik, O., Hampl, R., Hill, M., Wahala, K., Maharik, N.A. and Adlercreutz, H. (1998) *J. Steroid Biochem. Mol. Biol.*, 64, 261–268.

Lewis, P., Kaltia, S. and Wahala, K. (1998) The phase transfer catalysed synthesis of isoflavone-O-glucosides. *J. Chem. Soc. Perkin Trans.*, 1, 2481–2484.

Lewis, P. and Wahala, K. (1998) Regiospecific 4′-O-β-glucosidation of isoflavones. *Tetrahedron Lett.*, 39, 9559–9562.

Luk, K.C., Stern, L., Weigele, M., O'Brien, R.A. and Spirt, N. (1983) Isolation and identification of "diazepam-like" compounds from bovine urine. *J. Nat. Prod.*, 46, 852–861.

Mahal, H.S., Rai, H.S. and Venkataraman, K. (1934a) Synthetical experiments in the chromone group. Part XXI. Synthesis of 7-hydroxyisoflavone and of α- and β-napthaisoflavone. *J. Chem. Soc.*, 238, 1120–1122.

Mahal, H.S., Rai, H.S. and Venkataraman, K. (1934b) Synthetical experiments in the chromone group. Part XV. Synthesis of formononetin, daidzein and ψ-baptigenin. *J. Chem. Soc.*, 388, 1769–1771.

Messina, M., Barnes, S. and Setchell, K.D. (1997) Phyto-oestrogens and breast cancer. *Lancet*, 350, 23–27.

Messina, M.J. (1999) Legumes and soybeans: overview of their nutritional profiles and health effects. *Am. J. Clin. Nutr.*, 70, 439S–450S.

Mohan Roa, K.S.R., Rukmani Iyer, C.S. and Iyer, P.R. (1987) Synthesis of alpinum isoflavone, derrone and related pyranoisoflavones. *Tetrahedron Lett.*, 43, 3015–3019.

Nabaei-Bidhendi, G. and Bannerjee, N.J. (1990) Convenient synthesis of polyhydroxy flavonoids. *J. Indian Chem. Soc.*, 67, 43–45.

Nestel, P.J., Pomeroy, S., Kay, S., Komersaroff, P., Behrsing, J., Cameron, J.D. and West, L. (1999) Isoflavones from red clover improve systemic arterial compliance but not plasma lipids in menopausal women. *J. Clin. Endocrinol. Metab.*, 84, 895–898.

Ohsaki, A., Takashima, J., Chiba, N. and Kawamura, M. (1999) Microanalysis of a selective potent anti-*Helicobacter pylori* compound in a Brazilian medicinal plant, *Myroxylon peruiferum* and the activity of analogues. *Biorg. Med. Chem. Lett.*, 9, 1109–1112.

Park, H.J., Park, J.H., Moon, J.O., Lee, K.T., Jung, W.T., Oh, S.R. and Lee, H.K. (1999) Isoflavone glycosides from the flowers of *Pueraria thunbergiana*. *Phytochemistry*, 51, 147–151.

Parker, K.A. (1994) Novel methods for the synthesis of C-aryl glycoside natural products. *Pure App. Chem.*, 66(10–11), 2135–2138.

Parmar, V.S., Singh, S. and Jain, R. (1988) Synthesis of a new naturally occurring 3-phenyl-4H-1-benzopyran-4-one. *Synth. Commun.*, 18, 511–517.

Parmar, V.S., Singh, S. and Jain, R. (1987) Synthesis of isoflavone-H, occuring in *Tephrosia maxima*. *Indian J. Chem.*, 26B, 166–167.

Pelter, A. and Foot, S. (1976) New convenient synthesis of isoflavones. *Synthesis*, 326–327.

Pivovarenko, V.G. and Kihlya, V.P. (1993) Synthesis of 7-O-β-glucopyranozides of isoflavons and their heterocyclic analogs. *Chem. Nat. compd. (Engl. Transl.)*, 29, 181–186.

Powers, D.G., Casebier, D.S., Fokas, D., Ryan, W.J., Troth, J.R. and Coffen, D.L. (1998) Automated parallel synthesis of chalcone-based screening libraries. *Tetrahedron*, 54, 4085–4096.

Prakash, O., Pahuja, S., Goyal, S., Sawhney, S.N. and Moriaty, R.M. (1990) 1,2-Aryl shift in the hypervalent iodine oxidation of flavanones: A new useful synthesis of isoflavones. *Synlett.*, 337–338.

Rani, I. (1987) Two synthesis of 7,2'-dimethoxyisoflavone. *Indian J. Chem.*, 26B, 361–362.

Rasku, S., Wahala, K., Koskimies, J. and Hase, T. (1999) Synthesis of isoflavonoid deuterium labeled polyphenolic phytoestrogens. *Tetrahedron*, 55, 3445–3454.

Rasku, S. and Wahala, K. (1997) Synthesis of $D_4$-genistein, a stable deutero labeled isoflavone, by a perdeuteration-selective dedeuteration approach. *Tetrahedron Lett.*, 38, 7287–7290.

Rong, H., Stevens, J.F., Deinzer, M.L., De Cooman, L. and De Keukeleire, D. (1998) Identification of isoflavones in the roots of *Pueraria lobata*. *Planta Med.*, 64, 620–627.

Rooke, N., Li, D.-J., Li, J. and Keung, W.M. (2000) The mitochondrial monoamine oxidase-aldehyde dehydrogenase pathway: A potential site of action of daidzin. *J. Med. Chem.*, 43, 4169–4179.

Ruizlarrea, M.B., Mohan, A.R., Paganga, G., Miller, N.J., Bolwell, G.P. and Riceevans, C.A. (1997) Antioxidant activity of phytoestrogenic isoflavones. *Free Radic. Res.*, 26, 63–70.

Santhosh, K.C. and Balasubramanian, K.K. (1992) Ligand coupling route to isoflavanones and isoflavones. *J. Chem. Soc. Chem. Commun.*, 3, 224–225.

Sekizaki, H., Yokosawa, R., Chinen, C., Adachi, H. and Yamane, Y. (1993) Studies on zoospore attracting activity. II. Synthesis of isoflavones and their attracting activity to *Aphanomyces euteiches* zoospore. *Biol. Pharm. Bull.*, 16, 698–701.

Shao, G., Mo, R., Wang, C., Zhang, D., Yin, Z., Ouyang, R. and Xu, L. (1980) Studies on the synthesis and structure-antihypoxia relations of daidzein, an active principle of *Pueraria pseudohiruta*, and its derivatives. *Chin. Acad. Med. Sci.*, 15, 538–547.

Shibuya, Y., Tahara, S., Kimura, Y. and Mizutani, J. (1991) New isoflavone glucosides from white lupin (*Lupinus-albus L*). *Z. Naturforch.*, 46c, 513–518.

Shriner, R.L. and Hull, C.J. (1945) Isoflavones. III. The structure of prunetin and a new synthesis of genistein. *J. Org. Chem.*, 10, 288–291.

Singh, O.V. and Kapil, R.S. (1993) A general method for the synthesis of isoflavones by oxidative rearrangement of flavanones using thallium (III) perchlorate. *Indian J. Chem.*, Sect. B, 32, 911–915.

Steele, C.L., Gijzen, M., Qutob, D. and Dixon, R. (1999) Molecular characterization of the enzyme catalyzing the aryl migration reaction of the isoflavonoid biosynthesis in soybean. *Arch. Biochem. Biophys.*, 367, 146–150.

Suzuki, A. (1999) Recent advances in the cross-coupling reactions of organoboron derivatives with organic electrophiles, 1995–1998. *J. Organometallic Chem.*, 576(1–2), 147–168.

Thomsen, I. and Torssell, K.B.G. (1988) Use of nitrile oxides in synthesis. Novel synthesis of chalcone, flavanones, flavones and isoflavones. *Acta Chem. Scand.*, 42B, 303–308.

Traxler, P., Green, J., Mett, H., Sequin, U. and Furet, P. (1999) Use of a pharmacophore model for the design of EGFR tyrosine kinase inhibitors: Isoflavones and 3-phenyl-4(1H)-quinolones. *J. Med. Chem.*, 42, 1018–1026.

Tsukayama, M., Tsurumoto, K., Kishimoto, K. and Higuchi, D. (1994) Regioselective synthesis of prenylisoflavones. Syntheses of 2,3-dehydrokievitone and related compounds. *Chem. Lett.*, 11, 2101–2104.

Tsukayama, M., Li, H., Tsurumoto, K., Nishiuchi, M. and Kawamura, Y. (1998a) Regioselective synthesis of prenylisoflavones. Syntheses of allolicoisoflavone A, 2,3-dehydrokievitone, and related compounds. *Bull. Chem. Soc. Jpn.*, 71, 2673–2680.

Tsukayama, M., Li, H., Tsurumoto, K., Nishiuchi, M., Kawamura, Y., Takahashi, M. and Kawamura, Y. (1998b) Regioselective synthesis of prenylisoflavones. Syntheses of lupiwighteone, lupiwighteone hydrate and related compounds. *J. Chem. Res. Miniprint*, 5, 1181–1196.

Wahala, K., Makela, T., Backstrom, R., Brunow, G. and Hase, T. (1986) Synthesis of the ($^2$H)-labeled urinary lignans, enterolactone and enterodiol, and the phytoestrogen daidzein and its metabolites equol and O-demethyl-angolensin. *J. Chem. Soc., Perkin Trans.*, 1, 95–98.

Wahala, K. and Hase, T.A. (1991) Expedient synthesis of polyhydroxyisoflavones. *J. Chem. Soc., Perkin Trans.*, 1, 3005–3008.

Whalley, J.L., Oldfield, M.F. and Botting, N.P. (2000) Synthesis of (4-$^{13}$C)-isoflavonoid phytoestrogens. *Tetrahedron*, 56, 455–460.

Wu, E.S., Lock, J.T., Toder, B.H., Borelli, A.R., Gawlak, D., Rador, L.A. and Gensmantel, N.P. (1992) Flavones 3. synthesis, biological activities and conformational analysis of isoflavone derivatives and related compounds. *J. Med. Chem.*, 35, 3519–3525.

Xuan, B., Zhou, Y.H., Yang, R.L., Li, N., Min, Z.D. and Chiou, G.C. (1999) Improvement of ocular blood flow and retinal functions with puerarin analogs. *J. Ocular Pharmacol. Ther.*, 15, 207–216.

Yang, G., Zhang, L. and Fan, L. (1990) Anti-angina effect of puerarin and its effect on plasma thromboxane A2 and prostacyclin. *J. Mod. Dev. Tradit. Med.*, 10, 82–84.

Yokoe, I., Sugita, Y. and Shirataki, Y. (1989) Facile synthesis of isoflavones by the cross-coupling reaction of 3-iodochromone with arylboronic acids. *Chem. Pharm. Bull.*, 37, 529–530.

Zemplén, G. and Farkas, L. (1943) Synthese des genistins. *Chem. Ber.*, 76, 1110–1112.

Zemplén, G., Farkas, L. and Bien, A. (1944) Synthese des ononins. *Chem. Ber.*, 77, 452–457.

# 15 Research and development of *Pueraria* (Ge)-based medicinal products in China

*Guang-yao Gao and Wing Ming Keung*

## MEDICINAL USE OF GE IN CHINA – HISTORICAL PERSPECTIVE

### Traditional Chinese medicine

*Pueraria lobata* and *Pueraria thomsonii* are leguminous vines that can grow rapidly in almost every habitat where the weather is warm. They are known as Ge in China. Ge Gen, the root of Ge, is one of the most commonly used medicinal materials in traditional Chinese medicine (TCM) (The Pharmacopoeia Commission of People's Republic of China (PRC), 1995). Ge Gen was first described in Shen Nong Ben Cao Jing (Anonymous, *c.* 200 BC), the first Chinese materia medica, as sweet and acrid in taste, cool in nature and was used as an antipyretic, antidiarrhetic, diaphoretic and antiemetic agent. In Huang Di Nei Jing (Anonymous, *c.* 100 BC) and Shang Han Lun (Zhang, *c.* 200 BC), the two classic texts of TCM, Ge Gen was recommended for stiffness and pain of the neck, pain in the eye, febrile diseases, exhaustion and thirst, and for the induction of early measles eruption (Jiang Su Medical College, 1977). Today, it is still widely used in China, by laymen and medical professionals alike, for the treatment of "exo-pathogenic" diseases with symptoms such as fever, headache, stiffness and pain in the neck and back. It is also used for dire thirst in febrile disease, polydipsia in diabetes, and stiffness and pain in the neck resulted from hypertension.

Traditional Chinese medicine emphasizes a holistic approach to the treatment and prevention of diseases. As a norm than exception, composite formulas containing two or more medicinal ingredients are often used to accomplish this multiple task which includes rectification and maintenance of all normal bodily functions compromised by a disease. An effective composite formula should contain components that will remove the pathogens causing and all the symptoms of the disease, neutralize potential undesirable side effects caused by the essential components of the formula, and rectify compromised bodily function(s) that succumb a patient to the disease in the first place. For instance, for the induction of measles eruption, Ge Gen is often used with Cimicifuga root, peony root and liquorice as in Sheng Ma Ge Gen Tang:

Sheng Ma Ge Gen Tang (Jiang Su Medical college, 1977)
*Composition*: Pueraria root, 9 g; Cimicifuga root, 3 g; Peony root, 4 g; Liquorice, 3 g.
*Indications*: Dispels pathogenic factors from the superficial muscles; detoxicifies and induces measles eruption; relieves symptoms of fever, aversion to cold, headache, body pain, sneezing, cough, thirst and tearing and red eyes.
*Dosage form*: Water decoction.

In this formula, both Pueraria root and Cimicifuga root are regarded as the "Monarch Drugs" in TCM. The major action of Pueraria root is to enhance body fluid production whereas that of Cimicifuga is to detoxify and expel heat from the body. Together, they potentiate each other's therapeutic function in the induction of measles eruptions. Peony and liquorice are "Deputy Drugs" and are used to promote the therapeutic and modulate the extreme and/or side effects of the "Monarch Drugs."

Besides Sheng Ma Ge Gen Tang, two equally popular composite Ge Gen formulas have been used for centuries in China. They are Chai Ge Jie Ji Tang (a modified version of the classic Ge Gen Tang) and Ge Gen Qin Lian Tang.

> Chai Ge Jie Ji Tang
> *Composition*: Bupleurum root, 9 g; Pueraria root, 9 g; Liquorice, 3 g; Scutellaria root, 9 g; Peony root, 3 g; Notopterygium root, 3 g; Dahurican angelica root, 3 g; Platycodon root, 3 g.
> *Indication*: Induces perspiration; expels exterior pathogens and internal heat; relieves symptoms like fever, headache, anhidrosis, and pain and stiffness in the neck and back.
> *Dosage form*: Water decoction. In most cases, 3 g of gypsum, three pieces of ginger and two jujubes were added to make decoction. These items are regarded as "initiators" in TCM and are used to kick off the therapeutic effects of some herbal formulas which are slow to act on their own.

The effect of Pueraria root in Chai Ge Jie Ji Tang is to induce body fluid production, allay fever and relieve pain.

> Ge Gen Qin Lian Tang
> *Composition*: Pueraria root, 15 g; Liquorice, 3 g; Scutellaria root, 9 g; Coptis root, 9 g.
> *Indication*: Clears internal heat and dispels pathogenic factors; treats symptoms caused by trapped exopathgenic factors and unbalanced internal heat such as diarrhea with fever, vexation and fullness in the chest, dysphoria, and thrist.
> *Dosage form*: Water decoction.

The major action of Pueraria root in Ge Gen Qin Lian Tang is to clear internal heat; induce diaphoresis, and stop diarrhea.

The basic principles of TCM and their application in the use of Ge Gen containing composite formulas for various kinds of illnesses and symptoms have been discussed in greater detail else where (Zhu, 2002; Chapter 3, this volume).

## Chinese patent medicine

Since 1960s, studies on Ge Gen, and for that matter on other medicinal materials as well, have blossomed in China. Lessons learned from Western medicinal chemistry and pharmacology have led to the identification of active principles in and elucidation of the mechanism of actions of many medicinal materials used in TCM. New pharmacological activities of many ancient Chinese formulations have also been discovered. These not only have provided scientific information, based on these activity-based medications can be controlled and/or regulated but also led to the formulation, testing and development of new medications with efficacies that were not known to TCM before. Today, commercialization of medicinal products developed by and used according to non-traditional methods are controlled and regulated by The Ministry of Health in China. Only products that are manufactured to the standards set by the Ministry, and tested and proved effective and safe for human use would be certified. These medicinal preparations, usually sold over the counter as ready to use drugs are referred to as Chinese patent medicine (CPM) or simply as Medicine (Yao). The history of the making of Yu Feng Ning Xin Pian provides a typical example on how a new CPM is developed (Chapter 8, this volume). Yu Feng Ning Xin Pian, a tablet made of the powder and an isoflavone fraction of Ge Gen, was developed on the basis of an early finding that the isoflavone fraction is the active principle of Ge Gen and has since been used for the treatment of coronary heart disease, angina pectoris, and symptoms associated with hypertension such as dizziness, headache, pain and stiffness in the neck. Puerarin, the major isoflavone found in Ge Gen, was later identified as the major active principle and formulated for intravenous (i.v.) administration. Today, puerarin injection is widely used in hospitals in China for the treatment of coronary heart disease.

Composite formulas containing Ge Gen have also been tested and shown to be more effective in the treatment of patients with multiple symptoms of coronary heart disease. For example, Xin Ke Shu Tablet and Ge Tang Jiang Ya Tablet (Table 15.1–15.2) were developed in past two decades and are still used by some in the treatment of diseases of the cardiovascular system.

Besides cardiovascular diseases, modern pharmacological studies have also discovered that Ge Gen, used in combination with other medicinal herbs, is effective in improving cerebral circulation. Therefore, CPMs derived from these composite formulas have also been tested and certified for the treatment of cerebrovascular diseases such as apoplexy, cerebral embolism, acute and sudden deafness. In these instances however, Ge Gen plays a role as "Deputy Drug" and complements the pharmacological action(s) of one or more of the "Monarch Drug(s)" in the composite formula. Nao De Sheng Wan is a pill preparation that contains *Notoginseng*, *Chuanxiong* rhizome, Hawthorn fruit and Ge Gen and is used for the treatment of cerebral arteriosclerosis, ischemia, syncope caused by cerebral anemia and sequelae of cerebral hemorrhage. Yi Nao Fu Jian Jiao Nang is another CPM, a capsule preparation containing Ge Gen and a number of other medicinal materials (Table 15.2), developed for the treatment of acute cerebral embolism and thrombosis with symptoms such as facial paralysis, hemiplegia, involuntary drooling, retraction of the tongue and difficulties in speech. Because of its ability to promote salivation, Ge Gen has been used in combination with other herbs to treat patients suffering from diabetes. For instance, Yu Quan Pian, a tablet form of an ancient Chinese herbal medicine Yu Quan San, has been developed into a CPM for the clinical treatment of diabetes (The Ministry of Health of PRC, 1987–1998); and Xiao Ke Wan (Table 15.2), a pill that contains Ge Gen has been developed for the treatment of hyperglycemia and diabetes typified by the deficiency in both Qi and Yin.

## Nutriceuticals and dietary supplements

Herbal preparations that do not meet the standard of a CPM but that of a dietary supplement are certified as health food products and labeled as Jian (Health). These products, also regulated by The Ministry of Health in China, are developed and sold over the counter in similar fashions as the nutriceuticals and dietary supplements sold in the United States. Most, if not all, Ge Gen or Ge Hua (flower of Ge) preparations claimed to be effective for alcohol related health problems fall into this category (Table 15.3).

Table 15.1 Chinese patent medicines (Yao) derived from Ge Gen (Cao, 1983; The Pharmacopoeia Commission of PRC, 1995)

| Name | Composition | Actions and indications |
|---|---|---|
| Ge Gen Pian | Flavonoid fraction | Lifts yang and dispels pathogenic factors from superficial muscles, relieves dysphoria and quenches thirst. Indicated for headache; pain and stiffness in the neck and back; febrile disease with thirst and dysphoria; also for induction of measles eruption in children; hypertension with headache and dizziness; tinnitus; angina pectoris; numbness of limb. |
| Yu Feng Ning Xin Pian | Root powder and alcohol extract | Increases blood flow to cerebral and coronary arteries. Indicated for hypertension with dizziness, headache; pain and stiffness in the neck; angina pectoris; coronary heart disease; early stage acute deafness; migraine headache. |
| Ge Gen Jiu Jin Gao Pian | Alcohol extract | Promotes blood circulation. Indicated for pain; cardiovascular diseases. |
| Ge Gen Zong Huang Tong San | Flavonoid fraction | Same as Ge Gen Jiu Jin Gao Pian. |
| Ge Gen Huang Tong Pian | Flavonoid fraction | Reduces myocardial oxygen consumption and increases coronary and cerebral blood flow. Indicated for angina pectoris, arrhythmia, low immunity, hyperglycemia. |
| Ge Gen Zhu She Yie | Water extract | Indicated for fever; pain; diaphoresis resulted from cold; infection of the upper respiratory track. |
| Ge Gen Da Dou Dai Yuan Pian | Synthetic daidzein | Indicated for symptoms of hypertension such as headache, dizziness, tinnitus, stiffness in the neck; and for angina pectoris, acute sudden deafness and migraine headache. |
| Ge Gen Su Zhu She Yie | Puerarin | Increases blood flow to coronary arteries, dilates blood vessels and improves microcirculation. Indicated for coronary heart disease; angina pectoris; myocardial infarction; also for some eye diseases. |

*Table 15.2* Chinese patent medicines derived from Ge Gen composite formulas (The Pharmacopoeia Commission of PRC, 1995; The Ministry of Health of PRC, 1987–1998)

| Name | Composition | Actions and indications | Dosage form and doses |
|---|---|---|---|
| Tong Mai Chong Ji | Red sage root, Chuanxiong rhizome, Pueraria root | Promotes blood circulation. Indicated for ischemic cardiovascular and cerebrovascular diseases; arteriosclerosis; cerebral embolism; cerebral anemia; coronary heart disease; and angina pectoris. | Oral infusion, 10 g/dose, 2–3 doses/day |
| Yi Nao Fu Jian Jiao Nang | Notoginseing, Pueraria root, Red peony root, Herb of glandularstalk St Paulswort, Saffron crocus stigma, Chuanxiong rhizome, Earthworm, Dragon blood | Promotes blood circulation and removes blood stasis; dispels evil wind and removes obstruction of the channels. Indicated for acute cerebral embolism and cerebral thrombosis like facial paralysis; hemiplegia; involuntary drooling; retraction of the tongue and difficult in speech. | 3 g capsule, 3 capsules/dose, 3 doses/day |
| Xin An Ning Pian | Pueraria root, Hawthorn fruit, Fleece-flower root (processed), Pearl powder | Nourishes yin and calms the heart; removes blood stasis and obstructions of the channels. Indicated for hyperlipidemia; angina pectoris; hypertension with symptoms of headache, dizziness, tinnitus and palpitation. | Tablet, 0.25 g/tablet, 4–5 tablets/dose, 3 doses/day |
| Xin Xue Ning Pian | Extract of Pueraria root, Extract of hawthorn fruit, Starch | Indicated for hyperlipidemia; coronary heart disease; angina pectoris; hypertensive headache and stiffness in the neck. | 0.2 g Tablet, 4 tablets/dose, 3 doses/day |
| Xin Shu Ping Pian | Pubescent holly root, ginkgo leaf, Pueraria root, Mother wort, Herb of glandularstalk St Paulswort, persimmon leaf | Promotes blood circulation and removes blood stasis. Indicated for chest pain; coronary heart disease; angina pectoris; arteriosclerosis. | Tablet, 5–8 tablets/dose, 3 doses/day |
| Ge Tang Jiang Ya Pian | Uncaria stem with hooks, Pueraria root, Hydrochloro-thiazide | Calms the liver, keeps endogenous wind in check, dispels pathogenic factors from the superficial muscles to stop pain. Indicated for dizziness; headache; pain and stiffness in the neck; palpitation with insomnia; hypertension and coronary heart diseases. | Tablet dose: not listed |
| Gen Tong Ping Pian | Pueraria root, White peony root, Herb of common clubmoss, Saffron crocus stigma, Frankencense resin, Myrrh resin | Indicated for cervical vertebra disease and related symptoms such as pain and stiffness in the neck, shoulder, arm and sciatica. | Tablet, 0.5–1.2 mg puerarin/tablet |
| Ge Bang He Ji | Pueraria root, Arctium fruit Schizonepeta, Peppermint, Honeysuckle flower, Forsythia fruit, Cicada slough | Induces diaphoresis, removes heat, and counteracts toxicity. Indicated for diarrhea, dysentery with fever and thirst. | Liquid, oral 5–10 ml/dose |

Table 15.2 (Continued)

| Name | Composition l | Actions and indications | Dosage form and doses |
|------|---------------|-------------------------|----------------------|
| Mei Su Chong Ji | Black plum, Peppermint, Perilla leaf, Pueraria root | Removes summer-heat and promotes body fluid production and quenches thirst. Indicated for thirst; dryness in the throat; tightness in the chest; and dizziness resulted from summer heat. | Infusion, 10 g/dose, 3–4 doses/day |
| Jin Ju Wu Hua Cha Chong Ji | Honeysuckle flower, Kapok flower, Pueraria flower, Wild chrysanthemum flower, Sophora flower, Licorice root | Removes summer-damp and -heat, promotes diuresis to eliminate dampness and heat in the body, removes heat in blood and liver, cools the blood and improves the acuity of vision. Indicated for diarrhea; dysentery and bleeding syndromes resulted from dampness with heat in colon. | Infusion, 10 g/dose, 1–2 doses/day |
| Zhi Li Chong Ji | Pueraria root, Coptis root | Removes heat and toxic materials, stops dysentery and promotes the flow of qi. Indicated for dysentery and enteritis. | Infusion, 17 g/dose/day |
| Yi Qi Cong Ming Wang | Cimicifuga root, Pueraria root, Phellodendron bark, White peony root, Fruit of shrub chastetree, Pilose asiabell root, Astragalus root, Licorice root | Tonifies qi and lifts sunken yang, improves function of the eye and ear. Indicated for deafness; tinnitus and impaired eye-sight. | Pill, 4.2–5.2 mg puerarin/g, 9 g/dose/day |
| Xiao Ke Wan | Pueraria root, Scrophularia root, Astragalus root, Trichosanthes root, Schisandra fruit, Chinese yam | Improves kidney function, nourishes yin, tonifies qi and promotes production of body fluid. Indicated for diabetes; hyperglycemia. | Pill dose: not listed |
| Yang Yin Jiang Tang Pian | Astragalus root, Pilose, asiabell root, Pueraria root, Wolfberry fruit, Scrophularia root, Fragrant solomonseal rhizome, Scrophularia root, Anemarrhena rhizome, Moutan Bark, Chuanxiong rhizome, Giant knotweed rhizome, Schisandra fruit | Nourishes yin and replenishes qi, removes heat and promotes blood circulation. Indicated for diabetes. | Tablet, 0.8–1.0 mg puerarin/ tablet, 8 tablets/dose, 3 doses/day |

*Table 15.3* Ge-based health products for alcohol related diseases

| Name | Composition | Action and indication | Dosage form |
|---|---|---|---|
| Ge Hua Jie Xing Kou Fu Ye | Pueraria flower, Aucklandia root, White poria, Tangerine peel, White atractylodes rhizome, Dried ginger, Medicated leaven, Oriental water plantain rhizome, Immaturepericarp of bitter orange, Amomum fruit, Round cardamon fruit. | Warms the middle-jiao and invigorates spleen. Indicated for drunkenness and syndromes resulted from excessive drinking such as headache, dysphoria, dizziness and vomiting. | Liquid, oral |
| Zui Jiu Ling | Tangerine peel, Oriental water plantain rhizome, Licorice root, Agastache, Pueraria flower. | Indicated for drunkenness; alcoholism. | Liquid, oral |
| Ge Gen Zhi Yin Liao | Pueraria root, Germinated barley, Radish seed. | Used to treat drunkenness and the diseases due to excessive drinking (as an amethystic agent). | Beverage |
| Jie Jiu Bao Jian Cha | Pueraria root, tea. | Indicated for drunkenness. | Tea |

In TCM, both Ge Gen and Ge Hua are commonly used medicinal materials for the treatment of alcohol-related diseases or conditions (Luo *et al.*, 1988). Indeed, Ge Gen and Ge Hua are the major ingredients of almost all TCM formulas, ancient or contemporary, used for the treatment of alcohol abuse and acute and chronic alcohol intoxication. A modern nutriceutical preparation, Ge Hua Jie Xin Kou Fu Yie (Fang *et al.*, 1995), has been developed and sold over the counter in China for the treatment of alcohol intoxication and a series of symptoms resulted from excessive alcohol intake such as headache, dysphoria, dizziness, and stomach upset. A similar preparation, Ge Hua Jie Xin Kou Fu Yie (Table 15.3), was developed based on the classic formula Ge Hua Xin Jiu Tang of Li Dong-Yuan, a reputable herbalist of the thirteenth century China (Li, *c.* 1220). Zui Jiu Ling (Table 15.3), a simplified version of Ga Hua Xin Jiu Tang, was also shown to lower blood alcohol after drinking (Liu *et al.*, 1998) and was developed for the treatment of acute alcohol intoxication. However, despite their claims, The ministry of China certifies none of these preparation as CPM or Yao.

Ge, as a member of the leguminous family, is rich in isoflavones. Recently, plant products rich in isoflavones have attracted enormous attention in health, food and nutriceutical industries. Epidemiological findings have associated high dietary isoflavone intake with low incidence of a number of hormone-dependent diseases such as cancer of the breast, prostate, endometrium and postmenopausal symptoms including osteoporosis (Chapter 11, this volume). Recent approval by United States Food and Drug Administration of the claim that dietary isoflavone supplements can reduce the risk of coronary heart disease has provided additional impetus to the R&D of diets or dietary supplements containing these beneficial isoflavones. The fact that Ge contains a unique isoflavone, puerarin, which has been shown and used in China for the treatment of cardiovascular diseases renders this fast growing plant an attractive resource for the development of new product lines of dietary supplements and/or nutriceuticals. Readers are referred to Chapter 9 and 12 for further discussions on the pharmacological and chemopreventive actions of these isoflavones on cancer, osteoporosis and cardiovascular diseases.

## GE IN CHINESE PATENT MEDICINE

To date, more than 160 Ge-containing medicinal preparations have been certified by The Ministry of Health as CPMs. They can be classified into six groups on the basis of their indications:

1   For influenza and common cold, 34 preparations;
2   For cerebrovascular diseases, 21 preparations;
3   For cardiovascular diseases (hypertension, hyperlipidemia and coronary heart disease), 22 preparations;
4   For children's (pediatric) diseases; 14 preparations;
5   For symptoms related to diabetes, 9 preparations;
6   Miscellaneous, 60 preparations.

Among the CPMs of Ge, most were made of ingredients extracted from 2 to 10 different herbs and/or other medicinal materials (Table 15.2). The most elaborated formula contains 92 different medicinal materials or herbs. Yet there are a few of them contain only ingredients extracted from Ge Gen (Table 15.1). Seven most commonly used and Ge Gen containing CPMs are listed here in greater details to provide a glance on their composition, effects (as known in TCM), indications and contraindications, dosage and dosage form, potential side effects if any, and pharmacology. These information were obtained from the Pharmacopoeia and the Chinese patent Medicine Bulletin issued by the Ministry of Health of the People's Republic of China. Clinical data obtained during the research and development of these products are also included where available.

1.  Nao De Sheng (Cheng *et al.*, 1997)
*Composition*: Notoginseng root, Chuanxiong rhizome, Saffron crocus stigma, Pueraria root, Hawthorn fruit without seed.
*Effects*: Awaking; removing obstructions in main and subsidiary channels, promoting blood circulation and removing blood stasis.
*Indications*: It is indicated in the treatment of apoplexy with symptoms like difficulty in speech, hemiplegia, facial hemiparalysis and in the relief of sequelae caused by cerebral embolism and coronary heart disease. This medicine has been tested in 537 cases of cerebral embolism. Among them, 284 patients had headache and 131 of them felt. Effective rates were 61 per cent and 66 per cent among patients with headache and dizziness, respectively.
*Dosage form and dosage*: Tablet; taken orally 3 tablets/dose, 3 doses/day.
*Pharmacology*: Dilates coronary arteries, reduces obstruction and increases blood flow to coronary and cerebral arteries, and improve microcirculation. It could also lower serum cholesterol, $\beta$-lipoprotein and triglyceride levels and restrain atherosclerosis. It prevents thrombosis by decreasing blood viscosity and platelet aggregation and reduces and sustains blood pressure by lowering peripheral resistance.

2.  Jun Mai An (Cheng *et al.*, 1997)
*Composition*: Pueraria root, Uncaria stem with hooks.
*Effects*: Calming liver-wind; relieving pain by dispelling pathogenic factors from the superficial muscles; suppressing Liver-Yang and augmenting Liver- and Kidney-Yin.

*Indication*: It is indicated for hypertension and coronary heart disease and the associated symptoms like pain and stiffness in the neck and back, headache and dizziness, insomnia with palpitation, irritability, amnesia, tinnitus and numbness in the extremities. This medication has been tested in 210 cases of hypertensions in stage I, II and III together with 80 control subjects. Results indicate that it is very effective in 124 cases, somewhat effective in 67 cases and not effective in 19 cases. The overall effective rate was 91 per cent. In the same study, the effective rate of ginkgo leaves tablet, Hypertension Tea and Xin Tong Ding were 90 per cent, 90 per cent, and 80 per cent, respectively. Comparing to the controls, effect of Mai Jun An was mild, steady, safe and long lasting.

*Dosage form and dosage*: Tablet; taken orally 4–5 tablets/dose, 3–4 doses/day. Dosage should be decreased as symptoms recede.

*Pharmacology*: (1) It lowers blood pressure, regulates vasomotion, dilates capillaries, blocks sympathetic nerve control. It also controls neurotransmitters release from nerve endings and reduces heart rate and peripheral resistance. (2) It improves blood circulation in cerebral and coronary vasculature. (3) It is anxiolytic and antagonizes the anxiogenic effect of caffeine.

### 3. She Xiang Xin Nao Le Pian (Cheng *et al.*, 1997)

*Composition*: Red sage root, Pueraria root, Herb epimedii, Notoginseng root, Curcuma root, musk.

*Effects*: Promotes blood circulation and removes blood stasis; awakes patients from unconsciousness and relieves pain.

*Indication*: It is indicated for coronary heart disease, angina pectoris, myocardial infarction and cerebral embolism. It is also indicated for symptoms like hemiplegia, hemiparalysis, thready and irregular pulse.

*Dosage form and Dose*: Tablet (300 mg/tablet); 2–4 tablets/dose, 3 doses/day.

### 4. Yu Quan Wan (Cheng *et al.*, 1997)

*Composition*: Pueraria root, Trichosanthes root, Scrophularia root, Ophiopogon root, Schisandra fruit, Liquorice root.

*Effects*: Promotes body fluids production, clears heat and arrests irritability, nurtures middle-qi and balances yin and yang. Specifically it is for exhaustion, thirsty, deficiency in yin of the lung, stomach and kidney caused by fever.

*Indications*: It is indicated for diabetes with deficiency in body fluid of the lung, stomach, and deficiency in yin of the kidney. In a clinical study where 18 diabetic patients were treated with this medication, 6 showed significant improvement in two months, 7 showed some improvement and 5 showed no effect.

*Dosage form and dose*: Pill (150 mg/pill); 60 pills/dose, 4 doses/day.

*Pharmacology*: In general, it lowers blood sugar. In adrenaline induced hyperglycemic rats, administration of this medication at a dose of 25 g/kg for 4 days reduced blood sugar significantly. At higher doses (10 g/kg) and longer treatment (10 days), it reduced alloxan induced high blood sugar and increased glycogen storage in the liver.

### 5. Ge Gen Qin Lian Wei Wan (Cheng *et al.*, 1997)

*Composition*: Pueraria root, Scutellaria root, Coptis root, Liquorice root.

*Effects*: Clears heat and dispels pathogenic factors from exterior of the body.

*Indication*: It is indicated for diarrhea with fever caused by bacteria in children and adults and a variety of diarrheas and dysenteries. Clinical treatment of 114 cases of diarrheas achieved an effective rate of 75 per cent in 3 days and total effective rate of 96.4 per cent. The antipyretic effect was obvious. Its antiviral effect was tested in 35 cases. At the end of a 3 day treatment period, viral counts turned negative in 24 cases, an effective rate of 71.4 per cent.

*Dosage and administration*: Micropill; taken orally, 1 g/dose, 3 doses/day; 1 g/dose/day for children under 5.

*Pharmacology*: It boosts immune system in general and the B-cell activity in particular. In mice, it enhances antibody production from birth to maturation. It raises IgM, IgG and IgA level in stomach and intestinal track. In addition, it increases the number of *bacteriolyase* (*Rongjunmei*) and enhances the function of microphage. Also, it antagonizes the immunosuppressive effect of hydrocortisone.

## 6. Xin Ke Shu Pian (Cheng *et al.*, 1997)

*Composition*: Pueraria root, hawthorn fruit, Red sage root, Notoginseng, Aucklandia root.

*Effects*: It promotes blood circulation and disperses blood stasis, relieves chest pain and lowers blood pressure caused by stagnation of qi.

*Indication and contraindication*: It is indicated for hypertension and coronary heart disease due to decreased blood flow to coronary arteries and deficiency in the function of the heart. Clinical studies on 145 patients with coronary heart disease showed a 60 per cent effective rate. 30 per cent of the patients became normal and have improved electrocardiogram. Patients of angina pectoris in general felt better after using this medication for 7 to 8 days. Coronary blood flow was in general improved after treatment. It is not recommended for patients who are deficient in heart-yang.

*Dosage form and dose*: Tablet, 0.25 g; 4 tablets/dose, 3 doses/day.

*Pharmacology*: It increases tolerance to oxygen deficiency in animals. It has little effect on blood clotting time but protects heart muscle and lowers blood pressure. It also antagonizes the stimulating effect of adrenaline on the heart.

## 7. Xin Mai Tong Pian (Cheng *et al.*, 1997)

*Composition*: Chinese anglica root, Semen cassiae, Uncaria stem with hooks, Red sage root, Pueraria root, Flos sophorae, Notoginseng, Ilicis root, Achyranthis and Cyathula root, Spica prunellae.

*Effects*: It promotes blood circulation and disperses blood stasis, nourishes the heart and calms the liver. It also promotes the circulation of qi in the chest.

*Indications and contraindications*: It is indicated for high blood pressure, coronary heart disease, angina pectoris, hyperlipidemia, hypercholesterolemia and symptoms such as pain in the front area of the chest, feeling pressed and tight in the chest and difficult to breathe, dizziness, pain and stiffness in the neck. Clinical testing on 126 cases of hypertension showed an effective rate of 92.3 per cent. In cases of hyperlipidemia (38 cases), effective rate was 90 per cent. This medication is not recommended for pregnant women and women with excessive menses. It is also contraindicated for patients who are weak and deficient.

*Dosage and administration*: Tablet; 4 tablets/dose, 3 doses/day.

*Pharmacology*: This medication has been shown to have antihypertensive, antihyperlipidemia, antiarteriosclerotic, and vasodilating effects.

## QUALITY CONTROL OF MEDICINAL GE

Quality control has been one of the foremost important and difficult issues in the regulation of crude medicinal preparations. In the last decade, The Ministry of Health of The People's Republic of China has imposed increasingly stringent rules on the control of the qualities of medicinal preparations sold in China. Because Ge, and for that matter other medicinal plants as well, used for medicinal purposes in China is mainly collected from the wild, to maintain a standard composition of the various active ingredients contained in a crude medicinal preparation of Ge Gen is a major challenge to the CPM industry. Thus, to reveal the factor(s) that affect the contents of the active ingredients in Ge Gen has been a major subject of investigation in China.

### Flavones, isoflavones, trace elements and *Pueraria* species

While Ge Gen contains wide varieties of chemicals including flavones, isoflavones and their glycosides, saponins, coumarin etc. (for detail see Chapter 6, this volume), isoflavones are believed to be the major active principles in CPMs where Ge Gen acts as the "Monarch Drug." These isoflavones, which include puerarin, daidzin, daidzein, daidzein-4′,7-di-glucoside, formononetin and genistin, are under most circumstances used for the comparison and standardization of Ge Gen-based CPMs. The total flavonoid contents found in six common species of Ge varied by more than 10 folds with *P. lobata* being the highest followed by *P. thomsonii* Benth, *P. omeiensis* Tang et Wang, *P. phaseoloides*, and *P. montana* and *P. peduneularis* (see this volume Chapter 4, Table 4.5). The total isoflavone contents found in the root of these species follow similar order. Among the 3 common isoflavones determined, puerarin is by far the most abundant followed by daidzin and daidzein (Jiang Su Medical College, 1997).

Contents of trace elements in different species of Ge also exhibit qualitative and quantitative differences (Table 15.4). For instance, Zn, Cu, Mg, Ca, Sr, Cr contents found in *P. lobata* are much higher than those in *P. thomsonii* (Zeng and Zhang, 1996). Despite such differences, the root of both *P. lobata* and *P. thomsonii* are considered authentic species used in TCM in Chinese Pharmacopeia. Although *P. omeiensis*, *P. montana* and *P. edulis* are also used for medicinal purposes, they are considered more local and used mainly in areas where they are harvested. In this context, it is important to point out that species differences in chemical constituents are consistent with the fact that they are used for the treatment of different or slightly different disease states (Table 15.5). It is also important to emphasize that not all species of the genus *Pueraria* are safe for human consumption. For example, *P. peduneularis*, also known as bitter Ge to the locals, is toxic and used only for the production of pesticides in China. The geographical distribution and applications of various species of *Pueraria* are summarized in Table 15.5.

### Habitats, harvest time and isoflavone contents

Ge Gen of the same species but harvested from different geographical locations or at different time of a year differ significantly in isoflavone contents. The effects of habitats and harvest time on Ge Gen isoflavone contents have been discussed (Chapter 3, this volume) and, therefore, will not be detailed here.

*Table 15.4* Contents of elements and *Pueraria* species (µg/g)

| No.[a] | Zn | Cu | Fe | Ca | Mg | Mn | Pb | Sr | Ag | Cr | Co | Cd |
|---|---|---|---|---|---|---|---|---|---|---|---|---|
| 1 | 21.4 | 5.17 | 120 | 13100 | 305.3 | 2.93 | 0.34 | 32.12 | 0 | 0.160 | 0 | 14.0 |
| 2 | 19.6 | 2.97 | 120 | 9304 | 155.5 | 3.61 | 0.05 | 29.85 | 0 | 0.138 | 35.0 | 10.0 |
| 3 | 8.4 | 3.29 | 192 | 13156 | 19.2 | 8.7 | 0 | 45.07 | 0.1 | 0.102 | 55.0 | 13.0 |
| 4 | 12.4 | 3.98 | 18 | 11036 | 209.3 | 6.5 | 0 | 16.04 | 0 | 0.130 | 0 | 4.0 |
| 5 | 7.2 | 3.707 | 70 | 7620 | 255.4 | 10.9 | 1.78 | 23.17 | 0 | 0.118 | 47.0 | 39.0 |
| 6 | 11.2 | 35 | 80 | 9240 | 234.2 | 4.3 | 0 | 20.74 | 0 | 0.309 | 5.5 | 4.5 |
| 7 | 8.0 | 2.07 | 98 | 16968 | 351.4 | 23.5 | 0.80 | 36.67 | 0 | 0.098 | 62.0 | 25.0 |
| 8 | 10.6 | 6.62 | 210 | 17024 | 437.8 | 12.3 | 2.29 | 29.66 | 0 | 0.114 | 257.0 | 26.0 |
| 9 | 12.8 | 4.09 | 24 | 8276 | 158.4 | 10.8 | 0.10 | 31.65 | 0 | 0.050 | 9.0 | 1.0 |
| 10 | 12.0 | 2.65 | 38 | 18824 | 246.7 | 4.4 | 0.47 | 6.10 | 0.1 | 0.111 | 59.0 | 1.0 |
| 11 | 9.0 | 1.28 | 26 | 2000 | 96.0 | 4.8 | 0.70 | 1.69 | 0 | 0.069 | 4.5 | 1.0 |
| 12 | 8.6 | 3.01 | 30 | 5888 | 258.4 | 4.3 | 0 | 8.51 | 0 | 0.161 | 0 | 1.0 |
| 13 | 11.8 | 5.43 | 126 | 26704 | 263.0 | 10.8 | 0 | 20.00 | 0.1 | 0.135 | 51.0 | 2.0 |
| 14 | 58.0 | 12.3 | 96 | 18284 | 38.0 | 38.5 | 1.30 | 15.65 | 0 | 0.181 | 264.0 | 30.0 |
| 15 | 18.0 | 7.06 | 238 | 13016 | 168.0 | 22.3 | 0 | 19.67 | 0.02 | 0.713 | 275 | 5.3 |
| 16 | 14.0 | 4.63 | 65 | 29752 | 197.8 | 15.9 | 0.56 | 31.97 | 0.02 | 0.890 | 295.0 | 1.0 |

Notes

a  No. 1–9 were *P. lobata* (Willd.) Ohwi collected from different provinces (1. Liaoning, 2. Hebei, 3. Xian, 4. Shandong, 5. Guizhou, 6. Sichuan, 7. Yunnan, 8. Jiangxi, 9. Anhui). No. 10–16 were collected from different *Pueraria* species (10. *P. omeiensis* Tang et Wang, 11. *P. thomsonii* Benth., 12. *P. peduneularis* Grah., 13. *P. edulis* Pamp., 14. *P. phaseoloides* (Roxb.) Benth, 15. *P. alopercuroides* Craib, 16. *P. montana* (Lour.) Merr).

*Table 15.5* Actions and indications and Ge Gen species

| Ge Gen species | Distribution | Actions and indications |
|---|---|---|
| *P. lobata* (Willd.) Ohwi | All over China except Tibet, Xin-jiang, and Qing-hai provinces | PRC Pharmacopoeia: antipyretic; promotes salivation; induces measles eruption; antidiarrhea; raises yang |
| *P. thomsonii* Benth | Guangdong, Guangxi, Hainan, Sichuan, and Yunnan provinces | Same as *P. lobata* (Willd.) Ohwi |
| *P. omeiensis* Tang et Wang | Sichuan, Guizhou, Yunnan, and Tibet provinces | Used mostly by local people for diabetes, measles eruptions |
| *P. montana* (Lour.) Merr | Guangdong, Guangxi, Hainan, Taiwan, Yunnan, and Fujian provinces | Used mostly by Taiwanese, indicated for cough, hyperactive liver function |
| *P. peduneularis* Grah | Sichuan, Yunnan, Guizhou, and Tibet provinces | Toxic, used only for insecticides or pesticides |
| *P. edulis* Pamp | Yunnan province | Food supplements, also used in TCM as a substitute of *P. lobata* and *P. thomsonii* |

## Isoflavone contents in the root and vine of Ge

Virtually all parts of Ge such as the root, leaf, flower, vine and seed etc. have been used in TCM. Further, different parts of the plant were, in general, used for the treatment of different diseases. These empirical findings can now be explained by what we have learned from modern scientific studies that the chemical compositions in different parts of a plant can be very different. Indeed, the root and vine of Ge which share many

*Table 15.6* Chemical constituents of the root and vine of *P. lobata*

| Composition | Root (g/g, %) | Vine (g/g, %) |
| --- | --- | --- |
| Total flavonoids | 7.8 | 7.5 |
| Puerarin | 2.481 | 4.315 |
| Daidzin | 3.933 | 0.714 |
| Daidzein | 0.195 | 0.059 |
| Total amino acid | 2.4546 | 5.808 |
| Acetic amino acid | 0.4869 | 4.1357 |
| Alkaline amino acid | 0.4260 | 0.8345 |
| Neutral amino acid | 1.5417 | 3.8382 |
| Essiential amino acid | 1.2174 | 2.3550 |
| Aspartic acid (Asp) | 0.2449 | 0.5253 |
| Threonine (Thr) | 0.1396 | 0.2318 |
| Serine (Ser) | 0.1722 | 0.3058 |
| Glutamic acid (Glx) | 0.2420 | 0.6104 |
| Glycine (Gly) | 0.1519 | 0.2574 |
| Alanine (Ala) | 0.1575 | 0.2408 |
| Valine (Val) | 0.1952 | 0.3838 |
| Methionine (Met) | 0.0098 | 0.0588 |
| Isoleucine (Ile) | 0.1234 | 0.2451 |
| Leucine (Leu) | 0.1838 | 0.3605 |
| Tyrosine (Tyr) | 0.1105 | 0.2596 |
| Phenylalanine (Phe) | 0.1396 | 0.2405 |
| Lysine (Lys) | 0.1996 | 0.4303 |
| Histidine (His) | 0.0705 | 0.1709 |
| Arginine (Arg) | 0.1559 | 0.2333 |
| Proline (Pro) | 0.1582 | 0.2542 |

physical and chemical similarities have very chemical composition (Meng *et al.*, 1994; Li *et al.*, 1999). Their isoflavone contents are significantly different from each other (Table 15.6). Puerarin is most abundant in the vine whereas daidzin and daidzein are concentrated in the root. Moreover, their amino acid contents also differ, not only in total amounts but in types (Table 15.6). Since the appearance of processed root and vine are so similar, accurate identification of the exact part of the plant is, for obvious reason, extremely important.

## Processing methods and isoflavone contents

After harvest, medicinal plants are processed such as washing, cutting, debarking, breaching, drying etc. to be ready for the market. It is a well known fact that methods and steps of processing affect the chemical contents in the end products. In China, Ge Gen for medicinal used are harvested between October to the end of March. Roots dug out from the ground were washed, debark, cut into small pieces (1 to 2 cm thick and 3 to 4 cm long), and then dried under the sun or in an open fire place (Lian *et al.*, 1995). In some cases, herbalists practicing TCM use Ge Gen that were stirred fried with wheat bark powder to prepare customized TCM formulas catered for the specific needs of a patient. Stirred fried Ge Gen is called Wei Zhi Product and thought to be particularly effective in the treatment of diarrhea caused by yang-deficiency of the spleen and stomach. It stops diarrhea by lifting qi. It is usually used together with other ingredients such as Pilose asiabell root, Auklandia root and White atractylodes rhizome in Yi Wei Bai Zhu San.

*Table 15.7* Isoflavone contents in different root preparations of *P. thomsonii* Benth: with bark vs. without bark

| | Isoflavone contents (% dry weight) | | | | | |
|---|---|---|---|---|---|---|
| | Total isoflavone | Puerarin | 3-methoxy-puerarin | Daidzin | Daidzein | Daidzein-4,7-diglucoside |
| With bark | 1.13 | 0.45 | 0.094 | 0.069 | 0.041 | 0.019 |
| Without bark | 0.56 | 0.45 | 0.014 | 0.033 | 0.014 | 0.016 |

Studies on the effects of processing methods on isoflavone contents has shown that Ge Gen prepared with or without bark contain very different isoflavone contents. For instance, the total isoflavone contents in Ge Gen preparations without bark are only half of those found in preparations with bark attached. Such difference is attributed to extremely low contents of daidzin, daidzein and 3-methoxy-puerarin in the debarked Ge Gen (Table 15.7) (Lian *et al.*, 1995). Other processing methods have also been shown to change isoflavone contents in Ge Gen. Ge Gen treated with Rice water, stir-fried with wheat bark powder or with vinegar, stir-fried until it turns deep brown or becomes charcoal contain 3.9 per cent, 3.93 per cent, 5.13 per cent, 4.9 per cent and 2.42 per cent by weight, respectively (Liu *et al.*, 1998).

## Standardized medicinal Ge Gen

Except for special needs where individual herbalist may opt for specifically processed materials, Ge Gen used in China are mostly prepared according to the standardized methods listed in the PRC Pharmacopoeia (The Pharmacopoeia Commission of PRC, 1995) which are summarized below:

**Pueraria root** is the dried root of *Pueraria lobata* (Willd.) Ohwi or *Pueraria thomsonii* Benth.(Fam. Leguminosae). Roots are collected in autumn and winter. Fresh roots of *P. lobata* are often cut into thick slices or pieces and dried. Roots of *Pueraria thomsonii* Benth, also known as starchy *Radix Puerariae*, are debarked, fumigated with sulfur, cut into sections and then longitudinally into two parts, and dried.
*Description*: *Radix Puerariae Lobatae*: Longitudinally cut into rectangular, thick slices or small square pieces, 5–35 cm long, 0.5–1 cm thick. Outer bark is pale brown with longitudinal wrinkles; rough, cut surface is yellowish-white with indistinct striation. Texture pliable but with strong fibers. Odorless, tastes slightly sweet.
*Radix Puerariae thomsonii*: Cylindrical, subfusiform or semicylindrical, 12–15 cm long, 4–8 cm in diameter, some were cut longitudinally or obliquely into thick slices, varying in size. Externally yellowish-white or pale brown, or greyish-brown if the outer bark not removed. Transversely cut surfaces show pale brown concentric rings formed by fibers whereas longitudinally cut surfaces show striations. Heavy, texture hard and starchy.
*Identification*: (1) Powder: Pale brown, yellowish-white or pale-yellow. Starch granules abundant, simple granules spheroidal, semi-rounded or polygonal, 3–37 μm in diameter, hilum pointed, cleft or stellate, compound granules of 2–10 components. Fibers mostly in bundles, walls thickened and lignified, surrounded by parenchymatous cells mostly containing prisms of calcium oxalate, forming crystal fibers, crystal cells, with lignified and thickened walls. Stone cells some times visible, sub-rounded or

polygonal. 38–70 µm in diameter. Bordered pitted vessels relatively large, pits hexagonal or elliptical, arranged very densely. (2) Macerate 0.8 g of the powder in 10 ml of methanol for 2 h, filter and evaporate the filtrate to dryness. Dissolve residue in 0.5 ml of methanol as test solution. Dissolve puerarin CRS in methanol to produce a 1 mg/ml standard solution. Perform thin layer chromatography using silica gel H containing sodium *carboxymethyl-cellulose* as the coating substance and chloroform–methanol–water (7:2.5:0.25) as the mobile phase. Apply separately to the plate 10 µl of each of the two solutions. After development, examine plate under UV light (365 nm). The fluorescent spot of puerarin should be found in the test sample.

*Water*: Water content should be analyzed according to Method 1 in Appendix 9.3 and should not be more than 14.0 per cent.

*Total ash*: *Radix Puerariae Lobatae*: <7.0 per cent; *Radix Puerariae thomsonii*: <5.0 per cent.

*Processing*: Eliminate foreign matter, wash clean, soften thoroughly, cut into thick pieces and dry under the sun.

*Action*: relieves fever, promotes production of body fluid, facilitates measles eruption, and arrests diarrhea.

*Indications*: Fever, headache and stiffness of the nape in *exogenous affections*, thirst, diabetes, measles with inadequate eruption, acute dysentery or diarrhea, stiff and painful nape in hypertension.

*Dose*: 9–15 g.

*Storage*: Preserve in a ventilated dry place, protected from moth.


## PROSPECT

Ge, Ge Gen in particular, is one of the most ancient and popular medicinal materials used in TCM. Recent pharmacological and clinical studies have validated, at least to the satisfaction of The Ministry of Health in China, its efficacies in the treatment of traditional diseases claimed in TCM and its newly discovered efficacies in the treatment of cardiovascular, cerebrovascular diseases and diabetes. Significant progresses have been made in the understanding of the molecular mechanism underlying Ge Gen's therapeutic activities (See Chapter 5, 7, 9, 10 and 13, this volume) yet more work needed to be done. Results from more stringent, double-blind, placebo-controlled clinical trials are needed to convince modern treatment communities, especially health care providers practicing conventional medicine, that Ge Gen-based medications are safe and effective in the treatment of human diseases. Dietary isoflavones have been shown to have beneficial effects in preventing cardiovascular and a number of hormone dependent diseases. Ge Gen, rich in isoflavones, may become an attractive resource for the development of new product lines in the dietary supplement and health food industries (Yu and Shi, 1991).


## REFERENCES

Cheng, Z.-S., Wang, K.-G. and Lin, Z.-N. (1997) *Modern Chinese Patent Medicine*. Jiang Xi Science and Technology Publisher, Jiang, China.

Cao, C.-L. (1983) *Collections of Chinese Patent Medicine and Formulations*. People's Health Publisher, Beijing, China.

Fang, Y.-Z., Song, J.-Y. *et al.* (1995) Pharmacological study of orally administered Ge Hua Jie Xing Liquid. *Chin. Patent Med.*, 17(10), 30.

Jiang Su Medical College (1977) *Encyclopoedia of Chinese Materia Medica (Zhong yao da ci dian, in Chinese)*. Shanghai Scinence Publishers, Shanghai, China.

Li, S.-S., Deng, J.-Z., Liu, H. and Zhao, S.-X. (1999) Isoflavonoids from the vines of *Pueraria lobata*. *Natural Product Research And Development (Tian ran chan wu yan jiu yu kai fa)*, 11(1), 31–33.

Lian, W.-Y., Feng, R.-Z., Chen, B.-Z., Zhou, Y.-P., Su, X.-L., Zhong, Y., Gu, Z.-P., Xu, Q. and Fu, G.-X. (1995) Ge Gen. In Z.-Q. Lou and B. Qin (eds), *Species Systematization and Quality Evaluation of Commonly used Chinese Traditional Drugs (North-China edition)*, Peking Union Medical College & Beijing Medical University, Beijing, China, pp. 379–420.

Liu, L.-Q., Zhong, G.-S., Gao, W. *et al.* (1998) Experimental study of Zui Jiu Lin on acute alcohol intoxication. *J. Beijing Univ. Tradit. Chin. Med.*, 21(3), 24.

Liu, S.-P., Wang, J.-Z., Liu, C.-S., Wen, G.-Q. and Liu, Y.-H. (1998) Determination of puerarin contents in various processed Ge Gen product by HPLC. *China J. Chin. Mater. Med. (Zhong guo zhong yao za zhi)*, 23(12), 723–725.

Luo, Z.-Y., Hu, J., Ma, X.-H., Zhou, Z.-B. and Zou, L.-X. (1988) Experimental studies on amethystic agents. *China J. Ch. Mater. Med. (Zhong yao tong bao)*, 13(4), 28–30.

Meng, X.-Y., Li, X.-G., Wei, C.-Y. and Liu, S.-M. (1994) Analysis on the chemical constituents in the root and stem of *Pueraria lobata* (Wild) Ohwi. *J. Ji-lin Agric. Uni.*, 16(3), 47–58.

The Ministry of Health of PRC (1987–1998) *Pharmaceutical Bulletin: Chinese Patent Medicines*. The Ministry of Health, China.

The Pharmacopoeia Commission of PRC (1995) *Pharmacopoeia of the People Republic of China*. Guangdong Science and Technology Press, China.

Yu, Z.-Z. and Shi, J. (1991) Prospect on the development of Ge Gen food supplements. *Liang yiu shi pin ke ji*, 4, 2–3.

Zeng, M. and Zhang, H.-M. (1996) Analysis of trace element in *P. lobata* and other plants of *Pueraria* DC. *J. Chin. Med. Mater. (Zhong yao cai)*, 19(4), 190–191.

Zhu, Y.-P. (1998) *Chinese Materia Medica: Chemistry, Pharmacology and Applications*. Harwood, Amsterdam.

# 16  Kudzu (*Pueraria lobata*), a valuable potential commercial resource: food, paper, textiles and chemicals

*Llewellyn J. Parks, Robert D. Tanner, and Ales Prokop*

## INTRODUCTION

There are 15 recognized varieties of the Genus *Pueraria*, each having somewhat different specific characteristics. These include *P. phaseoloides, P. peduncularis, P. thomsonii, P. hirstuta, P. pseudo-hirstuta, P. javanica, P. elegans, P. tuberosa, P. omeiensis, P. triloba, P. mirifica, P. montana* and *P. lobata*, sometimes also known as *P. thunbergiana*.

Of these, *Pueraria lobata* is native to China where it grows primarily along the border with Vietnam. It is also found in the Yoshino Valley, Japan, in Korea, and throughout the Southeast United States. It was imported to the United States for the Philadelphia Exposition of 1876 where it was offered in the Japanese Pavilion as an ornamental vine. In mid 1920s, it began to be used in the Southeast United States as a cattle feed on a limited basis and by 1934 was being widely used there to control soil erosion. It is known as "kudzu" in North America, an American word derived from the Japanese "kuzu."

*P. lobata* is highly evolved. It is among the world's most opportunistic plants, choking out and excluding almost all other vegetation where it grows. Climatic, space and other considerations make the Southeast United States an ideal environment for *P. lobata*. Other regions where *P. lobata* grows may not share the combination of positive factors such as space, sunlight, humidity, and virtually free of natural enemies, necessary for successful development of a significant commercial crop.

*P. lobata* requires large growing space for cultivation. Its vines grow 60 to 100 feet in one growing season. Space for the production of this type of agricultural commodity is readily available on the existing farmland in the Southeast. The methodologies and technologies of the American Agricultural Industry have proven successful in the rapid production of soybean, requiring 3 years to deliver soybean in commodity quantities to the existing World Markets and merely 5 years to dominate the World Market for Soybeans. However, unlike soybean, an annual plant, *P. lobata* is perennial and requires 3 years to reach maturity. Seven to 10 years may be needed to develop a successful crop of *P. lobata* similar to that of soybean in the Southeast United States.

*P. lobata* became an "escaped plant" in the United States, considered by most a pest or weed, covering approximately 10 million acres at its peak. Despite its removal from large areas for real-estate and highway development as well as continual, determined efforts to control its spread, *P. lobata* remains entrenched on at least 6 million acres in the American South today. It is more prolific in Alabama than anywhere else in the World and ranges widely throughout the Southeast United States from North Carolina to Louisiana. By comparison, the Southeast cultivates as many as 10 million acres of cotton annually.

Notably, *P. lobata* has encountered no significant natural enemies in the Southeast United States for more than 60 years. Power, Railroad, Forestry Industries, State and Federal Government agencies in the Southeast are forced to spend millions of dollars, annually, to continually control its undesirable spread on their lands. Paradoxically, the hardiness of *P. lobata* suggests that it may be successfully grown on a commercial scale as a commodity field crop in the Southeast without facing serious negative commercial cultivation issues resulting from natural enemies, disease, fertilizer use, and irrigation problems.

All parts of *P. lobata* contain valuable components suitable for commercial production. Fine natural food and specialty starch is found in the rhizome; at least inch long, very strong, absorbent cellulose fiber suitable for paper-making and super absorbency is also found in the rhizome; very long, very strong fine fibers suitable for textiles are found in the vines, along with very long, very strong coarse fibers suitable for structural support or paper-making; at least 30 valuable chemicals and potential drugs have already been identified in *P. lobata* components and many more may be discovered as research proceeds. All parts of *P. lobata* are edible. Its leaves are an outstanding, readily harvested, readily renewable food, having similar properties to spinach, collards and other rich table greens.

## KUDZU IN COMMERCIAL AGRICULTURE

*P. lobata* is a nitrogen-fixing-legume, which means that it adds nitrogen to the soil and provides its own nitrogen fertilizer for its successful cultivation (Lynd and Ansman, 1990). For this reason, it may be successfully rotated with other commercial crops, including cotton which rapidly depletes the soil of nitrogen. However, unlike cotton, an annual plant that must be newly seeded each growing season, *P. lobata* is a perennial, deciduous dicotyledon.

*P. lobata* is flood and drought resistant, thermo-tolerant and is among the World's most photosensitive plants. It grows well in shade and prospers in full sunlight. It requires no irrigation, fertilization, insecticide or other expensive cultivation methods and techniques to make a good crop. These features make *P. lobata* potentially cheaper, easier and more profitable for farmers to produce than cotton and many other crops.

*P. lobata* grows rapidly, propagating by seed, nodes along its vines and shoots from its roots. Its seed has a very hard coating and the broadcast seeds have a germination rate of less than 20 per cent unless they are chemically treated to improve germination. This type of chemical treatment is generally undesirable for large scale commercial production. Instead, it is more reliably produced using cuttings from its vines, set in the row using standard water-wheel planters.

*P. lobata* may also be grown in a greenhouse and in aqua-culture for specific purposes. However, in aqua-culture and in the greenhouse, it will only produce a large mass of very fine feed roots to support the large mass of its vines and leaves. Under these conditions, *P. lobata* will not set corms or fleshy rhizomes and produces no flower nor seed.

Unlike many plants, *P. lobata* is readily produced and maintained in tissue culture. In the event it proves undesirable, difficult or impossible to chemically synthesize a substance found in low supply in *P. lobata* it may be induced to produce more of the specific substance in the field using specialized agricultural methods and techniques or grown rapidly in tissue culture to produce sufficient quantities of a rare substance for commercial production.

*P. lobata* prefers to grow thick vertical rhizomes which may be as deep as ten feet into the ground and weighing more than 5000 pounds. However, it is the smaller roots (one to three pound) that have great commercial value in the United States because they can readily be produced and processed according to the demands of modern American Agricultural and Industrial mass-production standards.

Lateral spreading of *P. lobata* is naturally encouraged in the Southeast United States where there is a table of hard earth that is often as little as two to three and no more than six feet below the surface. Where the soil has been packed by generations of use, this table is nearer the topsoil. For this reason, *P. lobata* found in this region tends to develop more lateral roots and more vines having more roots resulting from nodes along the vines.

*P. lobata* produces racemes of flowers that are a delicate magenta-red-purple, having a scent similar to that of grapes. The floral scent is volatile and begins to degrade within hours of harvesting unless flash freezing is employed in or very near the field of harvest (see also, Yokoyama, 1976). *P. lobata* does not begin to flower or set seed until it is 3 years old. It had been reported that *P. lobata* does not flower and set seed in the United States. It was subsequently discovered, that flowering and setting of seed are common in the United States where there is available pollination by bees. Honey made from the nectar of *P. lobata* flowers is very fine, having a faint red color and faintly grape-like scent.

*P. lobata* growing wild in forests and on farmland flowers profusely and sets seed in late summer-early fall, annually. Flowers are typically available for no more than six weeks in August–September. The flowers are fragile and therefore, must be hand-harvested. They deteriorate rapidly after harvesting and do not store well with controlled refrigeration methods. Thus they must be processed as rapidly after harvesting as possible.

Yields of *P. lobata* components are high in the Southeast United States. Wild plants typically produce approximately 2000–3000 pounds of roots and 3000–5000 pounds of vines per acre in the Southeast United States. Modern American agricultural methods and techniques may readily increase crop production to as many as 6000–10 000 pounds of roots and 6000–12 000 pounds of vines. Vines may be harvested twice annually in this region, producing as many as 18 000 pounds in one growing season.

Yields of starch, rhizome cellulose, and chemicals extracted from *P. lobata* do not appear to vary widely in the Southeast United States, regardless of the exact geographic location. However, seasonal effects are apparent: chemical content appears to be highest in late summer to early fall, whilst starch content is highest in late fall to early winter. Fine vine fibers are found only in the new growth at the crown of a plant, in the spring to early summer (Zhong *et al*., 1992; Achremowicz *et al*., 1994; Ululag *et al*., 1996a; An *et al*., 1999). By specific pruning, additional fine fiber may be produced in the late summer to fall .

Previously, it was reported that *P. lobata* may be a potentially suitable source of biomass for the production of renewable fuels such as ethanol. Subsequently, it was discovered that all parts of the plant have more valuable uses than as a biomass for renewable fuel production. Moreover, its rhizomes and vines contain approximately 70 per cent water, which is too high for economic renewable fuel production.

Commercial agricultural production of *P. lobata* will produce a valuable cash crop for farmers. In the Southeast United States there is adequate space for a new large commodity field crop for textiles, paper, and specialty and food starch markets. Moreover, a relatively small area of less than 100 000 acres under cultivation with *P. lobata* provides

*Table 16.1* Projected annual component yields. Wild (uncultivated) *Pueraria lobata* growing on farmland in the Southeast United States

| Component | Pounds/Acre | Pounds/100 000 Acres |
| --- | --- | --- |
| Starch<br>.20/rhizome pound<br>3000 pounds rhizomes/acre | 600 | 60 000 000 |
| Rhizome cellulose<br>.32/rhizome pound<br>3000 pounds rhizomes/acre suitable<br>for paper-making | 960 | 96 000 000 |
| Dry tannin<br>.01/rhizome pound(est.)<br>3000 pounds rhizomes/acre | 30 | 3 000 000 |
| Total vine fiber<br>.40/vine pounds<br>6000 pounds vines/acre | 2400 | 240 000 000 |
| Fine textile fiber<br>.33/total vine fiber | 792 | 79 000 000 |
| Coarse vine fiber<br>Remains after fine fiber<br>suitable for paper-making and<br>structural support | 1608 | 161 000 000 |

*Table 16.2* Current annual United states commercial acreage and yields*

| Crop | Acreage | Compound pounds |
| --- | --- | --- |
| Corn | 74 million | 4–5 billion pounds starch |
| Cotton | 13 million | 9 billion pounds lint |
| Pulpwood | 8.5 million acres of trees | 1 billion pounds pulpwood |

Note
* United States Department of Agriculture, 1995.

a sufficient crop for natural drug or other natural fine chemical extraction for World Markets.

The following tables demonstrate minimum extracts available from wild *P. lobata* found at a variety of sites on farmland in the Southeast United States (Table 16.1) as well as existing Corn, Cotton and Pulpwood acreage and yields (Table 16.2).

## *P. LOBATA* STARCH

*P. lobata* rhizomes contain approximately 25 per cent fine starch that is readily extractable using standard starch processing methods. Such starch has historic existing markets in China, Japan and Korea where pure and blended *P. lobata* starches have been prized for the finest cuisine and confections, as well as their superior digestive and medicinal properties. Limited amounts of *P. lobata* starch are available, imported to the United States from China and Japan. However, upon testing, these generally are found to be a blend of as little as 5 per cent *P. lobata* starch blended with other starches.

*Table 16.3* Comparison among several starches

| Starch | Average size micrometer | Amylose content, % | Gelatinization temperature, °C | Endothermic peak/ gelatinization, °C |
|---|---|---|---|---|
| *Pueraria lobata* | 12 | 19–24 | 60–72 | 60–68.5 |
| Corn | 15 | 26 | 62–74 | 80–84 |
| Tapioca | 20 | 17 | 52–64 | 68 |
| Wheat | 16 | 25 | 60–64 | 86–89 |
| Sweet potato | 19 | 21 | – | 70.5–74.5 |
| Arrowroot | 23 | – | – | – |
| White potato | 35 | 24 | 56–69 | 69 |

Some unique features of *P. lobata* starch make it valuable as a natural specialty food starch and to replace or improve upon other natural and certain synthetic gelling agents, emulsifiers and industrial products. Table 16.3 shows a comparison of *P. lobata* with other widely available table starches. Starch consists of two polysaccharide components: amylose and amylopectin. Amylose has straight chains (100–10 000 glucose units) while amylopectin is a branched molecule (20–30 glucose units). A visible difference is that amylose is more soluble and less viscous in water than amylopectin and facilitates gel formation. By looking at *P. lobata* starch vs. corn starch, there may be less amylose in *P. lobata* starch. Thus, it may appear to be less suitable as a gelling agent. However, other properties of *P. lobata* starch outweigh this feature, making it superior to other natural starches as a gelling agent. It is not only a superior gelling agent, but offers the valuable advantage of paste (gel) stability.

Starch granules derived from *P. lobata* have very sharp edges and more angular shapes than those from other natural sources. *P. lobata* starch has the finest (smallest) granule (starch particle size) known among starches (Figure 16.1). Among others, its granule is smaller than arrowroot, tapioca, potato, rice, wheat, pea, bean, and cornstarch. *P. lobata* starch also has different moisture absorption characteristics, different cross-linking of glucose and a different amylose:amylopectin ratio from other starches.

As a result of its fine granule size and other features, it is also the most absorbent natural starch. Unlike all other natural starches, *P. lobata* starch produces a stable, very clear, colorless, flavorless, soft, transparent gel, which has no cloudiness and no paper-like starch residue. Its gel is the most stable among natural starches and is superior to the gels of most natural and many synthetic gelling agents (Tanner and Hussain, 1979; Achremowicz and Tanner, 1996).

*P. lobata* starch is more cohesive in gel form than other starches. It is more elastic, resists crumbling and presents gel stability superior to all other natural starches due to its temperature–time–viscosity characteristics. *P. lobata* starch has a shelf life exceeding 15 years stored under ordinary conditions. Its gels may be held for indefinite periods of time at temperatures below 100 °C. *P. lobata* starch and gels may be sterilized at 121 °C under one atmospheric pressure. Upon boiling, *P. lobata* gels do not liquefy, only releasing water. Upon repeated freezing and thawing or warming and cooling, *P. lobata* starch gels do not permit the separation or crystallization of liquids. *P. lobata* starch and its gels impart a faint sweet taste to the tongue and dissolve rapidly.

Unlike most other gelling agents, *P. lobata* starch gels set up rapidly while hot and retain their shape and volume warm, at room temperature, chilled, or frozen for an indefinite period much longer than other gelling agents. Its gels can be composed for

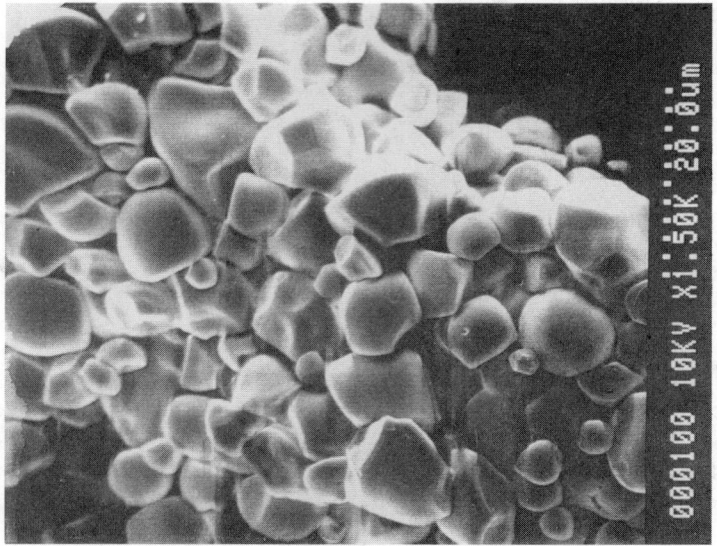

*Figure 16.1 Pueraria lobata* starch granules viewed by scanning electron microscopy (Copyright RHIZOMA CORP).

extrusion, cast and molded, used as a coating, and sprayed under pressure. *P. lobata* starch and its gels are shear resistant and lubricant.

*P. lobata* starch gels remain stable and will not dissolve in liquids under stress conditions for indefinite periods longer than other gelling agents. Its starch can be used to gel foods and medicines, including uncooked pineapple, kiwi, and papaya that contain high concentrations of proteases that prevent their gelling by proteins or with any other useful gelling agents until they have been cooked or the proteases otherwise broken down.

## BY-PRODUCT RHIZOME CELLULOSE FIBER

*P. lobata* rhizome cellulose fibers are as much as three inches long, which is unusual among natural fibers This cellulose has been used for centuries in the Orient as a source of raw material for making specialty papers. Because the fibers are so long and strong, they may be used to make very strong paper known as "acid free," requiring no chemical treatment to increase its strength. "Acid free" paper products are preferred for many uses because they last longer and do no damage to inks, dyes, paints, textiles and other materials with which they may be in contact over long periods. Moreover, they are cheaper and easier to manufacture and friendlier to the environment.

Typically, paper is made of pulpwood trees, often farmed specifically for papermaking purposes. Pulpwood tree farming is damaging to the environment and causes serious problems for human and animal life as a result. While it might not be economical to grow *P. lobata* as a field crop solely for papermaking, it will be extremely economical to use the by-product cellulose fibers from *P. lobata* rhizomes and vines from the production of starch, textiles, and chemicals for papermaking.

*Figure 16.2* Fine *Pueraria lobata* fibers viewed by light microscopy (Copyright RHIZOMA CORP).

*P. lobata* is a readily renewable crop, cheaper and easier to produce than trees. It is clearly a more economic crop for papermaking than trees. After starch, textile and chemical production, 100 000 acres of wild *P. lobata* growing in the Southeast United States, will produce approximately 250 million pounds (125 000 tons/year) of by-product cellulose fiber suitable for papermaking. The average modern paper mill operating in the United States consumes 800–1200 tons of wood pulp everyday. Clearly, *P. lobata* cellulose is a more valuable and less expensive raw material for papermaking than the usual wood pulp.

*P. lobata* rhizome cellulose is more absorbent than many other synthetics and natural products. It may be useful for cleaning up chemical and oil spills in industrial sites and/or in the environment. Alone or combined with other materials, it may be composed into materials to absorb specifically over time or to absorb rapidly and selectively.

## FINE FIBERS FOR TEXTILES

Approximately one-third of the new vines at the crown of wild *P. lobata* contain fine fibers (see Figure 16.2) useful for making a translucent cloth similar to silk but about the weight of linen. *P. lobata* fibers possess most of the properties of silk and nylon and may be used in most of the same ways. Further, *P. lobata* fibers have an advantage over silk and nylon because they are absorbents. Silk is a protein and has little absorbency. Nylon is synthetic material not composed for absorbency. Cultivated *P. lobata* plants may be engineered to produce high contents of these fine fibers and to produce two crops of fine fibers annually.

For at least 4000 years, the Chinese have woven *P. lobata* fibers into textile products, including their most important ceremonial garments. The first recorded use of *P. lobata* fibers for clothing was in China where a ceremonial robe made of these fibers was presented as a gift to the Emperor in the year 800 BC.

Today, a limited number of bobbins having four ounces of fine *P. lobata* thread ready for spinning are available for sale in Japan. These fibers are produced from vines hand-harvested in the spring from the wild. The fibers are then retted by hand using anti-quated methods of simmering the vines, soaking them in a caustic soda solution, and repeatedly washing them to wash away the sheath of the vine, leaving only fine fibers behind. This method results in approximately one pound of fine fiber retted from each fifty-two pounds of wild vines. The resulting fibers are spun on individual hand-operated spinning wheels and the bobbins of *P. lobata* thread are then sold for cottage type weaving to produce very fine cloth in small quantities.

*P. lobata* fibers are retted using industrial methods, including enzymatic hydrolysis and steam, or naturally field retting, similar to the retting of linen from flax (Eiamwat *et al.*, 1998; Uludag *et al.*, 1996b). *P. lobata* vines may grow as long as 100 feet in one growing season. The finest fibers for textiles are taken from the new growth at the crown of the plant in the spring, which comprises approximately one-third of the vines from a crown. Harvested at the optimum time, these fine fibers will be approximately thirty feet long. By comparison, the longest known cotton fibers are at best six inches long.

While *P. lobata* fibers are very long and fine, they have irregularities resulting from the nodes along the vine, similar to the irregularity found in silk fibers. This makes them valuable for certain textile fads and fashions, but difficult to weave using modern Textile Industry mass production methods and techniques. With the advent of nylon, it was expected that synthetic fibers would replace most natural fibers for most uses. As a result, little effort had been made to improve their agricultural production for textile markets. However, synthetics have not replaced natural fibers for most markets as had been anticipated. Instead presently the largest growth in many textile markets is in natural fibers, either alone or blended with synthetics.

In the past, fiber irregularity had made fine *P. lobata* fibers unsuitable for weaving on large modern power looms that operate at high speeds. Such looms depend upon fiber regularity for their speed and standardized mass production. However, modern cultivation and harvesting methods of American Agricultural Industry can be used to grow vines with very long sections between the nodes or with even spaces between the nodes. Such fibers may be cut and harvested in application-specific lengths at harvest, retted, spun and woven using the power equipment standard to the mass production demands of the American Textiles Industry.

Unlike cotton, which must be ginned and then spun for weaving, *P. lobata* fibers must be retted with methods similar to those used for linen. Next, *P. lobata* fibers must be combed and carded just as wool is combed and carded for spinning. There is no established system for retting harvested *P. lobata* vines in the Southeast United States similar to the system of cotton gins already in place. The modern spinning mills and looms of the South require short fibers such as cotton. They are not adaptable to the long *P. lobata* fibers. Instead, it is the great looms and mills for linen and wool in the Northeast United States, Canada and Europe that are readily adaptable to *P. lobata* textile production.

Fine *P. lobata* fibers are very strong and have great tensile strength. Blended with wool, silk, linen, cotton, synthetics and alone, they are most valuable for their strength and absorbency. Not only is the strength of fiber important to the strength of fabric, but also it is the length of fiber that imparts most of the strength to textiles. Using *P. lobata* fibers as the Warp (length) and other natural fibers as the Weft (width), it is likely possible to produce a strong natural fabric that may be expected to have as much as several times the tensile strength of most natural fiber textiles.

P. *lobata* fibers are more water absorbent than silk, nylon and similar fibers because they are composed of cellulose. Silk fibers are composed of proteins. Most proteins are not absorbent. *P. lobata* fibers are generally more absorbent than cotton and many other natural and synthetic fibers. By weaving *P. lobata* fibers very closely, it is possible to make a waterproof fabric similar to, but much stronger than, canvas. Such fabrics used for clothing may be woven to insulate, letting perspiration out without permitting water or air in. They may also be used for protection from the environment, for sails for ships, for parachutes and for similar products.

## COARSE FIBER OF KUDZU VINES

Coarser fibers can be extracted from the remains of *P. lobata* vines after fine fiber extraction. They are useful for woven floor matting, cordage, and for agricultural and structural support. They are an excellent source of cellulose for papermaking and can be cut into specific lengths to achieve application specific results. They may also be ground for rayon making. Because they have great tensile strength and are as much as 100 feet long, they had been used in the past to build suspension bridges in the Orient.

In Korea, combinations of *P. lobata* fibers are used to make small quantities of what is known as "grass cloth" wall covering. *P. lobata* grass cloth is made by hand, from hand-harvested and separated fibers, using *P. lobata* starch as an adhesive and textile finish to stiffen and glue the loosely woven split fibers and strips of woven outer sheath to rice paper. Grass cloth made of more traditional raw materials is widely exported to the United States and Europe from Korea and other eastern countries.

## FINE CHEMICALS FROM KUDZU

Among the members of the Genus *Pueraria*, it is *Pueraria lobata* that contains the largest amount of tannins and other fine chemicals. *Pueraria lobata* produces very large amounts of simple phenolics (tannins) and highly complex, condensed polyphenolics, having a very high molecular weight. Not all leguminous plants produce tannins. However, legumes that are highly evolved, perennial, deciduous, dicotyledons often contain large quantities of these types of chemicals relative to their biomass. This is the case with *Pueraria lobata*.

Generally, plants that are perennial and have been forced to develop their defenses against numerous enemies over long periods of evolution contain the greatest amounts and largest numbers of chemical substances. Phytochemicals are used by plants to enhance their survival. Tannins are simple phenols and polyphenols, which bind to proteins specifically. These chemicals are believed to protect the plant from digestion by herbivores, among their benefits to the host plant. They are also believed to direct the conversion of glucose, the primary product of photosynthesis. Tannins are found in all parts of plants and have been found to perform a variety of specific defensive tasks in plants.

*P. lobata* is among the most highly evolved plants, having developed a sophisticated defense system, permitting it to grow successfully under a wide variety of adverse conditions, including in soil contaminated with pollutants, in heavy air pollution, in acid rain, in bogs and semi-arid climatic conditions. It is difficult to eradicate using herbicides because it isolates and takes dormant rhizome segments affected by chemicals in the ground until the contaminants have dissipated in the soil before resprouting. *P. lobata* rhizomes are not damaged by nematodes and other pests found living in the soil in the Southeast United States. Few other plants can co-exist with *P. lobata* in the Southeast. Those that grow along with it on the same site are usually in limited amounts, generally along the fringes of a field.

All parts of a *P. lobata* plant contain valuable chemicals. Often, the same chemicals found in different parts of the plant exist in different forms. Whilst chemical extraction from *P. lobata* appears best made from its rhizomes, the leaves, vines, seeds and flowers may also be used for chemical extractions. Chemical content varies with seasons. Chemical content of *P. lobata* growing in the United States appears to be at its highest and most concentrated level during late summer to early fall.

The largest single chemical component of *P. lobata* rhizomes appears to be coumarin. This powerful anti-coagulant is the primary ingredient of human anti-coagulant drugs and also the primary ingredient of rat poison. Coumarin is also used as a newly mown hay scent in cheese, cosmetics and perfume and as a beige to tan and yellow dye, including as a laser marker dye. Among flavors found in *P. lobata*, the major component appears to be palmitate.

Chemicals found in *P. lobata* can be roughly separated into two major groups: (1) those that dissolve in water (also soluble in methanol, ethanol and/or butanol) are usually polar compounds of small molecular weights, and (2) those that dissolve in organic solvents such as acetone, chloroform, etc. are usually larger, featuring additional side groups or attached sugars. In general, the skeletons of the water-soluble compounds can also be found in a more complex form in the water-insoluble fraction of *P. lobata* extracts. Usually, smaller compounds can be obtained by chemical synthesis and mass-produced in factories at a cost much lower than that by the cultivation of and isolating from the natural plant. However, larger compounds with more complex structures can be difficult to make and isolation from natural source may be preferred.

*P. lobata* contains large amounts of polyphenolics. Unlike simple phenols, polyphenols have complex chemical structures, may be extremely condensed, and have high molecular weights. Thus, they may be difficult and costly to synthesize chemically. A natural substance may be preferred for that reason. Certain flavonoids can be chemically modified to yield more potent drugs. It is likely that extractions of natural chemicals from *P. lobata* may serve as the starting point for the discovery of more potent, commercially useful, therapeutics.

*Pueraria lobata* leaves can be separately used for the production of phytochemicals. Inherent chemicals in *P. lobata* leaves have been found to be unique and to differ from the chemicals in the rhizome. Among others, chemicals found in *P. lobata* leaves include a tobacco flavoring and others identical to certain food flavoring components. The leaves also contain a high level of chlorophyllin, a chlorophyll derivative that is able to inhibit the action of certain mutagens.

*P. lobata* contains volatile chemicals as indicated by the sweet grape-like scent of its flowers (Kinjo *et al.*, 1988a). Rhizome and other components of *P. lobata* contain volatile compounds as well. The chemicals in the flowers are not very different from those in the rhizome. The vines contain most of the same chemicals found in the rhizome and some chemicals that have not yet been found in the rhizome. It is likely that most, if not all, of the chemicals found in *P. lobata* will be found in its rhizomes.

Historically, *P. lobata* has been used in the Orient as a medicinal herb and food supplement. Koreans are fond of a *P. lobata* tannin tea as a healthful tonic. Small quantities of this tea are exported to the United States from Korea, usually flavored with orange or lemon and spices such as cinnamon and clove then highly sweetened with refined sugar. Japanese offerings exported to the United States tend to rely most upon starch-based recipes for foods, including a "kuzu crème" for dessert. Chinese most often describe the use of decoctions made by steeping the dried *P. lobata* rhizome for medicinal treatments, frequently describing it in mixtures also including peony and Siberian ginseng extracts.

Historic Chinese and Japanese medicine each reports extensive successful use of *P. lobata* extracts to treat a variety of ills. Among these are skin diseases, baldness, insect bites, digestive problems, strokes, drunkenness, and hangover after drinking alcoholic beverages, alcoholism, liver, and heart and kidney diseases. Their skin treatments from *P. lobata* are reported to be useful for moisturizing, skin-lightening, sun-screening, stimulating hair growth, preventing the formation of melanin deposits (age spots), treating and easing the symptoms of wounds, burns, insect bites, rashes, allergic skin reactions and similar conditions.

More recently, modern scientific methods have been used to study the pharmacological activity of these treatments, isolate and identify their active ingredients, and reveal their mechanism of actions. Among the biological activities reported in animals using *P. lobata* extracts are ones which are anti-spasmodic, (spasmolytic, muscle relaxing, musculatropic); anti-oxidative; anti-pyretic (hypothermic); cholinergic; anti-dyarhythmic; analgesic; anti-fertility; vasodilative (hypertension); anti-inflammatory (Kimura *et al.*, 1992); immune-stimulating; fungitoxic; herbicidal; anti-tuberculin; cardiovascular; anti-coronary artery disease; anti-mutagenic; anti-coagulant (Yu *et al.*, 1997); anti-alcohol abuse (Keung *et al.*, 1997); anti-liver intoxication/injuries (Nohara *et al.*, 1998; Shinho *et al.*, 1989); otolaryngologic (Xuan *et al.*, 1999); opthalmic; anti-hypoglycemic; anti-diabetic; preventive of melanin formation; edema preventive, urological (Yasuda *et al.*, 1995); anti-viral (Kumai, 1996; Nohara and Kinjo, 1999); neurological (Shen *et al.*, 1996; Oishi *et al.*, 1998); and others.

Among chemicals already identified in *Pueraria lobata* are the following: Chlorophyll; Chlorophyllin; Coumarin; Coumestrol; Puerarin; Biochanin-A; Genistin; Genistein; Daidzin; Daidzein; Formononetin; Oononin; Tectorigenin; Glycitein; Chryin; Puerarols 1, 2, 3, and 6 Puerasol; Glycosyl-tryptophan; Rutin; Robinin; Nicotiflorin; Kakalide; Kakatin; Palmitate; di-methyl-Sucherate; di-methyl-Azelate; Acetylcarbinol; Paenol; Furfural; Furfuryl alcohol; 2-Furfuryl-methyl-ketone; Isoliquirtigen; Allantoin; B-sitostrerol; 5-Methyl-hydantoin; Ephedra; 7-O-eanenne sapogenols, including *P. lobata* sapogenol, *P. lobata* sapogenol A, *P. lobata* sapogenol B, and *P. lobata* sapogenol methyl ester. Among sugar moieties in *Pueraria lobata* rhizomes, glucose, xylose and other sugars have been identified (Kubo *et al.*, 1973, 1975; Yokoyama, 1976; Sayed and Borisov, 1978; Shibata *et al.*, 1978; Inatomi *et al.*, 1979; Sajad *et al.*, 1979; Yamagishi and Houma, 1980; Chen and Zhang, 1985; Kinjo *et al.*, 1985, 1987, 1988b; Miyazawa and Kameoka, 1988; Oshima *et al.*, 1988; Hirakura *et al.*, 1989; Park *et al.*, 1992; Nohara *et al.*, 1993; Rong and Stevens, 1998).

## SUMMARY

All parts of the *P. lobata* plant contain valuable components that may be suitable for commercial application and production, making it a potentially valuable commercial crop as well. Fine food and specialty starch is found in the rhizome; very long, absorbent cellulose fiber suitable for paper-making is found in the rhizome; very long, very strong fine fibers suitable for textiles are found in the vines along with coarser fibers useful for structural support, cordage, flooring or paper-making; valuable fine chemicals are found in good supply and in all parts of *P. lobata*. All components of *P. lobata* plant are edible. Its leaves are an outstanding food, having similar properties to spinach, collards and other rich table greens.

*P. lobata* requires large space for commercial production in agriculture. Space is readily available in the Southeast United States. American Agriculture industry methods and techniques are the most sophisticated in the World. It is likely that America can rapidly produce a commodity type *P. lobata* crop for textiles, starch, chemicals and papermaking within no more than 7 years. Commercial agricultural and industrial production of *P. lobata* in the Southeast United States will produce a valuable cash crop for United States farmers to improve the balance of World Trade. Moreover, a relatively small area of less than 100 000 acres under cultivation is sufficient to produce a crop for natural drug and other natural chemical production. *P. lobata* is an under exploited natural resource. Its development as a crop for food, textiles, chemical and other industrial products as well as for papermaking should be encouraged.

## REFERENCES

Achremowicz, B., Tanner, R.D., Prokop, A. and Grisso, R.D. (1994) The effect of frost on the starch yield from kudzu (*Pueraria lobata*) roots grown in Northern Alabama. *Bioresour. Technol.*, 46, 149–151.

Achremowicz, B. and Tanner, R.D. (1996) Studies on starch isolated from kudzu roots (*Pueraria lobata* Willd.). *Pol. J. Food Nutr. Sci.*, 5(3), 63–71.

An, W., Xia, G. and Guo, R. (1999) Comparison study on total flavone contents in *Pueraria lobata* Ohwi and *P. thomsonii* Benth. grown in different areas. *Zhongguo Zhong. Za zhi*, 24(6), 339, 380.

Chen, M. and Zhang, S. (1985) Studies on the chemical constituents of *Pueraria lobata*. *Zhong. Tongbao*, 10(6), 274–276.

Eiamwat, J., Loha, V., Prokop, A. and Tanner, R.D. (1998) Batch foam fractionation of kudzu (*Pueraria lobata*) vine retting solution. *Appl. Biochem. Biotechnol.*, 70–72, 559–567.

Hirakura, K., Nakajima, K., Sato, S. and Mihashi, H. (1989) Isolation of 7-(6–0-malonyl-beta.-D-glucopyranosyloxy)-3-(4-hydroxyphenyl)-4H-1-benzopyran-4-one from *Pueraria lobata* Ohwi as aldose reductase inhibitors and pharmaceutical formulations. *Jpn. Kokai Tokkyo Koho* (*Japanese Patent*), 6 pp.

Inatomi, H., Ikawa, M., Kawamura, S. and Suyama, Y. (1979) The constituents of *Pueraria lobata* OHWI. II. Quantitative determination of salicylic acid. *Meiji Daigaku Nogakubu Kenkyu Hokoku*, 47, 5–10.

Keung, W.M., Klyosov, A.K. and Vallee, B.L. (1997) Daidzin inhibits mitochondrial aldehyde dehydrogenase and suppresses ethanol intake of Syrian golden hamsters. *Proc. Nat. Acad. Sci.*, *USA*, 94, 1675–1679.

Kimura, M., Kimura, I., Guo, X., Luo, B. and Kobayashi, S. (1992) Combined effects of Japanese-Sino medicine Kakkon-to-ka-senkyu-shin 'i' and its related combinations and component drugs on adjuvant-induced inflammation in mice. *Phytother. Res.*, 6(4), 209–216.

Kinjo, J., Furusawa, J. and Nohara, T. (1985) Two novel aromatic glycosides, pueroside-A and -B, from Puerariae radix. *Tetrahedron Lett.*, 26(49), 6101–6102.

Kinjo, J., Kurusawa, J., Baba, J., Takeshita, T., Yamasaki, M. and Nohara, T. (1987) Studies on the constituents of *Pueraria lobata*. III. Isoflavonoids and related compounds in the roots and the voluble stems. *Chem. Pharm. Bull.*, 35(12), 4846–4850.

Kinjo, J., Takeshita, T., Abe, Y., Terada, N., Yamashita, H., Yamasaki, M., Takeuchi, K., Murakami, K., Tomimatsu, T. and Nohara, T. (1988a) Studies on the constituents of *Pueraria lobata*. IV. Chemical constituents in the flowers and the leaves. *Chem. Pharm. Bull.*, 36(3), 1174–1179.

Kinjo, J., Takeshita, T. and Nohara, T. (1988b) Constituents of *Pueraria lobata*. V. A tryptophan derivative from *Puerariae flos. Chem. Pharm. Bull.*, 36(10), 4171–4173.

Kubo, M., Fujita, K., Nishimura, H., Naruto, S. and Namba, K. (1973) New irisolidone-7-O-glucoside and tectoridin from Pueraria species. *Phytochemistry*, 12(10), 2547–2548.

Kubo, M., Sasaki, M., Namba, K., Naruto, S. and Nishimura, H. (1975) Isolation of a new isoflavone from Chinese Pueraria flowers. *Chem. Pharm. Bull.*, 23(10), 2449–2451.

Kumai, S. (1996) Multilayered pharmaceutical troches for common cold. *Jpn. Kokai Tokkyo Koho (Japanes Patent)*, 3 pp.

Lynd, J.Q. and Ansman, T.R. (1990) Exceptional forage regrowth, nodulation and nitrogenase activity of kudzu (*Pueraria lobata* (Willd.) Ohivi) grown on eroded Dougherty loam subsoil. *J. Plant Nutr.*, 13(7), 861–885.

Miyazawa, M. and Kameoka, H. (1988) Volatile flavor components of crude drugs. Part IV. Volatile flavor components of *Puerariae radix* (*Pueraria lobata* Ohwi). *Agric. Biol. Chem.*, 52(4), 1053–1055.

Nohara, M. Takeshita, T., Kaneshiro, J., Ito, H., Niiura, Y. and Yamazaki, T. (1988) Therapeutic effects of chemical constituents of flos puerariae on experimental liver injuries. *Wakan Iyaku Gakkaishi*, 5(3), 408–409.

Nohara, T., Kinjo, J., Furusawa, J., Sakai, Y., Inoue, M., Shirataki, Y., Ishibashi, Y., Yokoe, I. and Komatsu, M. (1993) But-2-enolides from *Pueraria lobata* and revised structures of pueroosides A, B and sophoroside A. *Phytochemistry*, 33(5), 1207–1210.

Nohara, T. and Kinjo, J. (1999) Leguminous glycosides effective for hepatitis. *Studies in Plant Sciences*, 6 (Advances in Plant Glycosides, Chemistry and Biology), 131–145.

Ohshima, Y., Okuyama, T., Takahashi, K., Takizawa, T. and Shibata, S. (1988) Isolation and high performance liquid chromatography (HPLC) of isoflavonoids from the Pueraria root. *Planta Medica*, 54(3), 250–254.

Oishi, M., Mochizuki, Y., Takasu, T., Chao, E. and Nakamura, S. (1998) Effectiveness of traditional Chinese medicine in Alzheimer disease. *Alzheimer Dis. Assoc. Disord.*, 12(3), 247–250.

Park, H.H., Hakamatsuka, T., Noguchi, H., Sankawa, U. and Ebizuka, Y. (1992) Isoflavone glucosides exist as their 6″-O-malonyl esters in *Pueraria lobata* and its cell suspension cultures. *Chem. Pharm. Bull.*, 40(7), 1978–1980.

Rong, H. and Stevens, J.F. (1998) Identification of isoflavones in the roots of *Pueraria lobata*. *Belgica Planta Medica*, 64/7, 620–627.

Sajad, S.A., Borisov, M.I. and Kovalev, V.N. (1979) Isolation, identification and quantitative determination of robinin in the leaves of *Pueraria lobata*. *Farmakologhicheskii Zhurnal* (Kiev), 4, 52–55.

Sayed, S.A. and Borisov, M.I. (1978) Flavonoids of *Pueraria lobata* flowers. *Farmakologhicheskii Zhurnal* (Kiev), 6, 83–84.

Shen, X., Witt, M.R., Nielsen, M. and Sterner, O. (1996) Inhibition of [3H]flunitrazepam binding to rat brain membranes in vitro by puerarin and daidzein. *Yaoxue Xuebao*, 31(1), 59–62.

Shibata, S., Katsuyama, A. and Noguchi, M. (1978) On the constituents of an essential oil of Kudzu. *Agric. Biol. Chem.*, 42(1), 195–197.

Shinho, J., Yamazaki, R., Nohara, T., Kaneshiro, Y., Nakajima, K. and Ito, H. (1989) Flavonoids, saponins and glycoside thereof for improvement of urea nitrogen metabolism. *Jpn. Kokai Tokkyo Koho (Japanese Patent)*, 6 pp.

Tanner, R.D. and Hussain, S.S. (1979) Kudzu (*Pueraria lobata*) root starch as a substrate for the lysine-enriched baker's yeast and ethanol fermentation process. *J. Agric. Food Chem.*, 27(1), 22–27.

Uludag, S., Loha, V., Prokop, A. and Tanner, R.D. (1996a) The effect of fermentation (retting) time and harvest time of kudzu (*Pueraria lobata*) fiber strength. *Applied Biochemistry and Biotechnology*, 57/58 (Seventeenth Symposium on Biotechnology for Fuels and Chemicals, 1995), 75–84.

Uludag, S., Prokop, A. and Tanner, R.D. (1996b) A kinetic study of the kudzu (*Pueraria lobata*) retting process. *J. Sci. Ind. Res.*, 55(5 & 6), 381–387.

Xuan, B., Zhou, Y.-H., Yang, R.-L., Li, N., Min, Z.-D. and Chiou, G.C.Y. (1999) Improvement of ocular blood flow and retinal functions with puerarin analogs. *J. Ocul. Pharmacol. Ther.*, 15(3), 207–216.

Yamagishi, T. and Houma, S. (1980) Quantitative determination of Ephedra alkaloids in Chinese medicinal preparations by high-speed liquid chromatography. *Hokkaidoritsu Eisei Kenkyushoho*, (30), 6–9.

Yasuda, T., Ohtsubo, S., Kano, J. and Ohsawa, K. (1995) Fecal metabolites after oral administration of major components of Puerariae Radix in rats. *Annu. Rep. Tohoku Coll. Pharm.*, 42, 191–198.

Yokoyama, Y. (1976) On the flower color and anthocyanin of Pueraria lobata Ohwi. *Miyagi-Ken Nogyo Tanki Daigaku Gakujutsu Hokoku*, 23, 53–56.

Yu, Z., Zhang, G. and Zhao, H. (1997) Protection of Pueraria isoflavone against cerebral ischemia. *Zhongguo Yaoke Daxue Xuebao*, 28(5), 310–312.

Zhong, Y., Ding, X., Zuo, C. and Lan, J. (1992) Quantitative determination of puerarin in lobed kudzu vine (*Pueraria lobata*) collected at different seasons by HPLC. *Zhongcaoyao*, 23(6), 294–295.

# 17 Friend or foe? Changing cultural definitions of kudzu

*Kathleen S. Lowney*

## INTRODUCTION

When experts talk, people usually listen. In particular, scientists claim to offer the most authoritative interpretations of the natural world, and while scientific constructions can be challenged (e.g. the creationism–evolution debate), scientific claims usually receive very respectful hearings. This chapter tells the story of a time in America when scientific authorities spoke and lay people listened. But instead of a problem being solved, things went from bad to worse. This paper tells the ironic story of a plant becoming seen as a terrible foe, despised by most citizens who know it.

The story starts in the beginning of the twentieth century, when decades of cotton farming, sharecropping, natural disasters, and drought had caused severe soil erosion in the American South. Agricultural scientists advocated a simple solution: planting *Pueraria lobata*, better known as kudzu. The plant was lauded repeatedly as friendly vegetation and farmers, often despite doubts, listened to the experts and to government officials who championed the plant. Today, drivers on Southern highways can see the consequences of that scientific judgment – kudzu growing out of control, enveloping trees, telephone poles, even buildings. Kudzu the agricultural friend became kudzu the foe, an ecological problem which created severe economic burdens both for Southern farmers and foresters. In time, still more scientists spoke up, redefining kudzu as a severe problem demanding its own solution – the plant's eradication. But kudzu's history is even more complicated: in the last 15 years, new scientific claimsmakers have begun promoting kudzu as a solution to still other problems. In another fascinating historical twist, the green plant, once an ally, then an enemy, is on its way to becoming seen, yet again, as friendly and useful.

After a brief methods section, I discuss scientists' claims about kudzu during three periods: an enthusiastic period in which agronomists promoted kudzu planting (1917–1953); a period of disenchantment when foresters and other experts sought ways to eradicate kudzu (1954–1984); and a period of tempered enthusiasm when applied scientists began exploring new uses for kudzu (1985–present). Then I turn to analyzing how scientific claims have been heard by the American public, noting in particular how the public remains convinced that *Pueraria lobata* is still an enemy, invading more and more of the nation's territory. Thus there is a gap between scientific and lay perceptions of this plant.

## METHODS

I collected a large sample of English-language scientific and popular publications about kudzu.[1] Using several electronic databases – Agricola, Applied Science and Technology Abstracts, Biological and Agricultural Index, Biology Digest, Cambridge Scientific Abstracts, Current Contents, Dissertation Abstracts, Environment, General Science Abstracts, Geobase, CARL, and the General Academic Index – as well as such traditional print indices such as *Readers' Guide to Periodical Literature*, I searched under both the English and Latin names for the plant, and examined the bibliography in each located source for new references. I identified over 300 English-language sources published since 1917, and examined 110 (108 articles and two masters' theses). Since published research about kudzu has increased tremendously since 1985, it was impossible to read each article, but I have sampled approximately 31 per cent of this more recent literature (sixty articles).

I coded each source for whether it discussed kudzu in positive or negative terms, using five coding categories: (1) completely positive comments, (2) mostly positive but the article contained some negative comments, (3) mostly negative comments but with some positive ones as well, (4) completely negative comments, and (5) a residual category (articles which did not discuss kudzu in either positive or negative terms). As Table 17.1 demonstrates, assessments of kudzu varied over time: during the early (1917–1953) period, positive comments predominated; from 1954 to 1984, most comments were negative; while more recent commentary has been mixed.

*Table 17.1* Positive and negative assessments of kudzu in the scientific literature by time period

| Kind of comment | Time period | | |
| --- | --- | --- | --- |
| | *Enthusiastic*[a] (1917–1953; %) | *Disenchantment*[a] (1954–1984; %) | *Tempered enthusiasm*[b] (1985–current; %) |
| Completely positive | 73.0 | 0.0 | 17.5 |
| Mostly positive but some negative | 15.0 | 18.0 | 26.2 |
| Mostly negative but some positive | 3.0 | 6.0 | 19.3 |
| Completely negative | 0.0 | 65.0 | 32.0 |
| No comments | 9.0 | 11.0 | 5.0 |
| *n* | 33 | 17 | 60 |

Notes
a  All articles found in these two periods were read and categorized.
b  Sixty articles (31%) were sampled and categorized.

1  This chapter is adapted from Lowney, K.S. and Best, J. (1998) Floral entrepreneurs: kudzu as agricultural solution and ecological problem. *Sociological Spectrum*, 18, 89–110. When invited to write this chapter, I updated my research, focussing on recent articles on kudzu. That data have been added to the discussion about the period of tempered enthusiasm, 1985–present. I want to thank Joel Best for his work on the earlier article and his advice on this one and my husband, Frank Flaherty for his help and patience. I have also benefited from two other social scientific articles about kudzu: A vine for postmodern times: an update on kudzu at the close of the twentieth century. *Southeastern Geographer*, 37, 167–179 and Stewart (1997) Cultivating kudzu: the Soil Conservation Service and the kudzu distribution program. *The Georgia Historical Quarterly*, 81, 151–167.

## THE ENTHUSIASTIC PERIOD (1917–1953)

In 1876, the Japanese Pavilion at the Philadelphia Centennial Exposition exhibited kudzu, and it reappeared at the New Orleans Exposition in 1883 (Shurtleff and Aoyagi, 1985). Both exhibits depicted the plant as a decorative climbing vine appropriate for warm climates, and a few householders throughout the American South began cultivating kudzu. Over time, its use shifted from ornamental shading, to being used primarily for pasturage (Winberry and Jones, 1973). In 1917, the first published article gave kudzu a mixed review:

> ...the optimistic tone [that] has been adopted by a number of Southern seed dealers in their advertising matter, [shows] some lack of restraint and a resulting tendency to exaggerate the possible advantages of the crop and minimize its probable... disadvantages...the practical farmer tends to class kudzu as a worse pest than Johnson or Bermuda grass...the far-sighted farmer will probably leave it alone until it has been more thoroughly acclimated and reformed so that it can economically be harvested for hay or ensilage (Dacy, 1917).

However, such caution did not last. Three years later, the first agricultural circular about kudzu was almost completely enthusiastic:

> It grows with remarkable rapidity...Kudzu is a most excellent vine for arbors and porches...Kudzu is excellent for soiling, as was shown by the experience of the Louisiana Agricultural Experiment Station. During an extremely dry period the only green forage available was furnished by the kudzu fields...Kudzu is very nutritious...Horses, cows, and sheep eat the green leaves readily, as well as the hay (Piper, 1920).

Although the circular recommended first "...make an experimental planting of a small area" (Piper 1920: 6), it did not specify kudzu's troublesome properties. Four years later, only one other negative comment about kudzu had appeared in print: "The most serious disadvantage of the crop is the difficulty of cutting and handling the hay" (Funchess and Tisdale, 1924). Instead, agricultural publications repeatedly highlighted the plant's positive benefits – both agricultural and financial:

> There is a need for a perennial forage plant which will produce large yields of hay, which is adapted to grazing by livestock, and which is sufficiently drouth-resistant to produce high yields when other crops fail. Kudzu has shown more promise of meeting these requirements than any other plant now being grown in Alabama.... The remarkable growth of kudzu on land too poor and rough for other crops, together with its high feeding value, shows beyond a doubt that there is a place for this crop on Alabama farms.... The possibilities of kudzu on thousands of acres of similar land in Alabama are almost unlimited (Bailey & Mayton, 1931; see also Pierre & Bertram, 1929).

In the 1930s, articles began noting farmers' growing concern about kudzu. The experts' articles did not address the substance of these concerns; rather, they attacked the farmers' anxieties. Several articles published during the 1930s contained – almost word for word – a final section, entitled "Kudzu as a Pest," designed to calm the lay public. This wording first appeared in an Agricultural Experiment Station of Alabama circular, co-authored by R.Y. Bailey (who is better known as the "Father of Kudzu"):

> One of the chief reasons why kudzu has not been grown more generally in Alabama was the prevailing idea that this crop was a dangerous pest. Farmers were told by some agricultural workers that if this plant were allowed to become established in cultivated fields it would be impossible to eradicate it. They were also warned against it on account of the possibility of its spreading to fields where it was not desired. As evidence that this warning is unfounded, kudzu on a small area at Auburn gave no trouble on the adjoining cultivated fields over a period of more than 25 years... close grazing followed by plowing will eradicate kudzu. (Bailey & Mayton, 1931).

Obviously, the authors – as the state's leading agricultural experts – expected that their words would assuage worries. Their reports carried the weight of scientific research, of over two decades of studying kudzu. Bailey continued to promote the plant:

> ...most farmers were rather strongly prejudiced against [kudzu]. They thought of it as an aggressive plant that would spread very rapidly and become a dangerous pest if planted near cultivated fields.... Agronomists in the [Soil Conservation] Service and at agricultural colleges, knew from direct experience with the plant that its habits of growth made it easy to control. They knew that, where kudzu was planted on land adjacent to cultivated areas, the plowing necessary for the production of row crops would prevent any undesirable spread to cultivated areas (Bailey, 1939).

Nor was Bailey the only agronomist attempting to debunk "misinformation" about kudzu. A Mississippi Agricultural Experiment Station Bulletin criticized the belief that kudzu:

> spreads and covers the place and can not be killed. It is true that it is an aggressive plant, but this Station keeps it confined to individual plots without any serious trouble. It is easily killed by grazing or frequent and thorough plowing (Miles & Gross, 1939; see also Sturkie & Grimes, 1939).

These experts not only criticized less-educated farmers for worrying about kudzu, but they presented extensive experimental results illustrating the beneficial roles kudzu could play in Southern agriculture: building soil; replenishing nitrogen; producing moisture-retaining mulch; solving erosion problems; offering protection from silting; stabilizing gullies and banks; producing higher hay yields; unaffected by drought, and rarely susceptible to disease or killed by frost, kudzu did not require cutting at specific times, and cows, chicks, and other animals fed kudzu gained more body weight than those raised on grass pastures or other feed, while corn and oats grown on land previously planted with kudzu produced significantly higher yields, often for several years (Kudzu in South Carolina, 1938; Miles and Gross, 1939; Sturkie and Grimes, 1939; McKee and Stephens, 1943; Polk and Gieger, 1945).

By the 1940s, the agronomists' campaign had been successful. Kudzu was being planted by private farmers and, more importantly, it was being planted at the behest of government agencies, such as the soil conservation service (SCS). "This past planting season the [SCS] nurseries supplied more than 30 million seedlings to farmers in 82 soil conservation districts in the south-east – enough to plant 60 000 acres. The previous year 23 million seedlings were produced, and in 1938 a total of 13 million" (Tabor and Susott, 1941). By the mid-forties, 500 000 acres of kudzu had been planted (McKee and Stephens, 1943). Government fiscal policy fostered planting; planting kudzu for conservation purposes, "entitled the farmer to assistance payments as high as $6.00 to $8.00 per acre" (Winbery and Jones, 1973).

By all accounts, kudzu had become a worthwhile addition to the Southern landscape (Alexander, 1950; Davis and Young, 1951; Porter, 1953; Stephens, 1953). It had proven its economic value, especially during drought conditions in the early 1950s: "... while other crops, notably corn, withered and died under the blistering sun, kudzu was almost startling in its greenness and rank growth. It was entirely unaffected! That one fact has pushed many kudzu doubters over to the side of 'progress'" (Stephens, 1953). No longer a rare, decorative plant, kudzu had become a multi-purpose component of Southern agriculture. Its popularity was largely due to positive claims about kudzu made by Southern agronomists. Early cautions were largely forgotten or explained away as the result of inexperience with kudzu. Still, as the period of enthusiasm drew to a close, even kudzu's most fervent advocates began to note – however subtly – the plant's problematic side. Often the subject was broached through comedy; one article ended: "If you must plant it, put it on the backside of your farm and maybe you'll have a chance to escape with your family before it reaches your home!" (Stephens, 1953). But kudzu's growth potential was not something to joke about; it was becoming a serious problem, recognized even by the "Father of Kudzu." While insisting that "The habits of growth of kudzu make its control very simple," Bailey acknowledged that: "Where a vigorous stand of kudzu adjoins a plantation of young trees, *serious damage to the trees may result* unless steps are taken to prevent its spread..." (Bailey, 1944, emphasis added). Yet these pro-kudzu scientists continued to claim that kudzu's growth could be controlled; for example, "[t]he telephone company maintains a special 'kudzu brigade' in kudzu territory to periodically cut the vines away from poles and switch boxes" (Stephens, 1953). This confidence in humanity's ability to manage kudzu changed during the next decades, and kudzu became redefined as a serious ecological problem for the American South.

## THE PERIOD OF DISENCHANTMENT (1954–1984)

By the mid-1950s, many scientists and farmers had shifted their attention from methods of planting or managing kudzu, to finding ways to eradicate it. A letter to the editor of *Crops and Soils* put the problem succinctly; "Elmer Paulk of Ocilla, Ga., inquires about weed killers that will kill kudzu vines" (Kill Kudzu, 1955). There was also a shift in claimsmakers (see Table 17.2). During the period of disenchantment, a new set of expert claimsmakers – foresters – emerged who were especially critical of the dangers of unchecked kudzu growth:

> Where honeysuckle and kudzu are free to grow, establishment of tree seedlings, either pine or hardwoods, is seriously inhibited.... Kudzu can smother trees 80 feet tall. This is not a Paul Bunyan story; it's a true tale of the South.... We have a problem on our hands, one that can be as serious as wild fires, insects and diseases.... We applied this herbicide ... as often as four times without killing it.... For kudzu, complete eradication is necessary, since a single surviving plant will soon cover the ground and wrap up the tallest tree.... (Brender, 1960).

*Table 17.2* Authorship of scientific literature by time period (Discipline of articles' lead authors)

| Type of scientist | Time period | | |
| --- | --- | --- | --- |
| | Enthusiastic (1917–1953; %) | Disenchantment (1954–1984; %) | Tempered enthusiasm[a] (1985–current; %) |
| Agricultural, Biology, Botany, or Soil Expert | 82 | 41 | 33 |
| Forestry | 0 | 24 | 18 |
| Other Scientists[b] | 0 | 29 | 39 |
| Unknown/Lay[c] | 18 | 6 | 10 |
| *n* | 33 | 17 | 60[a] |

Notes
a  Sample of articles published in this time period.
b  Such as chemical engineers, health physicists, and pharmaceutical chemists.
c  But still appeared in a scientific journal.

Five years later, this same expert took the unusual step of releasing preliminary data, explaining: "[t]he problem of kudzu control in the south-east is so urgent that even interim results of new findings should be released promptly" (Brender and Moyer, 1965).

In 1970, the United States Department of Agriculture declared kudzu a common weed, and some authors claimed to have developed successful eradication methods (Winberry and Jones, 1973; Fears and Frederick, 1977). But success often came at a high cost to the timber crop: "Standing timber can be a detriment to efficient kudzu control.... Pine mortality will be excessive if two successive annual broadcast treatments of TORDON 10K pellets at 64 lb/A are required" (Fears and Frederick, 1977). Timber was a primary concern for authors writing during the period of disenchantment, for "[a]n estimated two million acres of forest land in the south-east are infested with kudzu" (Rosen, 1982). Timber killed by either kudzu or the herbicides used to eradicate it, as well as timber that could not be harvested due to kudzu infestation, posed significant losses to the region's economy. Yet, by 1985, opinions about kudzu's role in the South would change again.

## THE PERIOD OF TEMPERED ENTHUSIASM (1985–PRESENT)

This period has produced an explosion of research about *Pueraria lobata* (Table 17.3). During the first two periods of scientific research about kudzu, there were approximately 60 articles written about the plant. But during this third period, approximately 200 articles have been published. A significant number of them emphasize the plant's potential usefulness.

This second shift in opinion could first be discerned in a 1985 *Crops and Soils* column on weeds: "Kudzu, under control, could be valuable. That rapidly growing leafy vine that covers the ground, trees, power poles, and just about anything else that does not move in the south-east has a good side" (Crop Protection, 1985). Experts were not necessarily recommending "additional planting of the vines for these uses, but rather turning what is now considered to be a problem into an opportunity by making use of the plant which is already present in abundance" (Crop Protection, 1985). A few articles

*Table 17.3* Evidence of recent scholarship about kudzu 1985–1999

| Year article published | Number of articles published |
| --- | --- |
| 1985 | 6 |
| 1986 | 7 |
| 1987 | 3 |
| 1988 | 9 |
| 1989 | 9 |
| 1990 | 14 |
| 1991 | 10 |
| 1992 | 6 |
| 1993 | 34 |
| 1994 | 15 |
| 1995 | 13 |
| 1996 | 24 |
| 1997 | 21 |
| 1998 | 19 |
| 1999 | 4 |
| Total | 194 |

Sources: Compiled from the following bibliographical databases: Agricola, Applied Science and Technology Abstracts, Biological and Agricultural Index, Biology Digest, Cambridge Scientific Abstracts, Current Contents, Dissertation Abstracts, Environment, General Science Abstracts, and Geobase (Notes: there exists the possibility that a few of these may be duplicate entries; nevertheless it is clear that the study of kudzu has significantly increased in recent years).

even *encouraged* planting more kudzu; it is "greatly underutilized for improved pasturage" (Lynd and Ansman, 1990, see also Rhoden *et al.*, 1991; Shields *et al.*, 1995).

The *Crops and Soils* column suggested at least two possible industrial applications for the plant that was once thought to be such a pest; kudzu could

> be made into a very high quality paper... kudzu has about half the heat value of coal and a low sulfur content, which means that one potential use could be to burn it with coal in power plants for cheaper energy and less pollution. With its rapid growth, it is certainly a renewable energy source ("Crop Protection", 1985; see also Hart *et al.*, 1993).

But before kudzu can be technologically useful, it must be broken down. Studies have proposed various retting methods (Hart *et al.*, 1993; McLees, 1993; Tanner *et al.*, 1993; Uladaag *et al.*, 1996; Eiamwat *et al.*, 1998) that will allow the plant to be used more efficiently.

Still more publications stress kudzu's potential medicinal value. The period began with popular books promoted kudzu as a cure for hangovers (Shurtleff and Aoyagi, 1985) and as a nutritional supplement ("the 'green menace' is in fact one of Japan's most honored wild plants, [and]... could become America's finest-quality cooking starch and natural herbal medicine" (Shurtleff and Aoyagi, 1985; see also Zurbel and Zurbel, 1984). Scientific studies which supported these claims soon followed. Kudzu's antidipsotropic properties have been well researched (Keung, 1993; Keung and Vallee,

1993a,b;  Miller, 1993; Health Report, 1995; Heyman *et al.*, 1996; Keung and Vallee, 1998; Overstreet, 1998). Kudzu is a main ingredient in some blood pressure medications (Hintz, 1993) and is a substance rich in isoflavonoids, which are potentially effective in the fight against cancer (Hakamatsuka *et al.*, 1994; Gu *et al.*, 1996; Wong and Keung, 1997, 1999).

Praise for kudzu in this period, however, was not universal. Forestry scientists, among others, still sought more efficient methods of eradicating kudzu from forests without also killing the trees ( Edwards and Gonzalez, 1986; Miller and True, 1986; Miller, 1986, 1988a,b; Smith, 1990; Zidack and Backman, 1996; Hipkins, 1998; Kay and Yelverton, 1998; Rader *et al.*, 1998). They called for "zero-percent kudzu . . . [as] the uncompromising goal of any herbicide treatment program for this pest vine" (Miller, 1986). Such eradication programs were costly; "removal at several south-eastern parks in the early 1980s cost about $228 per acre" (Hester, 1991).

## SCIENTIFIC INFLUENCES ON THE LAY PUBLIC

Scientific advances involves sharing of information. For pure scientists, knowledge is exchanged in the pages of technical papers, in journals not available in most public libraries, and at annual meetings of professional associations. But applied scientists, while still primarily talking among themselves, also have other potential audiences. These disciplines seek to *use* scientific knowledge; their discourse involves what Aronson (1984) calls "interpretive" scientific claimsmaking, claims that "assert the existence of a social problem which a particular scientific specialty is uniquely equipped to solve." Their scientific successes often become the basis for social policy promulgated by all levels of government. The media will publicize their findings to a lay public anxious for reports about how science is conquering dreaded diseases, attacking environmental problems, and so on. Only when these other audiences have become convinced of the applied scientists' claims can there be significant cultural change. How successful, then, have scientists been in bringing claims about kudzu to the attention of the lay public? How influential have they been in shaping American opinion about *Pueraria lobata*?

During the enthusiastic period, certain members of the media were active in promoting the applied scientists' claims that kudzu could be useful. Channing Cope, who wrote for *The Atlanta Constitution*, also gave two daily radio broadcasts about farming, and was a farmer himself. Paralleling the scientific claims of the agronomists, Cope stressed that soil erosion was the South's foremost problem:

> 'It isn't just topsoil that is rushing along here under the bridge; it's children's shoes and clothes and school books; it's the washing machine and the refrigerator that the family was planning to buy; it's the labor of the past and the hopes of the future.' . . . Soil erosion is not merely topsoil being moved off the land. It is school erosion, church erosion, and family erosion (Cope, 1949a).

And kudzu was the answer:

> The South believes the Almighty had its cottoned-out gullies and hillsides in mind when He designed the wonder crop, Kudzu. We call it 'Earth's Best Friend.' . . . We believe it is the best land-holder now known to farmers from Florida to Maryland;

the best moisture-holder; the best land-builder; the quickest soil-maker; and the best insurance against summer and fall drought (Cope, 1949b).

Cope began a "Kudzu Club" for listeners who were following the scientists' advice. By the mid-1940s, it had some 20000 members (Shurtleff and Aoyagi, 1985). Between Cope's broadcasts and the agronomist Bailey's circulars, claims about kudzu's promise for Southern agriculture reached most of the region's farmers. When the *Reader's Digest* advocated its use (Lord, 1945), an even larger audience became convinced of the plant's fantastic powers.

So by the early 1950s, it seemed that many Southern farmers had accepted the claims of the scientists. Stories appeared in print in which farmer after farmer bore witness to kudzu's potential. Many of these testimonies came from uneducated individuals, highlighting the trust that they placed in those who knew more about kudzu – the scientists. These stories were employed to allay others' worries. One such article appeared in *Soil Conservation*; it quoted farmer John Woody, telling how his mule had significantly improved with a diet of kudzu:

> When I fust got the mule she got down and I would have ta help her up. When I was feeding her some would say not to feed her too much it mo kill her. I didn't see her git no better 'till I fust started to cutting kudzu and feeding it to the cow. On the 15th day of April was the fust day I started to feed the mule on kudzu. It was not over 4 or 5 days that before she got where she could git up and den eat by herself and has been picking up ever since. If I feed her sweet feed and kudzu, she will eat the kudzu 'fore she will the sweet feed. They say she has picked up 200 pounds, so they tell me, I don't know (Rowalt, 1936).

Woody's story was repeated by Brink (1942), who introduced it with the words, "Kudzu willingly and capably repairs the hurts of stricken agriculture wherever they are found. It makes no distinction of race or creed when rescue and rehabilitation are the stakes." All that mattered is that one believed in the healing power of kudzu – and kept on planting it.

In large measure, scientists in the period of enthusiasm were able to convince the public that they knew how to master the plant. One SCS scientist wrote that "the vine is an obedient child, amenable to forms of discipline easily exercised by the farmer" (Brink, 1942). Farmers were repeatedly told that acreage planted with kudzu could be controlled by mowing or other means. Those reassurances, coupled with governmental incentive programs, were enough to assuage most lingering worries. But the public was never entirely convinced of kudzu's efficacy; there had always been some skeptical farmers, concerned about how uncontrollable the plant seemed. So when scientific confidence about human mastery over kudzu diminished in the period of disenchantment, the public readily embraced this new definition of the situation. They accepted that kudzu had become the "South's Frankenstein monster" (Robertson, 1971). Southern folklore mirrored scientific concern about kudzu:

> An Arkansas farmer once quipped that, 'When you plant kudzu, drop it and run.' One legend tells of a man who planted kudzu behind his barn, and its branches grew fast enough to beat him back to the house. Another asserts that the tale of *Jack and the Beanstalk* was but a slightly exaggerated account of a boy who was careless

with kudzu. In some rural areas children are warned that if they are naughty they will be thrown into the kudzu patch and swallowed up....In Georgia, legend says that you must close your windows at night to keep kudzu from coming into the house (Shurtleff & Aoyagi, 1985).

Similar imagery appeared in Southern literature. James Dickey's poem "Kudzu" calls the plants "Green, mindless, unkillable ghosts" (Dickey, quoted in Shurtleff and Aoyagi, 1985), while the villains' bodies in Pat Conroy's novel *The Prince of Tides* (1987) are thrown into a kudzu stand, never to be discovered.

This view of kudzu – a commanding foe, out of control, and unable to be eradicated – has remained popular with the public; they seem less ready to embrace the renewed scientific enthusiasm for kudzu. Despite some fifteen years of scientific claims about kudzu's industrial and medicinal usefulness, much of the public still considers the plant to be a pest, a foe worthy of annihilation. A recent article in *The Washington Post* refers to kudzu this way:

> What is green, can grow a foot a day, kills trees, and is prepared to take over the Southeast? The answer is kudzu, the Green Menace. Kudzu, as practically any southerner knows, is that vine that rises each spring and summer to coat tall pines, cascade down embankments, drape itself over abandoned houses and cars, and give the landscape the look of a lumpy green blanket. Unchecked, it will cover everything in its path, depriving trees of needed sunlight, choking the life out of ferns and undergrowth, and encroaching on farmlands and yards. As a perennial, it always springs right back to life the next year, stronger than ever, naturally drought-resistant. To say kudzu is hardy is like saying summer is hot (Presley, 1998).

In fact, much of the literature read by the lay public still discusses kudzu as the pre-eminent model for biological invaders which need to be stopped before they devastate American flora and fauna. Walter Reeves, a columnist for Atlanta, Georgia's largest newspaper, wrote of his troubles with yellow archangel with these words, "I decided that the archangel was a bit too similar to kudzu for my taste. A tank of weedkiller was employed on my first attempt to wipe out this noxious plant....Kudzu was originally presented as an ornamental trellis cover...[it is the subject] of harrowing tales that begin with, 'I just had a little bit over there...and now it's in my neighbor's yard!'" (Reeves, 1999; see also Bragg, 1997; Johnson, 1997; Fenyvesi, 1998; Tompkins, 1998; Viney, 1998; Howe, 1999; May, 1999).

There appears to be a gap between scientists' renewed enthusiasm about kudzu in the last 15 years and the public's skepticism. In part this discrepancy is a result of the fact that recent scientific discoveries have not been covered well in the lay press. In fact, of the last 25 articles about kudzu published in newspapers or popular magazines, only two discussed *any* of kudzu's positive properties. One tells the story of Dr Jim Duke and his garden. "'My neighbors think I'm crazy for putting in kudzu, but it has more genistein than soy,' said Dr Duke,...Genistein is what everybody's promoting soy for, to help prevent breast cancer'" (Raver, 1998). The other article briefly mentions kudzu's value in the treatment of alcoholism and as a heart stimulant (Rising, 1997). Two other stories discussed the plant's potential nutritional uses – as both *haute cuisine* (Hurst and Hoots, 1995) and kudzu roots as a substitute for other starches, "You can bake it just like sweet potatoes" (Presley, 1998). The only other positive use of the plant was as the main ingredient in wreath- and basket-making (May, 1998; Reinolds, 1999).

Instead, what the public hears, over and over again, is scientific confidence that kudzu can be controlled, albeit through innovative means. One such technique, frequently mentioned in the lay press, was invented by David Orr and his colleagues, who are:

> working on a hybrid kudzu-eating bug that gorges on the greens, then obligingly allows itself to be eaten alive....Orr's strategy involves the soybean looper, a caterpillar found on soybean plants that really likes kudzu, and stingless wasps that are used to inject the caterpillars with wasp eggs. The effect of the parasite eggs on the looper is something like a tapeworm, multiplying hundreds, even thousands, of times within the caterpillar's body and make it eat much more kudzu than usual. But just before the caterpillar gets ready to spin a cocoon, the parasites, receiving a chemical signal, devour it, leaving a husk of wasp larvae (Presley, 1998, see also Bragg, 1997; Johnson, 1997; Rising, 1997; Kudzu might, 1998).

But while these stories offer a glimmer of scientific control of the promiscuous plant, they also unfailingly reinforce the public's negative opinion of kudzu.

And the popular press continues to incite worry that the plant is not yet under control and is still causing significant problems. A television station in Atlanta, Georgia reported that the Atlanta Police Department employs a full-time police officer who is assigned to handle kudzu complaints between neighbors in an attempt to prevent further infestation by the plant (it is illegal to grow kudzu within city limits) (WAGA TV, 1995), while an article in *Garden Magazine* warned: "the controversial vine that established a love–hate relationship with the South is creeping North....With inadequate vigilance, the weed may yet be known as 'the plant that ate the North'" (Watson, 1989, see also Kudzu attempts, 1998).

Thus while the scientific community has shifted from seeing kudzu as the Southern savior, to one of its most ignoble pests, to an industrial and medicinal solution to other problems, the lay public has not followed suite. The public still sees the weed as a blight on the Southern landscape, one that is stealing ever northward. Until scientists are better able to inform the public about their current opinions about the plant, there will remain a knowledge gap, a cultural lacuna, between those who work with kudzu in laboratories and those who just want the troublesome vine gone from their backyard.

## REFERENCES

Alexander, E.D. (1950) *Kudzu*. Agricultural Extension Service Circular, Athens, Georgia.

Aronson, N. (1984) Science as a claims-making activity: implications for social problems research. In J. Schneider and J. Kitsuse (eds), *Studies in the Sociology of Social Problems*, Ablex, Norwood, New Jersey, pp. 1–30.

Bailey, R.Y. (1939) The use of kudzu on critical slopes. *Soil Conservation*, 5, 48–50.

Bailey, R.Y. (1944) Kudzu for erosion control in the Southeast. U.S. Department of Agriculture. Farmers' Bulletin Number 1840, Washington, D.C.

Bailey, R.Y. and Mayton, E.L. (1931) *Kudzu in Alabama* Agricultural Experiment Station of the Alabama Polytechnic Institute, Circular 57, Auburn, Alabama.

Bragg, R. (1997) Experts use strategy on kudzu weed. *The Plain Dealer*, September 7, 11A.

Brender, E.V. (1960) Progress report on control of honeysuckle and kudzu. *Southern Weed Conference*, 187–193.

Brender, E.V. and Moyer, E.L. (1965) Further progress in the control of kudzu. *Down Earth*, 20, 16–17.

Brink, W. (1942) Kudzu – a mender of tattered lands. *Better Crops Plant Food*, 26, 10–13, 38–40.

Conroy, P. (1987) *The Prince of Tides*. Bantam, New York.

Cope, C. (1949a) *Front Porch Farmer*. Turner E. Smith and Company, Atlanta, Georgia.

Cope, C. (1949b) Kudzu. *The Farm Quarterly*, 5, 54–57, 88, 90.

Crop protection: weeds. (1985) *Crops and Soils*, 37, 22.

Dacy, G.H. (1917) Be cautious with kudzu. *Country Life*, 31, 100–101.

Davis, R.L. and Young, W.C. (1951) Kudzu-23: a new fine-textured variety. *Soil Conservation*, 16, 279–280.

Edwards, M.B. and Gonzalez, F.E. (1986) Forestry herbicide control of kudzu and Japanese honeysuckle in loblolly pine sites in central Georgia. *Proc. South. Weed Soc.*, 39, 272–275.

Eiamwat, J., Loha, V., Prokop, A. and Tanner, R.D. (1998) Batch foam fractionation of kudzu (*Pueraria lobata*) vine retting solution. *Appl. Biochem. Biotechnol.*, 70–72, 559.

Fears, R.D. and Frederick, D.M. (1977) Kudzu control with TORDON 10K Pellets brush killer for timber site preparation in central Alabama. *Down Earth*, 33, 6–9.

Fenyvesi, C. (1998) When prolific plants are just what you need. *The Washington Post*, July 30, T21.

Funchess, M.J. and Tisdale, H.B. (1924) Kudzu for Hay. In The Thirty-fifth Annual Report of the Agricultural Experiment Station of the Alabama Polytechnic Institute, Auburn, Alabama, pp. 5–6.

Gu, Z.P., Chen, B.Z., Feng, R.Z. and Dong, X. (1996) The source utilization and evaluation of medical kudzu and roots from the genus plants *pueraria DC* in China. *Yao hseuh hsueh pao*, 31, 386.

Hakamatsuka, T., Ebizuka, Y. and Sankawa, U. (1994) *Pueraria lobata* (kudzu vine): *in vitro* culture and the production of isoflavonoids. *Biochemistry Agriculture Forestry*, 28, 386.

Hart, P.W., Brodgon, B.N. and Hsieh, J.S. (1993) Anthraquinone pulping of kudzu (*Pueraria lobata*). *Tappei J.*, 76, 162–166.

Health Report. (1995) *Time*, 146, September 25, 24.

Hester, F.E. (1991) The US National Park Service experience with exotic species. *Nat. Areas J.*, 11, 127–128.

Heyman, G.M., Keung, W.M. and Vallee, B.L. (1996) Daidzin decreases ethanon-consumption in rats. *Alcohol. Clin. Exp. Res.*, 20, 1083–1087.

Hintz, H.F. (1993) Kudzu. *Equine Practice*, 15, 5–6.

Hipkins, P.L. (1998) Control of kudzu during the dormant season. *Proc. South. Weed Sci. Soc.*, 51, 191–192.

Howe, P.J. (1999) US sets move on invasive sea organisms. *The Boston Globe*, January 27, A3.

Hurst, A.S. and Hoots, D. (1995) Consuming kudzu. *South. Living*, 30, August, 20ga.

Johnson, R.G. (1997) Another plot against kudzu. *The Atlanta Journal and Constitution*, August 8, 1C.

Kay, S.H. and Yelverton, F.H. (1998) Dormant season herbicide treatments for kudzu control. *Proc. South. Weed Sci. Soc.*, 51, 190–191.

Keung, W.M. (1993) Biochemical studies of a new class of alcohol-dehydrogenase inhibitors from *Radix puerariae. Alcohol. – Clin. Exp. Res.*, 17, 149–151.

Keung, W.M. and Valle, B.L. (1993a) Daidzin: A potent, selective inhibitor or human mitochondrial aldehyde dehydrogenase. *Proc. Nat. Acad. Sci.*, 90, 1247–1251.

Keung, W.M. and Valle, B.L. (1993b) Daidzin and daidzein suppress free-choice ethanol intake by Syrian Golden hamsters. *Proc. Nat. Acad. Sci.*, 90, 10008–10012.

Keung, W.M. and Valle, B.L. (1998) Kudzu root: an ancient Chinese source of modern anti-dipsotropic agents. *Phytochemistry*, 47, 499–506.

Kill kudzu (1955) *Crops and Soils*, p. 9.

Kudzu attempts northern invasions – maintaining the integrity of a 160-acre arboretum and 900 miles of roadside rights-of-way meant controlling a kudzu invation without harming compatible species (1998) *Public Works*, 129, 40.

Kudzu in South Carolina (1938) The Extension Service of Clemson Agricultural College, Circular 164, Clemson, South Carolina.

Kudzu might meet its match in hybrid bug (1998) *The Times-Picayune*, July 5, B3.

Lord, R. (1945) Kudzu: Another agricultural miracle. *Readers Digest*, January.

Lynd, J.Q. and Ansman, T.R. (1990) Exceptional forage regrowth, nodulation and nitrogenase activity of kudzu [*pueraria lobata* (willd.) Ohivi] grown on eroded dougherty loam subsoil. *J. Plant Nutr.*, 13, 861–865.

May, L. (1998) Kudzu putting a new face on Christmas spirit. *Atlanta J. Constitution*, December 19, 1G.

May, L. (1999) My nandina an invader? Say it ain't so! *Atlanta J. Constitution*, March 6, 1G.

McKee, R. and Stephens, J.L. (1943) *Kudzu as a Farm Crop*, US Department of Agriculture. Farmers' Bulletin 1923, Washington D.C.

McLees, L. (1993) Pulping kudzu. *Res. Horizons*, 11, 20–22.

Miles, I.E. and Gross, E.E. (1939) *A Compilation of Information on Kudzu*. Mississippi Agricultural Experiment Station, Bulletin 326, State College, Mississippi.

Miller, J.H. (1986) Kudzu eradication trials testing fifteen herbicides. *Proc. South. Weed Soc.*, 39, 276–281.

Miller, J.H. (1988a) Guidelines for Kudzu Eradication Treatments. In J.H. Miller and R.J. Mitchell (eds), *A Manual on Ground Applications of Forestry Herbicides*, US Department of Agriculture. Forest Service Management Bulletin R8-MB21, Washington D.C., pp. 6.1–6.7.

Miller, J.H. (1988b) Kudzu eradication with new herbicides. *Pamphlet*, pp. 220–225.

Miller, J.H. and True, R.E. (1986) Herbicide tests for kudzu eradication. *Georgia Forest Res. Pap.*, 65, 3–11.

Miller, S.K. (1993) Hamsters take the pledge after ancient Chinese cure. *New Scientist*, 14, 4–5.

Overstreet, D.H. (1998) The Chinese herbal medicine NPI-028 suppresses alcohol intake in alcohol-preferring rates and monkeys without inducing taste aversion. *Perfusion*, 11, 381.

Pierre, W.H. and Bertram, F.E. (1929) Kudzu production with special reference to influence of frequency of cutting on yields and formation of root reserves. *J. Am. Soc. Agronomy*, 21, 1079–1101.

Piper, C.V. (1920) *Kudzu*. United States Department of Agriculture, Department Circular 89, Washington, D.C.

Polk, H.D. and Gieger, M. (1945) *Kudzu in the Ration of Growing Chicks*. Agricultural Experiment Station, Bulletin 414, State College, Mississippi.

Porter, H.L., Jr. (1953) Kudzu weak . . . . *Crops and Soils*, pp. 6.

Pressley, S.A. (1998) Plan of attack for an attacking plant. *The Washington Post*, July 29, A03.

Rader, L.T., Harrington, T.B. and Taylor, J.W., Jr. (1998) First-year response of Kudzu and associated vegetation to five herbicides. *Proc. South. Weed Soc.*, 51, 141–142.

Raver, A. (1998) A man with a garden that's a medicine cabinet. *The New York Times*, October 15, F1.

Reeves, W. (1999) Start now to trump bamboo, kudzu. *Atlanta J. Constitution*, February 20, 4G.

Reinolds, C. (1999) Cable TV discovers Ball Ground artist's vine. *Atlanta J. Constitution*, January 28, 12JQ.

Rhoden, E.G., Woldeghebriel, A. and Small, T. (1991) Kudzu as a feed for Angora goats. *Tuskegee Horizons*, 2, 23.

Rising, G. (1997) Kudzu vine grows a foot a day, chokes its host. *The Buffalo News*, September 29, 7B.

Robertson, W.J. (1971) How we controlled kudzu. *Forest Farmer*, 30, 8–9, 18.

Rosen, A.E. (1982) *Feasibility Study: Eradication of Kudzu with Herbicides and Revegetation with Native Tree Species in Two National Parks*. Masters Thesis, University of Tennessee, Knoxville.

Rowalt, E.M. (1936) Kudzu is this farmer's friend. *Soil Conservation*, 1, 12–14.

Shields, F.D., Jr, Bowie, A.J. and Cooper, C.M. (1995) Control of streambank erosion due to bed degradation with vegetation and structure. *Water Resource Bulletin*, 31, 475–490.

Shurtleff, W. and Aoyagi, A. (1985) *The Book of Kudzu: A Culinary and Healing Guide*. Avery, Wayne, New Jersey.

Smith, A.E. (1990) *Kudzu Control in Nonforested Areas with Herbicides*. Georgia Agricultural Experiment Stations, Research Report 591, Athens.

Stephens, R.W. (1953) Kudzu: versatile wonder-bean. *Organic Farmer*, 4, 32–35.

Stewart, M.A. (1997) Cultivating kudzu: the Soil Conservation Service and the kudzu distribution program. *Georgia Historical Quarterly*, 81, 151–167.

Sturkie, D.G. and Grimes, J.C. (1939) *Kudzu: Its Value and Use in Alabama*. Agricultural Experiment Station of the Alabama Polytechnic Institute, Circular 83, Auburn, Alabama.

Tabor, P. and Susott, A.W. (1941) Zero to thirty million mile-a-minute seedlings. *Soil Conservation*, 7, 61–65.

Tanner, R.D., Prokop, A. and Bajpai, R.K. (1993) Removal of fiber from vines by solid-state fermentation enzymatic degradation – a comparison of flax and kudzu retting. *Biotechnol. Adv.*, 11, 646–643.

Tompkins, S. (1998) 'World's worst weed' picks Houston for invation landing. *The Houston Chronicle*, December 4, 13.

Uladag, S., Prokop, A. and Tanner, R.D. (1996) A kinetic study of the kudzu (*Pueraria lobata*) retting process. *J. Sci. Ind. Res.*, 55, 381.

Viney, M. (1998) Special forces for tackling aliens. *The Irish Times*, August 1, 68.

WAGA TV (1995) Newscast on the "kudzu cop." Aired several times in September.

Watson, R. (1989) The green menace creeps north. *Garden* 13 (March–April), 8–9, 11.

Winberry, J.J. and Jones, D.M. (1973) Rise and decline of the 'miracle Vine' kudzu in the southern landscape. *Southeastern Geographer*, 13, 61–70.

Wong, C.K. and Keung, W.M. (1997) Daidzein Sulfoconjugates are potent inhibitors of sterol sulfatase (EC 3.1.6.2). *Biochem. Biophys. Res. Commun.*, 233, 579–583.

Wong, C.K. and Keung, W.M. (1999) Bovine adrenal 3$\beta$-hydroxysteroid dehydrogenase (E.C. 1.1.1.145)/5-ene-4-ene isomerase (E.C. 5.3.3.1): characterization and its inhibition by isoflavones. *J. Steroid Biochem. Mol. Biol.*, 71, 191–202.

Zidack, N.K. and Backman, P.A. (1996) Biological control of kudzu (*Pueraria lobata*) with the plant pathogen *Pseudomonas syringae pv. phaseolicola. Weed Sci.*, 44, 381.

Zurbel, R. and Zurbel, V. (1984) *The Natural Lunchbox*. Holt, Rinehart, and Winston, New York.

# Index

abortificient effect 71–4
abrisapogenol 200
acamprosate 147, 161
acetaldehyde 146, 151–2, 156, 162, 168, 173–4
acetylkaikasaponin III 92–3
acetylsoyasaponin I 92–3
aconitine 136–7
adenylate cyclase 138
ADME 190, 222–3
adrenal 112–14
$\beta$-adrenergic receptor 132–3, 137–9, 161
agronomy 29, 276
alcohol 59, 144–8, 156, 159, 173–5
alcohol addiction 145–6, 159, 228; *see also* alcohol
    craving, alcoholism/alcohol abuse
alcohol craving 147, 159, 161, 173; *see also* alcohol
    addiction, alcoholism/alcohol abuse
alcohol dehydrogenase, ADH 146, 153, 162,
    174, 227–8
alcoholism/alcohol abuse 42, 144, 161, 228, 249;
    *see also* alcohol addiction, alcohol craving
aldehyde dehydrogenase, ALDH 146, 152–6,
    162, 174, 228
amino acid 32, 123–4, 126, 255
angina pectoris 60, 62, 132, 141, 245, 246–8;
    *see also* coronary heart disease
anhydrotuberosin 72
antabuse 145, 174; *see also* disulfiram, calcium
    carbimide
anthocyanin 32, 120
anthropology 29, 40–9
anti-angiogenic activity 184, 226; *see also*
    cancer
anti-atherosclerosis 227; *see also* heart diseases
anti-coagulant 268–9
anti-diabetic 42, 62–4, 269
antidiarrhetic 42, 243
antidipsotropic 144–50, 174, 182, 243, 279;
    mechanism of action 151–6, 174–5; *see also*
    alcoholism/alcohol abuse
anti-drunkenness 41, 62, 249
antiemetic 43, 243

antiestrogenic 74, 180, 183–4, 191; *see also*
    estrogen
anti-fertility 70, 72, 78, 80, 108
anti-*Helicobacter pylori* 227
antihypertensive 62, 133, 141, 252; *see also*
    hypertension
antihypoxia 228
anti-implantation 71–2
anti-inflammatory 62, 227
antimicrobial 61, 127
antiosteoporosis 227–8; *see also* ipriflavone,
    osteoporosis
antioxidant 161, 184, 226–7, 269
antipyretic 61–3, 243, 252; *see also* febrile diseases
antispasmodic 71, 93, 198, 212, 269
aromatase 184, 188
arrhythmia 136–7, 246
aversion therapy 144, 146–7; *see also* disulfiram,
    calcium carbimide

$\psi$-baptigenin 230
biliary excretion 214–16, 219
bioavailability 150–1, 165–7
biochanin A 83, 89, 180, 182, 226, 234–6
biogenic aldehyde 152, 156, 174, 228
biosynthesis 119, 121–2, 124, 126–7, 188, 231
bisdesmoside 94, 201
botany 1, 29–30, 51
but-2-enolide 85, 90, 92

cabreuvin 228
calcium carbimide 146–7, 152
cancer; anti 70, 227, 280; breast 103, 114, 180,
    184–5, 188–9, 282; endometrium 184, 186,
    188; prostate 184–6, 188, 191; skin 227
carbohydrate 31, 78
cardiovascular diseases 60, 132, 141, 243–50
cell culture 119–22
central reward pathway 147
cerebrovascular diseases 245, 247–50; *see also*
    cardiovascular diseases
chalcones 119–20, 122–5, 231

(Continued)